Artificial Intelligence for Air Quality Monitoring and Prediction

This book is a comprehensive overview of advancements in artificial intelligence (AI) and how it can be applied in the field of air quality management. It explains the linkage between conventional approaches used in air quality monitoring and AI techniques such as data collection and preprocessing, deep learning, machine vision, natural language processing, and ensemble methods. The integration of climate models and AI enables readers to understand the relationship between air quality and climate change. Different case studies demonstrate the application of various air monitoring and prediction methodologies and their effectiveness in addressing real-world air quality challenges.

Features:

- A thorough coverage of air quality monitoring and prediction techniques.
- In-depth evaluation of cutting-edge AI techniques such as machine learning and deep learning.
- Diverse global perspectives and approaches in air quality monitoring and prediction.
- Practical insights and real-world case studies from different monitoring and prediction techniques.
- Future directions and emerging trends in AI-driven air quality monitoring.

This is a great resource for professionals, researchers, and students interested in air quality management and control in the fields of environmental science and engineering, atmospheric science and meteorology, data science, and AI.

AI Applications in Earth Science

Series Editor – Prof. Dr. Gaurav Dhiman

Geoinformatics is one of the areas where Artificial Intelligence (AI) is already making a significant impact. Satellites and other sensors are collecting vast amounts of data on the Earth's surface, its atmosphere, and its oceans. AI algorithms are being trained to recognize patterns in remote sensing data to help researchers quickly identify changes in land use, vegetation cover, ocean currents, climate modeling, and predictions. It has great potential in hydrology for predicting extreme weather events and for disaster planning and response. This book series provides a comprehensive view of how AI is transforming the field of Earth Science, including the latest research and techniques, and explains the potential of AI in improving our understanding of Earth's natural systems. It addresses some of the most pressing environmental challenges of our time. Books in this series serve and benefit professionals, academics, and students in the Earth Science community and other related fields.

Artificial Intelligence for Air Quality Monitoring and Prediction
by Amit Awasthi, Kanhu Charan Pattnayak, Gaurav Dhiman, and Pushp Raj Tiwari

For more information about this series, please visit: https://www.routledge.com/AI-Applications-in-Earth-Science/book-series/CRCAIES

Artificial Intelligence for Air Quality Monitoring and Prediction

Edited by
Amit Awasthi, Kanhu Charan Pattnayak,
Gaurav Dhiman, and Pushp Raj Tiwari

CRC Press
Taylor & Francis Group
Boca Raton London New York

CRC Press is an imprint of the
Taylor & Francis Group, an **informa** business

Designed cover image: © maeching chaiwongwatthana, Shutterstock

First edition published 2025
by CRC Press
2385 NW Executive Center Drive, Suite 320, Boca Raton FL 33431

and by CRC Press
4 Park Square, Milton Park, Abingdon, Oxon, OX14 4RN

CRC Press is an imprint of Taylor & Francis Group, LLC

ISBN: 978-1-032-68379-9 (hbk)
ISBN: 978-1-032-68382-9 (pbk)
ISBN: 978-1-032-68380-5 (ebk)

DOI: 10.1201/9781032683805

Typeset in Times
by Newgen Publishing UK

Contents

Preface

In the present era, the quality of air that we breathe has become a major concern for the well-being of our society. The air quality can be influenced by both natural and anthropogenic reasons, as it is a crucial aspect of our environment that has a profound impact on our health and well-being. Hence accurate monitoring and prediction of air quality are important as it has a lot of negative impact on society. The objective of this book is to provide a comprehensive overview of the transition from conventional methods to AI-driven approaches in air quality monitoring and prediction. It explores the integration of AI techniques, such as Machine Learning (ML) and deep learning (DL), with existing monitoring systems to enhance accuracy, efficiency, and real-time analysis. This book consist of 15 chapters is structured into 10 reviews and 5 case studies with the main aims to make educators understand how to monitor and predict air quality using cutting-edge technology like AI and its subdomains like DL, ML, etc.

Chapter 1 "Air Quality Monitoring (AQM) and Prediction: Transitioning from Conventional to AI Techniques" is the introductory chapter, in which the journey of conventional to AI techniques for air quality measurement will be discussed along with its fundamentals and principles behind it. The current state-of-the-art in air quality measurement utilizing ML and DL techniques will be interpreted. A comparative analysis of conventional techniques concerning AI techniques will be discussed by considering major challenges and the future direction of Air Quality Monitoring (AQM). In the concluding remark, balanced solutions between the strengths and weaknesses of both methodologies are proposed in the present scenario for better air quality measurement and prediction.

Chapter 2 "Temporal Variations of Sulphur Dioxide Levels across India: A Biennial Assessment (2020–2021)" is a case study which examines the levels of sulfur dioxide (SO_2) in different states of India' to understand ambient air quality. the study concentrates on five cities designated by the national government as ecologically sensitive namely Agra, Firozabad, Mathura, Alwar, and Dehradun. Dehradun was found to have the highest SO_2 concentration among the 378 cities analyzed in 2020 and 2021.

Chapter 3 "The Effectiveness of Machine Learning Techniques in Enhancing Air Quality Prediction" is a case study of ML in enhancing the air quality prediction. In this chapter, a case study of United Kingdom is presented to highlight the importance of ML model in improving the concentration of ammonia (NH_3) as well as the computing time. Intercomparison analysis is carried out between WRF-chem and the ML model in simulating spatial distribution of NH_3 during the month of April 2016. Results indicate the ML model significantly reduces the half of the computing time compared to WRF-chem. The results of this test case study encourage the possible usage of ML models with proper realizations to enhance the prediction of air quality. This will consequently help in framing climate policies.

Chapter 4 "Enhancing Environmental Resilience: Precision in Air Quality Monitoring through AI-Driven Real-Time Systems" discusses the use of artificial intelligence (AI) in real-time air quality monitoring systems. It explores how AI can integrate weather model data to improve air quality assessments. This chapter reviews about the effectiveness of AI in analyzing data, predicting pollution patterns, and providing insights for decision-makers. This chapter highlights the practicality and adaptability of AI-based monitoring systems and emphasizes their potential impact on community health and policy development.

Chapter 5 "Forecasting Air Pollution with Artificial Intelligence: Recent Advancements at Global Scale and Future Perspectives": To address the issue of rapid urbanization, it is important to develop advanced methods for monitoring and predicting air pollution. Hence this chapter discusses the use of AI and ML algorithms to interpret data from sensors. In this chapter, it is proposed that these types of models can forecast air pollutants such as $PM_{2.5}$, PM_{10}, O_3, CO, SO_2, NO_2, and CO_2. Hybrid models that combine different AI tools can improve performance in forecasting and warning approaches. Performance evaluation measures such as R^2, RMSE, MAE, and MAPE are used to assess the accuracy of these models.

Chapter 6 "Integrating AI into Air Quality Monitoring: Precision and Progress" provides an overview of AI applications in air quality monitoring, encompassing sensor data processing, pollutant concentration estimation, and forecasting models. Discussion on various AI techniques such as ML, DL, and ensemble methods are explored for their effectiveness in handling diverse data sources and improving prediction accuracy.

Chapter 7 "Application of AI-Based Tools in Air Pollution Study" gives a summary of techniques for forecasting and identifying the sources of air pollution that are based on AI and how they contribute to solving this pressing problem. Complex air quality data, such as atmospheric measurements, satellite images, and meteorological data, have been progressively analyzed using AI approaches like ML algorithms and neural networks (NN). It explores several AI techniques that are frequently used in source detection, such as DL models, support vector machines, and random forests. It emphasizes how crucial it is for researchers, environmental organizations, and local populations to work together when deploying and validating AI models for source detection.

Chapter 8 "Study of Extreme Weather Events in the Central Himalayan Region through Machine Learning and Artificial Intelligence: A Case Study" is a case study that focuses on various ML approaches for analyzing extreme weather events in in the central Himalayan region of Uttarakhand, India, based on meteorological data and aerosol properties. Out of different cutting-edge methodologies, including Support

Vector Machines (SVM), Artificial Neural Networks (ANN), and General Regression Neural Networks (GRNN), etc. a Random Forest (RF) model is used to predict rainfall events in the Srinagar Garhwal region, utilizing parameters such as temperature, humidity, wind direction, and wind speed. The model's efficacy was assessed through rigorous evaluation metrics, including the Mean Absolute Error (MAE), Coefficient of Determination (R^2), and Root Mean Square Error (RMSE). The Srinagar rainfall model, with 100 trees, demonstrated a delicate balance between complexity and performance, excelling with RMSE of 2.74, MAE of 1.39, and R^2 of 0.57.

Chapter 9 "Machine Learning Applications in Air Quality Management and Policies" advocates the significance of employing ML methods in generating air quality datasets. The chapter explores the growing role of ML algorithms in predicting air quality distributions and trends, forecasting pollutant levels, and optimizing policy interventions. The application of several ML algorithms, including SVM, Multiple Linear Regression (MLR), RF, NN, and cluster analysis, is observed across diverse air quality contexts. In conclusion, this chapter acknowledges the significant role of accurate data in the realm of air quality management and policymaking and emphasizes that ML methods are instrumental in providing decision support and hold unparalleled promise for the future.

Chapter 10 "A Glimpse into Tomorrow's Air: Leveraging PM 2.5 with FP Prophet as a Forecasting Model" is a case study that assesses the degrading quality of air, by monitoring the concentration of several pollutants and subsequently classified by the Air Quality Index (AQI). Solid particulate matter, nitrogen dioxide (NO_2), sulfur dioxide (SO_2), carbon monoxide (CO), ozone depletion, toluene, ammonia (NH3), benzene, and nitrite (NO), etc., are monitored. This use-case model investigates the last six years' data from ten cities in India for air quality analysis and predicts the future conditions in these cities. The predictive analysis showed that nine out of ten cities show a defined pattern of particular matter having 2.5 microns ($PM_{2.5}$) levels over the years (Jan 2013–Dec 2023).

Chapter 11 "Air Quality Forecast Using Machine Learning Algorithms" examines the key gaps, challenges, usage, and improvements, as well as the important trends and research needs for the continued development of air quality forecasting and ML techniques including gradient boosting (GB), MLR, principal component analysis (PCA), RF, support vector regression (SVR) etc., based on published reviews and recent analyses.

Chapter 12 "Deep Learning Approaches in Air Quality Prediction" explores the review on the application of DL techniques in the realm of air quality prediction. It provides an in-depth analysis of various DL models and methodologies employed to enhance the performance and efficiency of air quality predictions. The state-of-the-art techniques are discussed meticulously. A performance analysis of these methods is also presented for analyzing these methods quantitatively. A subsection is also dedicated to the details of datasets for simulating different models for prediction/classification of air quality. The chapter also discusses the challenges faced by existing techniques, and future directions in the field of air quality research using DL approaches.

Chapter 13 "Incorporation of AI with Conventional Monitoring Systems" reviews the recent improvements on the Internet of Things (IoT) and big data that raised the capacity of prediction and alleviated air quality efficiently and effectively. This review will try to shed light on the precise use of AI along with conventional methods and also to check the merits and demerits of combining this high-end technique to monitor or lessen air pollution in the near future.

Chapter 14 "A Comparative Evaluation of AI-Based Methods and Traditional Approaches for Air Quality Monitoring: Analyzing Pros and Cons": This chapter investigates into the innovative realm of real-time air quality monitoring systems, harnessing the potential of AI to provide both conceptual frameworks and practical implementations. It explores the integration of weather model data, enhancing real-time air quality assessments. This chapter comprehensively reviews how AI, real-time data, and weather models enhance the air quality monitoring system.

Chapter 15 "Machine Learning-Driven Hydrogen Yield Prediction for Sustainable Environment" is the case study in which a ML-based approach is used to investigate the effect of various operating parameters like steam-to-ethanol ratio, temperature, and pressure on the hydrogen yield. The performance of ML algorithms are analyzed in various categories using the test data generated from conventional mathematical model. The ML models MLR, SVR, decision trees (DT), and RF algorithms are analyzed and tested to suitably predict the yield of hydrogen under different operating conditions.

Contributors

Satadal Adhikary
Post Graduate Department of Zoology
A.B.N. Seal College
Cooch Behar, West Bengal, India

Amit Awasthi
Department of Physics
School of Advanced Engineering
UPES
Dehradun, Uttarakhand, India

Mamta Bansal
Department of Zoology and Environmental Sciences
Maharaja Agrasen University
Baddi, India

Suchandra Bhattacharya
Department of Chemistry
A.B.N. Seal College
Cooch Behar, West Bengal, India

Tania Ghatak (Chakraborty)
Department of Botany
Shibpur Dinobundhoo Institution (College)
Howrah, West Bengal, India

Soni Chaurasia
Department of CSE
SGT University
Gurugram, Haryana, India

Sunil Kumar Chhikara
University Institute of Engineering and Technology
Maharshi Dayanand University
Rohtak, Haryana, India

Tanupriya Choudhury
Graphic Era
Deemed to be University Dehradun
Uttarakhand, India

Sourangsu Chowdhury
CICERO Center for International Climate Research
Oslo, Norway

Sushant Das
Department of Earth and Atmospheric Sciences
National Institute of Technology Rourkela
Odisha, India

Abhratanu Ganguly
Department of Animal Science
Kazi Nazrul University
Asansol, West Bengal, India

Alok Sagar Gautam
Atmospheric and Space Physics Laboratory
Department of Physics
Hemvati Nandan Bahuguna Garhwal University (A Central University)
Srinagar Garhwal, Uttarakhand, India

Sneha Gautam
Department of Civil Engineering
Karunya Institute of Technology and Sciences
Deemed University
Karunya Nagar, Tamil Nadu, India

Gunjan Goyal
Department of Botany
University of Rajsthan
Jaipur, Rajsthan, India

Juhi Gupta
Amity Institute of Environmental Sciences (AIES)
Amity University
Uttar Pradesh, India

Mukul Kumar Gupta
Department of Electronics and Instrumentation Engineering
Faculty of Engineering and Technology
MJP Rohilkhand University
Bareilly, Uttar Pradesh, India

Navjot Hothi
Department of Physics
School of Advanced Engineering
UPES
Dehradun, Uttarakhand, India

Sarfraz Hussain
ECE
REVA University
Bengaluru, Karnataka, India

Rashi Jain
Department of Physics
School of Advanced Engineering
UPES, Dehradun, India

Prabhjot Kaur
School of Computing
DIT University Dehradun
UK, India

Rakesh Kumar
Uttaranchal University
Dehradun, India

Sanjeev Kumar
Atmospheric and Space Physics Laboratory
Department of Physics
Hemvati Nandan Bahuguna Garhwal University (A Central University)
Srinagar Garhwal, Uttarakhand, India

Saurabh Kumar
University of Hertfordshire
Hatfield, UK

Sagar Lachure
Department of Computer Science and Engineering
YCCE
Nagpur, India

Ankit Mahule
Department of Computer Science and Engineering (Cyber Security)
Shri Ramdeobaba College of Engineering and Management (RCOEM)
Nagpur, India

Rahul Malik
Amity Institute of Environmental Sciences (AIES)
Amity University
Uttar Pradesh, India

Bisma Nadeem
Amity Institute of Environmental Sciences (AIES)
Amity University
Uttar Pradesh, India

Yasti Panchbhaiya
Atmospheric and Space Physics Laboratory
Department of Physics
Hemvati Nandan Bahuguna Garhwal University (A Central University)
Srinagar Garhwal, Uttarakhand, India

Subrat Kumar Panda
Department of Atmospheric Science
School of Earth Sciences
Central University of Rajasthan
Kishangarh, Ajmer, India

Kanhu Charan Pattnayak
Leeds University Business School
University of Leeds
United Kingdom

Kumargaurao D. Punase
Solar Industries Private Limited
Nagpur, India

Prem Rajak
Department of Animal Science
Kazi Nazrul University
Asansol, West Bengal, India
0000-0002-1693-8090

Kaushik Roy
Department of Computer Science and Engineering (Cyber Security)
Shri Ramdeobaba College of Engineering and Management (RCOEM)
Nagpur, India

Anita Sagwan
NIMS
Jaipur, Rajsthan, India

Ayan Sar
School of Computer Science
UPES
Dehradun, Uttarakhand, India

Abhijit Sarkar
Department of Botany
Kaliachak College
Malda, West Bengal, India

Ankush D. Sawarkar
Department of Information Technology
Shri Guru Gobind Singhji Institute of Engineering and Technology
Nanded, India

Nzeyimana Bahati Shabani
Department of Environmental Science
Bishop Heber College
Trichy, India

Abhinav Sharma
Department of Electrical and Electronics Engineering
University of Petroleum and Energy Studies
Dehradun, Uttarakhand, India

Ayushi Sharma
Uttaranchal University
Dehradun, India

Puneet Sharma
School of Interdisciplinary Research
Indian Institute of Technology Delhi
New Delhi, India

Asha Singh
Department of Environmental Sciences
Maharshi Dayanand University
Rohtak, Haryana, India

Karan Singh
Atmospheric and Space Physics Laboratory
Department of Physics
Hemvati Nandan Bahuguna Garhwal University (A Central University)
Srinagar Garhwal, Uttarakhand, India

Kush Singla
Grey B
Mohali, India

Love Singla
Department of Microbiology
Maharaja Agrasen University
Baddi, India

Siddharth Swami
Uttaranchal University
Dehradun, India

Pushp Raj Tiwari
Centre for Atmospheric and Climate Physics Research
University of Hertfordshire
United Kingdom and Centre for Climate Change Research
University of Hertfordshire
United Kingdom

Abhishek Upadhyay
Laboratory of Atmospheric Chemistry
Paul Scherrer Institute
Villigen, Switzerland

Aman Deep Vishwkarma
Atmospheric and Space Physics Laboratory
Department of Physics
Hemvati Nandan Bahuguna Garhwal University (A Central University)
Srinagar Garhwal, Uttarakhand, India

Abhilasha Yadav
Centre of Excellence for Energy and Environmental Studies
Deenbandhu Chhotu Ram University of Science and Technology
Murthal Sonipat, Haryana, India

Sashi Yadav
University Institute of Engineering and Technology
Maharshi Dayanand University
Haryana, India

About the Editors

Amit Awasthi is Associate Professor at the University of Petroleum and Energy Studies, Dehradun, India. He completed his PhD at Thapar University, Patiala, in 2011, and has over 18 years of professional experience in the areas of research and development, teaching, and allied functions with reputed educational institutions and universities, offering research experience in atmosphere and environmental sciences including air monitoring, exposure studies, extreme events, water quality, climate change. He has the merit of receiving International Fellowship from National Science Council-Taiwan as Postdoc Fellow. He has published 4 textbooks and more than 60 research papers in reputed journals of atmospheric/environment sciences/climate change with an h-index 17 and citation ~ 1300 and impact factor ~200. He has received several research grants from national and international organizations in the areas of aerosol and water quality measurements. He has edited several books with reputed publishers such as Taylor & Francis, Elsevier, Springer.

Kanhu Charan Pattnayak is Senior Climate Impact Scientist at the National Environmental Agency, Singapore, with over 17 years of experience in climate impact research and climate modeling. He has a Ph.D. in Climate Science from the Indian Institute of Technology Delhi and has held research positions at the University of Leeds, the National Environment Agency of Singapore, and the NCMRWF. He has published over 25 research papers in top-tier journals and has served as a reviewer for the Sixth Assessment Report of the United Nations Intergovernmental Panel on Climate Change.

Gaurav Dhiman is Assistant Professor in the School of Sciences and Emerging Technologies at Jagat Guru Nanak Dev Punjab State Open University, Patiala, India. He holds a Ph.D. in Computer Engineering from Thapar Institute of Engineering & Technology, Patiala. He is recognized as one of the world's top researchers in Stanford University list of world's top 2% of scientists, prepared by Elsevier. He is a Senior Member at IEEE; Research Faculty at Lebanese American University, Lebanon; and Research Scientist at Universidad Internacional Iberoamericana, Mexico. He has authored over 300 peer-reviewed research papers, and 10 books. He is currently serving as a guest editor for more than 40 special issues in various peer-reviewed journals. He is an Editor in Chief of the *International Journal of Modern Research* (IJMORE), and Co-Editor-in-Chief of the *International Journal of Electronics and Communications Systems* (IJECS) and *International Journal of Ubiquitous Technology and Management* (IJUTM). He is an Associate Editor of *IEEE Transactions on Industrial Informatics*, *IET Software* (Wiley), *Expert Systems* (Wiley), *IEEE Systems, Man, and Cybernetics Magazine*, *Spatial Information Research* (Springer), and more.

Pushp Raj Tiwari is a climate scientist and Fellow of UK's Higher Education Academy (FHEA), and specializes in climate change, big data, and earth system modelling. A former RCUK-ECR Fellow, he now leads a research group on Climate Change Modelling and Applications. His work focuses on aerosol-cloud–climate interaction and reducing the associated uncertainties related to them in climate models.

1 Air Quality Monitoring (AQM) and Prediction

Transitioning from Conventional to AI Techniques

Amit Awasthi, Kanhu Charan Pattnayak, Pushp Raj Tiwari, Subrat Kumar Panda, Sneha Gautam, Tanupriya Choudhury, and Ayan Sar

1.1 INTRODUCTION

Air quality is deteriorating due to many natural and anthropogenic reasons. Industrial emissions, vehicle emissions, deforestation, open agriculture residue burning (Agarwal et al, 2014; Singh et al., 2010), construction activities, etc. are some of the common anthropogenic activities that are no doubt deteriorating air quality, but these are somewhat essential due to modernization. The impact of this decline in air quality is significant and far-reaching. It has been linked to an increase in different types of health problems like respiratory problems, heart disease, cancer, cardiovascular problems, and even neurological disorders. The direct relationship of public health with air quality is one of the key motivations and reasons for researchers and academicians to research its accurate measurement. Susceptible populations, including children, the elderly, and individuals with pre-existing health conditions, are particularly at more risk due to poor air quality (Maung et al., 2022). Consequently, the need for accurate and real-time air quality data has never been more urgent, as it serves as a foundation for informed decision-making aimed at safeguarding public health.

Along with health outcomes, air quality has a significant effect on the ecosystems and overall ecological balance. Elevated levels of air pollutants, such as nitrogen oxides (NOx), Volatile Organic Compounds (VOCs), and Particulate Matter (PM), etc. cause several environmental concerns like climate change, acid rain formation, ozone depletion, agricultural impacts, etc. Many air pollutants not only contribute to local health issues but also play a role in climate change by influencing the radiative balance of the Earth's atmosphere. The shifts in the radiative balance of the Earth's atmosphere due to the increase in the concentration of air pollutants are significantly responsible for climate change (Kabir et al., 2023).

DOI: 10.1201/9781032683805-1

In developing countries, these issues are worse, because developing countries face challenges in maintaining environmental hygiene as they focus on developing industries and infrastructure, which may lead to accidentally or purposely violating necessary actions. To address these types of issues, globally governments have executed several monitoring programs to trace the concentration of pollutants in the air. These monitoring programs use a variety of technologies to collect data and forecast future air quality levels, a variety of technologies such as ground-based sensors, satellite imaging, and air quality models.

Hence it is important to measure the quality of air we breathe and identify the level of deterioration so that we understand the air quality situation based on the different quantitative and qualitative analysis (Awasthi et al., 2017). For a long time, we have been using different traditional methods, but new technologies have been developed which enhance the available methods of measurement. For any investigation or designing any strategies, we must have proper channels or options for the prediction and monitoring of air quality. Moreover, deterioration of air quality is one of the important environmental factors that affect the health of populations globally. Accurate air quality predictions are a needed at present so that communities and individuals can make plans to protect themselves from the harmful effects of air pollution. Artificial intelligence (AI) has appeared as a crucial technology with more accuracy and efficiency for the prediction of air quality. There are some challenges like an absence of normalization in Air Quality Monitoring (AQM) systems, due to which comparison of data and maintaining a global understanding of air quality trends are difficult. The use of AI in air quality prediction is a promising development that can help us understand and address air pollution (Essamlali et al., 2024). Ongoing research and development in this area will lead to more accurate predictions, benefiting people worldwide. So in this chapter effort will be made to discuss the different perspectives of air quality measurement and prediction by conventional and AI techniques. In this introductory chapter, after the brief introduction, the fundamentals of conventional AQM are discussed. Subsequently, a discussion on the transition to AI technologies will be presented followed by a dialogue on the different types of AI techniques used for AQM and prediction. The current state-of-the-art in air quality measurement utilizing Machine Learning (ML) and Deep Learning (DL) techniques will be elucidated from the year 2022. Before the chapter's conclusion or summary, challenges and prospective directions for the application of AI techniques in AQM and predictions will also be addressed.

1.2 FUNDAMENTAL OF CONVENTIONAL AQM TECHNIQUES

As discussed, AQM plays an important role in assessing and managing the environmental impact of pollutants on human health and ecosystems. Conventional AQM techniques serve as the foundation for understanding the concentrations and sources of various pollutants in the atmosphere. In this chapter, techniques that do not employ AI are regarded as traditional or conventional approaches to AQM. Conventional AQM employs diverse methods to gauge concentrations of pollutants in the air. Determining the sampling location is crucial in choosing the appropriate instrument for measuring air pollutant concentrations or meteorological parameters. Once the sampling location and target pollutant are selected, various instruments and techniques, as outlined in Table 1.1, become available for use. Dispersion and transformation of pollutants in the atmosphere vary because of different metrological parameters like temperature, humidity, wind speed, direction, etc., hence meteorological conditions are important to understand for accurate interpretation of air quality data. Different metrological parameters along with the different instruments used to measure it are tabulated in Table 1.1. Instruments like anemometers, thermometers, and barometers are employed to measure wind speed, temperature, and atmospheric pressure. Meteorological data assists in identifying factors influencing air pollutant dispersion.

If the pollutants are solid in nature then the gravimetric method has been used for a long time in which a different type of filter media is used to collect the PM of different sizes. In, the gravimetric method proper desiccation of filter papers is required before and after the sampling to avoid the different types of over and underestimation errors or artifacts due to different metrological factors especially relative humidity, temperature, etc. (Liu et al., 2014). Beta attenuation methods and sensors are also used to measure the PM of different sizes to reduce the time and cost of measurement or where we need instant preliminary information. Instruments like Beta Attenuation Monitors (BAM), High-Volume Samplers (HVS), Respirable Dust Sampler (RDS), and Cascade Impactors, etc. capture particulates for subsequent analysis, providing insights into PM of different size (Idrees & Zhang, 2020). To measure the different types of gases like SO_X, NO_X, CO, O_3, Volatile Organic Carbon (VOC), etc. methods like Chemiluminescence, Ultraviolet Absorption, Infrared (IR) Absorption, Chromatography, etc. are used (Rahman, 2023). Gas analyzers are fundamental instruments for measuring concentrations of various gaseous pollutants. Different sensors are also used to measure the different pollutant concentrations

TABLE 1.1
Common Principles Techniques and Applications Used in Conventional AQM

Parameters monitored	Principle of measurement techniques	Applications
Particulate matter (PM_{10}, $PM_{2.5}$ & PM_1)	Gravimetric, beta attenuation, optical scattering, sensor	Dust and particulate matter concentration at construction sites, industrial areas, urban air quality.
Gases (SO_X, NO_X, CO, O_3, VOC, etc.)	Chemiluminescence, ultraviolet absorption, infrared (IR) absorption, Non-dispersive Infrared (NDIR), chromatography, sensor	Industrial emissions, vehicular exhaust, ambient air quality.
Meteorological parameters (wind speed, wind direction, temperature, humidity, etc.)	Sensors	Weather forecasting station, air quality dispersion modeling, environmental studies.

in the atmosphere (Wang et al., 2023). These analyzers employ techniques such as Non-dispersive Infrared (NDIR) absorption and chemiluminescence to quantify gas concentrations accurately as shown in Table 1.1.

Along with the gas monitor analysers, particulate monitors, etc. discussed in Table 1.1, Continuous Emission Monitoring Systems (CEMS) are also used to record and measure pollutant emissions directly from exhaust or stack sources of industrial plants. Passive samplers are also used to absorb pollutants for a specific period which will be analysed later on for the measurement of the concentration of different types of pollutants. Automated AQM stations are installed at fixed locations to constantly monitor multiple pollutants and meteorological parameters, providing real-time data using various sensors.

The remote-sensing techniques like Light Detection and Ranging (LIDAR) use laser light to measure the movement and concentrations and movement of pollutants and PM in the atmosphere. Satellite observations are another example based on remote sensing techniques that provide comprehensive statistics on air quality, especially for pollutants like trace gases and aerosols. These conventional methods are foundational in providing valuable data, but one of the main drawbacks of conventional air monitoring methods is that they can be time-consuming and expensive. All these methods are often demanded for their usage in various expensive and bulky equipment, along with trained personnel in order to operate them. As a result, the air monitoring using all the bulky methods could also be limited to certain types of areas or other periods, which might not be able to provide an overall and comprehensive picture of overall air quality. Another major concern with the other conventional air monitoring methods was that they are often measured along with some of the few pollutants including particulate matter, ozone nitrogen dioxide, etc. In order to address all these constraints, many scientists and researchers are interested in the exploration of more advanced air monitoring methods, which would lead to the emergence and application of AI for improved AQM, along with its prediction.

1.3 TRANSITION TO AI TECHNIQUES

Many traditional and conventional methods related to the measurement of air quality, which consists of weighing particles or measuring gases, were being replaced or updated with some of the available state-of-the-art smart methods, which employ the potential of AI. Traditional methods of inspecting air quality, such as weighing particles or measuring gases, are being replaced by new, smart methods that utilize AI. This change is analogous to upgrading to a high-tech electric bike from an old bicycle, as it allows for faster and more economical AQM. This does not mean that the old bike is not fine, but the new one is smarter, and faster, and enables you to get where you want to go more efficiently. Similarly, with the use of the power of AI, measurement, and understanding of the air we breathe in a much better and quicker way. The introduction of AI does not imply any evidence suggesting that conventional methods are less accurate and reliable.

Figure 1 shows that traditional AQM techniques have limitations in spatial and temporal resolution, hindering the identification of localized pollution hotspots and

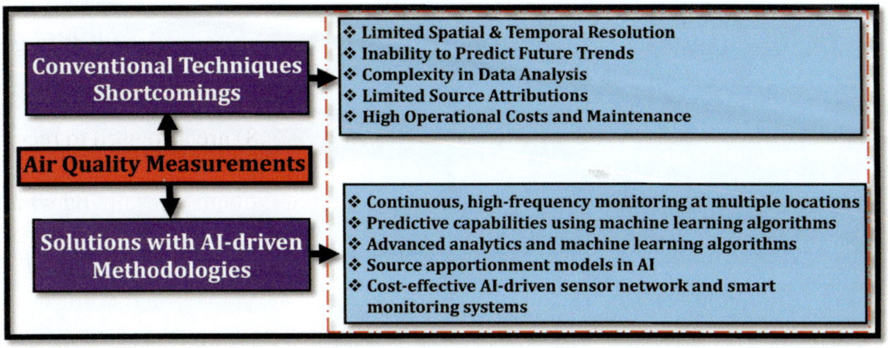

FIGURE 1.1 Bridging gaps: limitations in traditional air quality methods and the AI-driven methods.

proactive measures. Conventional techniques generate large datasets that are difficult to analyze comprehensively and struggle with accurate source attribution. The integration of AI-driven methodologies offers continuous monitoring, predictive analytics, advanced data analysis, accurate source attribution, and cost-effective sensor networks, providing transformative solutions for efficient air quality management (Figure 1.1). There are several review articles available in which the basics of AI and its application in the measurements of air quality parameters are discussed (Kang et al., 2018; Kaginalkar et al., 2021; Li et al., 2022; Masood & Ahmad, 2021). In this chapter, a concise overview of AI and its different key techniques are discussed for basic understanding. For in-depth information, readers are encouraged to consult different reviews and research articles available in the literature (Subbiah, & Paramasivan, 2024), or refer to the references covered in this chapter.

1.4 AI APPLICATION IN AIR QUALITY PREDICTION

Conventional methods have long been used to perform various tasks. However, these methods have limitations that have led to the adoption of AI for the measurement and prediction of the concentration of air pollutants. AI can be trained to recognize patterns and make predictions based on historical observations, making it more effective in controlling complex tasks. AI can identify patterns and correlations in data that may not be visible to humans, making it a valuable tool in decision-making. Basic AI techniques are shown in Figure 1.2 which are used to perform complex tasks and make informed decisions.

ML is one of the most basic used AI techniques in AQM. Algorithms and statistical models are used in this technique to examine a large amount of data and classify the different patterns. This technique is used to examine the data collected from sensors installed in various locations for air quality measurements. The data collected includes information about the concentration of pollutants in the air and meteorological parameters such as temperature, humidity, etc. As per the information on Science Direct, by using the keywords "machine learning and air pollution"

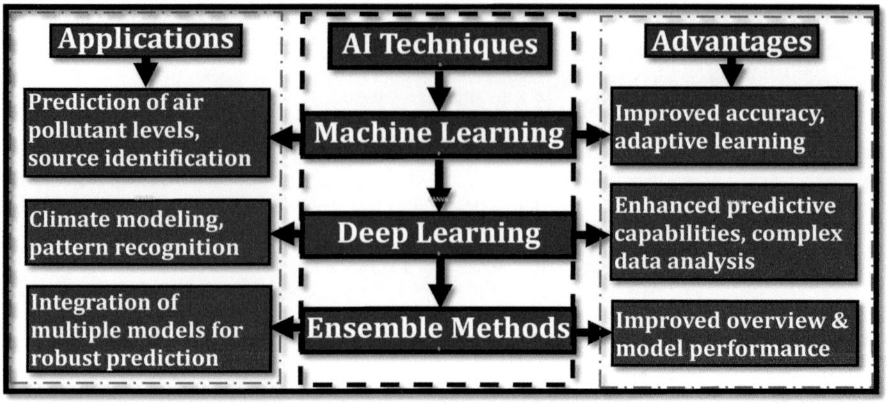

FIGURE 1.2 Basic AI techniques in Air Quality Monitoring (AQM).

in the advanced search of "title", the first publication is observed in 2017 (Wang et al., 2017), and now there are 37 publications. For information related to air quality and ML, a review article was recently published in 2024 in which important issues related to ML are discussed in detail (Tang et al., 2024). Various studies have shown the use of ML approaches in the field of air quality and meteorological parameters (Tandon et al., 2023). ML approaches have been increasingly utilized in the realm of air quality research, as evidenced by a multitude of studies. In Table 1.2, studies related to the prediction of the air quality parameters with the help of ML from 2022 to date are only presented. In 2024 Wang et al. introduced a promising and optimized technique with the integration of PALM along with CNN models, thus aiming for the enhancement of the accuracy and performance of the prediction of urban air quality. The fusion and integration of such techniques capitalized on PALM's fine-grained spatial resolution and CNNs in connection with the extraction of intricate features from different complex datasets, thus providing a more robust framework for the forecasting of air pollutant concentrations in various urban environments. Moreover, Wu et al. (2024) demonstrated the real-time feasibility of employing ML to address the data gaps in various networks which monitor air pollution. This would, in turn, revolutionize the monitoring of air quality along with offering timely and comprehensive insights into different pollutant levels.

Miao et al. (2024) came up with the idea by shedding light on the real influence of different types of roads on the concentrations of air pollutants in different urban areas, thus paving the way for more targeted strategies in mitigation. Anggraini et al. in 2024 also showcased the real superiority of global land area where multiple models were proved to be better than single models, which in turn emphasized the fact that the consideration of diverse factors such as socio-economic and environmental data play a major role in forecasting. Specific models in the field of ML have also shown promising results, with Matthaios et al. (2024) reporting improvements in the prediction of pollutant concentrations, particularly for NO_2. In 2023 Livingston et al. also

TABLE 1.2
Studies Related to Air Quality Measurement by Using ML from 2022 to 2024

SN	References	Key variables measured	Study area	Key points
1	Wang et al., 2024	CO concentration	Nanjing, China	Integrating PALM with CNN models, enhances accuracy and performance in predicting urban air quality.
2	Wu et al., 2024	O_3, NO_2, $PM_{2.5}$, PM_{10} and SO_2	Pearl River Delta region in South China	Study showed that ML is feasible for filling data gaps in air pollution monitoring networks.
3	Miao et al., 2024	PM, SO_2, NO_2, CO, and O_3	Shenyang, China,	Study addresses the past research limitations and offering an understanding of how road types influence air pollutant concentrations in urban areas.
4	Anggraini et al., 2024	AQI, CO, NO_2, SO_2, $PM_{2.5}$, and PM_{10}	Global land area	Multiple models are more accurate than the single models, showing that the addition of socioeconomic and environmental data can enhance accuracy.
5	Matthaios et al., 2024	PM_{10}, $PM_{2.5}$, PM_1, NO_2, NOx, aerosol lung surface deposited area and ultrafine particles	Birmingham, UK	ML model predictions matched with the performance of the first model MB for most species and verified improvement for NO_2.
6	Livingston et al., 2023	NO_2, CO, O_3, $PM_{2.5}$, PM_{10}, SO_2, Temperature, Pressure, Rain, Dewpoint, Wind direction,	Beijing, China	ML approach AirStackNet is recommended to improve accuracy which has significant application in commercial and industry settings with minimal hardware.
7	Ravindiran et al., 2023	AQI. PM, gaseous pollutants, and meteorological parameters	Tamil Nadu, India	XGBoost model surpassed other models, achieving a notable R^2 value of 0.9935.

(continued)

TABLE 1.2 (Continued)
Studies Related to Air Quality Measurement by Using ML from 2022 to 2024

SN	References	Key variables measured	Study area	Key points
8	Zhang et al., 2023	$PM_{2.5}$, CO_2	China	ML framework provide accurate estimations of annual average $PM_{2.5}$ directly from a high-resolution dataset on fossil energy use.
9	Liao et al., 2023	Meteorological, spatial and temporal factors on Air Quality Prediction.	Shandong Province, Yangtze River, China	Observed that mechanism model and ML show the appealing performance over competitive baselines.
10	Ravindiran et al., 2023	AQI with 12 contaminants and 10 meteorological parameters	Visakhapatnam, Andhra Pradesh, India	Used Catboost model show low RMSE (0.76) and high prediction accuracy (0.9998).
11	Calatayud et al., 2023	NO_2, PM, O_3	Valencia, Spain, Europe	Based on ML model, NO_2 concentration increases and rise in O3 concentration is observed.
12	Maltare & Vahora, 2023	AQI, PM_{10}, $PM_{2.5}$, NO_2, SO_2, CO, O_3, NH_3, Pb, Ni, As, Benzo(a)pyrene, and Benzene.	7 AQM station of Ahmedabad city, Gujarat, India	Various ML model like SARIMA, SVM and LSTM are compared and SVM with RBF kernel outperformed in comparison to other model.
13	Wang et al., 2023	Health parameters and exposure to heat and air quality	Australia	Based on ML, V-shape relationship is observed temperature and health and air quality.
14	Mehmood et al., 2022	Review on the ML approaches	Studies of any area in between 1992 to 2021	ML approaches are used extensively, as a lot of publications have been observed in the last 5 years.
15	Lv et al., 2023	$PM_{2.5}$, SO_2 during COVID-19	40 cities of North China	The algorithm in ML and the Theil-Sen regression technique were used, and it was also observed that during the lockdown, air pollutants were overestimated twofold with the raw observation data.

TABLE 1.2 (Continued)
Studies Related to Air Quality Measurement by Using ML from 2022 to 2024

SN	References	Key variables measured	Study area	Key points
16	Majdi et al., 2022	Indoor air quality, temperature, air humidity, and carbon dioxide.	Kian Centre, Iran	Proposed models check the pattern of temperature, humidity, and CO_2 data on air quality, with a predictions error of 3%, considering the importance of smart buildings and provide better solutions for smart management
17	Fabregat et al., 2022	Pollutant levels, meteorology and transport intensity	Barcelona, Spain	Prediction based on ML, a reduction of CO and NOx is observed whereas O_3, PM_{10} and SO_2 are found to be unaffected.
18	Munir et al., 2022	Traffic pollutant like NO_2, NO, O_3, PM_1, $PM_{2.5}$ and PM_{10}.	Thatcham, West Berkshire, UK	Results shows that observed concentration is higher than the average predicted concentrations.
19	Li et al. 2022	AQI	Shanghai, Beijing and Xi'an, China	New hybrid prediction model with the optimization of VMD & KELM parameters is proposed.
20	Dai et al., 2022	$PM_{2.5}$, PM_{10}, O_3, SO_2, NO_2, and CO	Fenwei Plain, China	During 2017–2020 air quality is improved in the Fenwei support by the residential and industrial sector
21	Guo et al., 2022	$PM_{2.5}$, O_3, NO_2, meteorological paramters	12 China megacities	During clean air action periods, except for O_3, Beijing's meteorological conditions positively amplified reductions in major pollutants.
22	Gladkova, & Saychenko, 2022	$PM_{2.5}$, PM_{10}, O_3, meteorological data	7 cities of Russia	Insufficient data is presented in the public domain for Russian cities that hinders the further research.

recommended the use of a specific ML approach known as AirStackNet in order to improve accuracy, especially in the field of commercial and industrial settings.

Additionally, in 2023 Ravindran et al. also highlighted the relative effectiveness of the XGBoost model (Ravindran et al., 2023), while on the other hand, in 2023 Zhang et al. demonstrated the main accuracy of frameworks which fall under ML in the estimation of $PM_{2.5}$ concentrations directly from the datasets which have high-resolution data (Zhang et al., 2023). Calatayud et al. in 2023 provided many valuable insights into the relationship which falls between ML-predicted NO_2 concentrations and its rising O_3 concentrations, which were very crucial in informing the policymakers and urban planners. In 2023, Maltare and Vahora conducted a study of the identification of SVM along with RBG kernel, which proved to be a top performer among the various ML models, ultimately offering a totally robust methodology for the prediction of air quality. Wang et al. (2023) also further emphasized the multi-dimensional nature of the quality of air dynamics through any type of modelling through ML. Despite all these advancements, many challenges persisted, as highlighted by Gladkova and Saychenko (2022); they also pointed out the lack of sufficient data on different Russian cities (Gladkova and Saychenko, 2022). Addressing these challenges required greater transparency in data and their accessibility for further propelling in research in their field.

DL is also another notable AQM power that falls under AI. Among these techniques, Artificial Neural Network (ANN) is one of the most widely used which is in about human thinking that simulates to analyze and comprehend data. In local air quality monitors, neural networks are employed for forecasting the concentration of pollutants in the air using historic data. Our monitor can contribute to the identification of pollution focus areas and actions needed to tackle their implications before they even become a problem. As per the information on Science Direct, by using the keywords "deep learning and air pollution" in the advanced search of "title," first publications were observed in 2019 (Heo et al., 2019; Ma et al., 2019; Zhou et al., 2019), and there are total number of 27 publications till the date of writing this chapter in January 2024. For a comprehensive approach to how DL is used to estimate PM, refer to the recently article by Zhou et al. (2024); the article is especially informative on the core parameters often used by DL in making such predictions.

Recent studies concerning air quality parameters based on DL techniques have been portrayed in Table 1.3 from the last year 2022 until now. The DL technique in modeling has recently become acknowledged as a valid research tool for air quality studies in a global context. According to Yadav et al. (2024), the trained model presented an explanation of up to 54% of air quality variation, hence it is only understandable to be amazed by the accuracy of the predictive capability of the model. Prado-Rujas et al. (2024) presented an efficient AI-based forecasting model Spatio-Temporal Air Quality Forecaster (ST-AQF), which is capable of managing multiple pollutants simultaneously with as good as fine spatio-temporal resolutions that help in forecasting comprehensive air quality scenarios. In their recent work, Li et al. (2023) have identified the linked impact of factor pairs on air quality dynamics, demonstrating an attribute that is nonlinear and is characterized by initial declines and then by increases once inflection points are reached. Guo et al. (2023) developed a novel model with outstanding exploration accuracy of the spatial distribution and the

TABLE 1.3
Studies Related to Air Quality Measurement by Using DL from 2022 to 2024

S.N	References	Key Variables measured	Study area	Key Points
1	Yadav et al., 2024	Urban Air quality (AQ)	Accra in Ghana, Africa, Los Angeles and New York City	Trained models explain up to 54% of the variation in the AQ distribution
2	Prado-Rujas et al., 2024	Concentrations of eleven pollutants	Madrid (Spain)	Spatio-Temporal Air Quality Forecaster (ST-AQF) is capable of simultaneously handling multiple pollutants at once with fine spatio-temporal resolutions.
3	Ahmed et al., 2024	Air quality index	Seven sites of Bangladesh	The CLSTM-BiGRU model, which combines a Convolutional Neural Network (CNN), a long-short term memory (LSTM), and bi-directional gated recurrent unit (BiGRU) network, outperforms benchmark models in air quality forecasting.
4	Li et al., 2023	$PM_{2.5}$ and O_3 and other factors on the AQI	China	The combined impact of factor pairs decreases initially, then increases after reaching inflection points.
5	Guo et al., 2023	fine-granular $PM_{2.5}$ concentrations	417 micro monitoring stations in Lanzhou City, China,	Method accurately predicts the spatial distribution and fluctuations of $PM_{2.5}$ with low RMSE for 1 or 6-h forecasts.
6	Jamei et al., 2023	$PM_{2.5}$ and PM_{10} forecasting	Queensland, Australia.	The proposed two-stage filtering framework is favourable for AQM.
7	Shin et al., 2023	Indoor air quality	Dataset of 2132 different cases of location.	Even in untrained condition, FCNR-H0.5 showed strong prediction performance.

(*continued*)

TABLE 1.3 (Continued)
Studies Related to Air Quality Measurement by Using DL from 2022 to 2024

S.N	References	Key Variables measured	Study area	Key Points
8	Wu et al., 2023	NO_2 and O_3	51 AQM stations in Shanghai	Incorporating topological information from the monitoring network has the potential to enhance model performance.
9	Sarkar et al., 2022	$PM_{2.5}$	Delhi, India	The LSTM-GRU model outperforms other existing approaches in terms of MAE and R2 metrics
10	Du et al., 2023	Surface-level $PM_{2.5}$ mainly focus on Traffic pollution	Shanghai, China	Proposed deep learning model "iDeepair" provide hight-efficient and accurate traffic management and pollution control
11	Iskandaryan et al., 2023	Meteorological and traffic data along with NO_2	Madrid, Spain	The proposed method with preceding steps can be applied not only to air quality forecasting but also to any data with the purpose of predictive analysis.
12	Das et al., 2022	PM_{10} and SO_2	Basaksehir district, Istanbul	Used Long Short-Term Memory Networks (LSTM) performed well in comparison to similar studies
13	Jamei et al., 2022	$PM_{2.5}$ and PM_{10}	Miles Airport, Queensland, Australia	Long short-term memory (LSTM) performed well in comparison to similar studies.
14	Kanmani et al., 2022	Air quality management, Cryptogams	South zone of India	VGG-16 architecture generates the training and validation accuracy of 92.6% and 93.1%
15	Hu et al., 2022	$PM_{2.5}$ and PM_{10}	22 stations in Beijing City, China	The proposed model shows superiority over the baseline models.
16	Yang et al., 2022	Indoor air quality	Subway station Seol, South Korea	Based on the mean square error and mean absolute error, the proposed model is improved in comparison to the conventional neural network and LSTM, respectively.

TABLE 1.3 (Continued)
Studies Related to Air Quality Measurement by Using DL from 2022 to 2024

S.N	References	Key Variables measured	Study area	Key Points
17	Midday, & Roy, 2022	$PM_{2.5}$, PM_{10}, SO_2, NO_2, NO, and O_3	Kolkata, India	Based on the MAE and RMS, Holt-Winters, Bi-LSTM, and ConvLSTM achieve superior prediction precision.
18	Wang et al., 2022	$PM_{2.5}$ and O_3	South-central Beijing-Tianjin-Hebei	The APTR model excels in proximity forecasting (24-h), while the DeepPM model demonstrates superior performance in optimizing short- and medium-term forecasts (144-h).
19	Wood, 2022.	Air quality index based on six key pollutants	Dallas County (USA)	Distinctive prediction results are observed for the annual and quarterly timeframes in 2019 and 2020.
20	Kow et al., 2022	$PM_{2.5}$, PM_{10}, photos, and the air quality index	Kaohsiung, Taiwan	Solution for rapid and reliable multi-pollutant estimation based on captured images.

turbulence of $PM_{2.5}$ in low value of Root Mean Square Error (RMSE) for both 1-hour and 6-hour forecasts, respectively. Among a range of options Jamei et al. (2023) have suggested a two-step approach for checking data quality and reliability which may be providing the more perfect solution in AQM. In related work, Shin et al. (2023) achieved high efficiency and resilience with the Few-Shot Concept Neural Rerank-H0.5, showing its effectiveness even in cases where it is not trained for. In their study, Wu et al. (2023) examined the use of spatial connectivity from networks of monitors in modelling systems of this kind with an emphasis on enhancing predictability using topological information. This points to the promising effects of considering connectivity on modelled spatial systems for better methodologies of forecasting and thus predicting. Sarkar et al. (2022) reported that the LSTM-GRU model outperformed existing approaches in terms of Mean Absolute Error (MAE) and R-squared (R2) metrics, highlighting its superior predictive accuracy. The former means that it predicts future behavior more accurately. Additionally, Du et al. (2023) introduced the “iDeepair” deep learning model for efficient and accurate traffic management and pollution control, demonstrating its practical potential. Iskandaryan et al. (2023) proposed a versatile method applicable not only to air quality forecasting but also to various predictive analysis tasks, emphasizing its broad utility.

In addition, LSTM was found to outperform similar studies by Das et al. (2022) in carrying out air pollution prediction based on past atmospheric observations and greenhouse gas emissions. Wood (2022) showed that in Dallas County (USA) the air quality index consisting of six key pollutants (the index is annually specific for one quarter) was the best method to predict air cleanliness with differing results but pronounced for both time lengths. Therefore, they proved that prediction of air quality for different times increases significantly with different granularity used. First and foremost, authors such as Kow, et al (2022) devised another novel perspective of air quality evaluation, which is timely and dependable in bringing out the intensity of multiple pollutants, through the photos we are taking.

The next key method is that of ensemble, which requires the use of several ML or DL models to increase the accuracy of prediction. In simple terms, an approach known as the ensemble method can combine different ML algorithms. The training of the algorithms has been done through different sections related to the quality of air. In faking this blending, the ensemble combination method can deliver predictions that are more accurate compared to those made by one model.

Likewise, in addition to AI procedures, Natural Language Processing (NLP) is being applied in AQM. NLP is another approach, and its function is an algorithm that enables the interpretation of the spoken language. In the AQM, NLP is employed to study the data from social media platforms like Twitter, Facebook etc. The aim is to assess/rate the level of air quality in every given locality. Lastly, AI in AQM has computer vision, which is a technology that uses pixels to break down objects and assign features to them. This way of using computers can be illustrated as a process in which visual information is interpreted using computer algorithms. The computer vision system employed in AQM can assess air quality from satellite photographs and possibly cameras positioned on the ground to determine the places where the air is contaminated. Thus, the data collected is further used to help implement AQM programs in critical locations.

1.5 COMPARATIVE ANALYSIS

In this section, a thorough comparative assessment of AI-driven approaches and conventional AQM techniques will be done. Strengths and weaknesses are identified and also how to put those factors into a desktop, mobile, or tablet format. This section will cover the air quality assessment topics by devoting appropriate time to them. Significant comparisons will be made between conventional methods and AI techniques by considering influential factors like accuracy, real-time monitoring, cost, adaptability, and scalability as shown in Figure 1.3.

Figure 1.3 visualizes the analytical comparison, from which the readers could easily grasp what specific distinctions can be achieved through the traditional approach and AI-based approach to AQM. Although the existing paradigm does suffer from inadequate accuracy because it relies on calibration, AI methods can act as a substitute capable of achieving a constant enhancement by utilizing ML. Timely reports are extended since well-established ways usually don't offer this, unlike advanced technology means, which surpass their passive and immediate responses. Both are

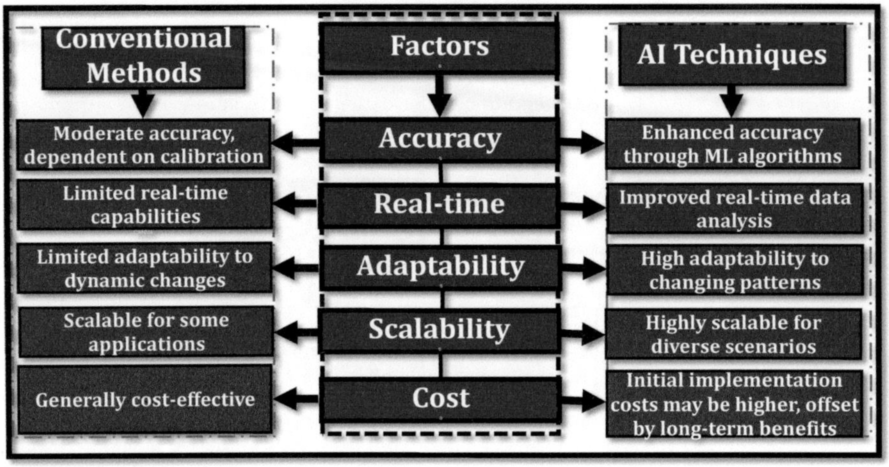

FIGURE 1.3 Comparative analysis of conventional vs. AI techniques.

necessary to take care of pollution issues. Regardless of cost, traditionally priced methods often outcompete the new techniques in the long term. Nonetheless, AI gains are mostly in resilience due to fewer mistakes and enhanced operational efficacy, which would be at the expense of the initial investment but pay off handsomely. AI is adaptive as its systems can easily work with dynamic air quality data as compared to a few conventional methods, which are notoriously monotonous and thus cannot adjust to changing air patterns. Lastly, the scalability of these systems is of the utmost importance because AI may become used in a mixture of monitoring applications. It may, thus, be applicable to the challenges that are usually associated with the rigid tools of traditional approaches as well as being adaptable in the scope. Although AI and conventional approaches have strengths and weaknesses, they can be used effectively in the AQM. Also, the utilization of the best parts of each approach will lead to the obtaining of the most efficient and productive result.

1.6 CHALLENGES OF AIR QUALITY MEASUREMENT BY USING AI

In order to take the most advantage of AI methods for AQ prediction and forecast, awareness of specific difficulties and their effect should be taken into consideration. Then proper planning and action should made to minimize or remove these shortcomings (Figure 1.4). Ensuring data quality and reach is one of the problems when applying an AI model for measuring air quality levels. It is really not an easy task to get 100% flawless and factual datasets, especially in cases when the data cover various geographic resources related to different types of pollutants. The quality and the quantity of data used for training AI algorithms to gain more precision and efficiency are crucial factors. AI models are meant to be well-trained to become an accurate solution to the data being processed. Still, data should be collected and arranged according to the problem so it can achieve perfection. This

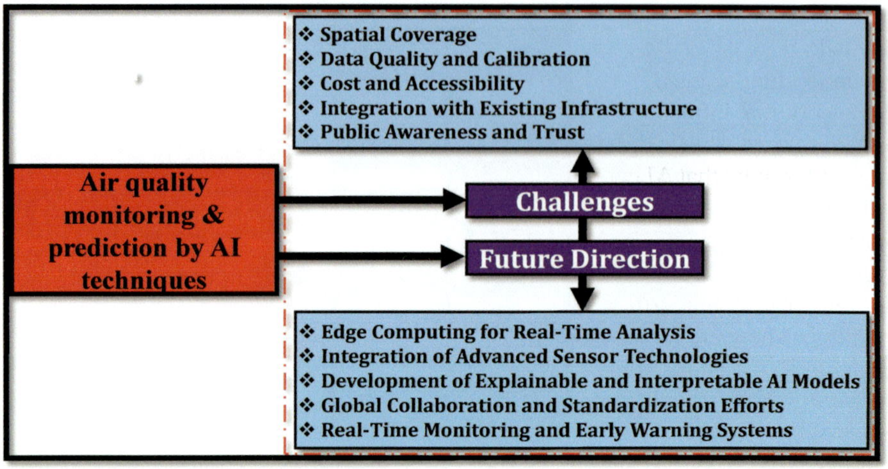

FIGURE 1.4 Challenges and future direction of AQM by using AI.

contains high-quality, relevant, and comprehensive data and is the nucleus of all AI model training. Incomplete or inaccurate data makes it impossible that the results are fully reliable, whereas the data that is of no importance negatively affects the process of problem-solving. In light of that, it becomes fundamental to scrutinize whether the data used for training an AI model is reliable, relevant, and comprehensive. The degree of information is also among the key matters affecting the accuracy and correctness of AI models. The accuracy and reliability of a model will be determined by the amount of data and the diversity of the data it's been trained on. A multitude and broad-based database is very useful in driving the model to a variety of situations, which are characterized as the capability of making accurate predictions and decisions. At the core of the model reliability issue is the amount and quality of data being used to train AI models, which are major factors in this regard. To enable such an AI model that is flexible and fully featured, it must be applied in a dataset in which the data is complete, true, and relevant. Moreover, training sample diversity is essential. The reliability, accuracy, and efficiency of AI in the resolution of complex issues can be achieved through that condition fulfilment, which means that the AI systems should be able to perform accurate, efficient, and reliable resolutions.

Model complexity and interpretability are other challenges for some advanced AI techniques, like DL models that can be extremely complicated and difficult to comprehend (Figure 1.4). Artificial neural networks, which draw inspiration from the structure and functions of the human brain, are used to build these DL models. Although DL models have demonstrated impressive results in several applications, such as voice and image recognition, they are associated with challenges of complexity, which makes it difficult to understand how they arrive at their conclusions. Hence, to explore the potential of AI, we must address its complexity, which makes it challenging to interpret.

Establishing effective collaboration between AI systems and human experts in air quality management is crucial (Figure 1.4). It requires a deep understanding of the underlying science, transparent decision-making processes, and a collaborative framework that influences the strengths of both AI and human expertise. AI systems can provide valuable insights and predictions, and human experts bring a level of intuition and context that AI cannot replicate; hence the right balance between AI systems and human expertise is required. Another important challenge is the cybersecurity threat. Cybersecurity risks are coming into existence when we are dealing with sensitive air quality data. Ensuring robust cybersecurity measures to protect data integrity and privacy is a critical challenge that requires a proactive and comprehensive approach. These are important challenges that need to be addressed for the successful implementation and advancement of AI techniques in AQM and prediction that contribute to improved public health outcomes and environmental management. Hence, it is crucial to focus on these areas as they establish the future scope, as they are vital for advancing the potential of AI in assessing air quality and its prediction.

1.7 FUTURE DIRECTION FOR AQM AND PREDICTION BY AI TECHNIQUES

In the future, there could be a shift towards utilizing AI algorithms directly on edge devices like sensors and monitoring equipment (Figure 1.4). This would enable the analysis of data in real-time without relying too much on centralized processing. By doing so, we can minimize delays in interpreting data and respond faster to changes in air quality conditions. Hence, we can say that real-time analysis with edge computing is the future outcome of AI techniques. In the future, there is a possibility of creating sensor networks that can adjust themselves according to the environment. These networks would have AI algorithms that can constantly optimize the placement of sensors to ensure efficient coverage and accurate data collection. This adaptability would greatly improve the effectiveness of AQM systems.

When the demand for AI decision-making transparency is rising, there is more and more chance that we will need to choose explainable and interpretable AI methods. Once in the future, instead of just air quality prediction models to create the work can be shifted to the ones which will not only predict incoming air pollution but also provide the logical explanations about the factors of these predictions. Such a campaign shall mobilize and activate the will of the user and confidently make decisive decisions. AI-guided clean air management of tomorrow calls for a heavy investment of big corporations and laying down the characteristics criteria across all sectors. Cooperation cannot only facilitate the sharing of data but also contribute to ensuring the consistency of approaches and the standardization of related documentation pertinent to all situations. Standardization is essential to make an internationally applicable methodology for air quality that suits the different territories as well as the changeability of the AI prediction. Current and forthcoming technologies, like real-time monitoring and early warning systems, shape the future of AQM. The environmental pollution and the health of humans will be neutered by this technology through pollution detection that occurs in real time and reporting so that the air quality can be improved and the air quality regulations and policies that are in place can be improved. The future looks bright for air quality and regularly timed routed monitoring and early warning systems are certain tools that will be critical for the provision of clean air.

The utilization of edge computing and real-time processing in intelligent AI will be a highly probable cycle of improvement (Figure 1.4). Explainable AI, adaptive sensor networks, and closer integration with urban planning and policy decision-making will be the most influential developments in the future of AI in air quality management and forecasting. These innovations strive to achieve timely and accurate readings through the use of air quality sensors that are cheaper to deploy and readily available for the general public. AI algorithms are easy to comprehend and optimize sensor placement in changing and mobile environments. Urban planning strategies are developed that can be implemented to effectively combat air pollution problems.

1.8 CONCLUSION

In the chapter where we move from AQM through the conventional techniques to the use of AI for AQM, we will discuss how AQM has changed over time, from traditional techniques to AI advancements. Traditional techniques fundamentally are explained depicting the fact that these methods were largely effective in dealing with air pollution. Although they remained cost-effective and the most available remedies during that time. The outfit of AI methods uproots the previous traditions, as we are in a new era that provides us with a chance of novel possibilities that make the work more accurate, real-time, and doable. Technologies like ML and DL are progressively being used to effectively forecast air quality and pollution sources pin-pointing, which are a plus to tackling pollution. Recognizing the differences between traditional methods and AI techniques, the most inexpensive and preferred ones by the people are facing the dilemma of adopting the newest innovations, and the limitations are caused by the large areas of their job. On the other hand, AI approaches to this issue offer flexibility and the capability of mass production. However, they are data-driven and often even require additional investments in the beginning.

To be able to address the range of problems which have been mentioned in this chapter, the opportunity is now open to make further advances. Reaching more land area can be done by having sensor networks deployed systematically covering bigger areas, and ongoing tech administrations can introduce solutions to the problems of data quality and calibration like humidity and sensitivity. As AI is becoming easier to use, the initial costs for setting it up should be decreased, promoting a balance between the cost-effectiveness of conventional methods and the transformative potential of AI. What we learn through working together or by combining the traditional and AI methods becomes crucial as these two key take-homes. The inter-relationships between these strategies can culminate in highly effective AQM, which can provide the audience with timely decision-making in environmental management and public health. Ahead of the developments are more sensor technologies and algorithm selections, whereas the people's participation in the process improvement is of the utmost importance to achieve a universal and decent perception of the whole issue of AQM.

In summing up all these, the rapid transition from traditional methods to that of AI in AQM was the main need of the present scenario as it was a very important step towards ensuring a healthy environment for the betterment of society. The strengths and weaknesses, along with the pros and cons of both the methodologies and finding the right balanced solutions, every scientist is quite poised in leading the new era of AQM with accuracy and instant widely available measures of air quality. This built the base for a healthier future for every individual, which would further lead to a better and more long-lasting sustainable future for everyone.

REFERENCES

Ahmed, A. A. M., Jui, S. J. J., Sharma, E., Ahmed, M. H., Raj, N., & Bose, A. (2024). An advanced deep learning predictive model for air quality index forecasting with remote satellite-derived hydro-climatological variables. *Science of The Total Environment*, 906, 167234.

Agarwal, R., Awasthi, A., Mital, S. K., Singh, N., & Gupta, P. K. (2014). Statistical model to study the effect of agriculture crop residue burning on healthy subjects. *Mapan*, 29, 57–65.

Anggraini, T. S., Irie, H., Sakti, A. D., & Wikantika, K. (2024). Machine learning-based global air quality index development using remote sensing and ground-based stations. *Environmental Advances*, 15, 100456.

Awasthi, A., Hothi, N., Kaur, P., Singh, N., Chakraborty, M., & Bansal, S. (2017). Elucidative analysis and sequencing of two respiratory health monitoring methods to study the impact of varying atmospheric composition on human health. *Atmospheric Environment*, 171, 32–37.

Calatayud, V., Diéguez, J. J., Agathokleous, E., & Sicard, P. (2023). Machine learning model to predict vehicle electrification impacts on urban air quality and related human health effects. *Environmental Research*, 228, 115835.

Dai, X., Zhang, B., Jiang, X., Liu, L., Fang, D., & Long, Z. (2022). Has the three-year action plan improved the air quality in the Fenwei Plain of China? Assessment based on a machine learning technique. *Atmospheric Environment*, 286, 119204.

Das, B., Dursun, Ö. O., & Toraman, S. (2022). Prediction of air pollutants for air quality using deep learning methods in a metropolitan city. *Urban Climate*, 46, 101291.

Du, W., Chen, L., Wang, H., Shan, Z., Zhou, Z., Li, W., & Wang, Y. (2023). Deciphering urban traffic impacts on air quality by deep learning and emission inventory. *Journal of Environmental Sciences*, 124, 745–757.

Essamlali, I., Nhaila, H., & El Khaili, M. (2024). Supervised machine learning approaches for predicting key pollutants and for the sustainable enhancement of urban air quality: A systematic review. *Sustainability*, 16(3), 976.

Fabregat, A., Vernet, A., Vernet, M., Vázquez, L., & Ferré, J. A. (2022). Using machine learning to estimate the impact of different modes of transport and traffic restriction strategies on urban air quality. *Urban Climate*, 45, 101284.

Gladkova, E., & Saychenko, L. (2022). Applying machine learning techniques in air quality prediction. *Transportation Research Procedia*, 63, 1999–2006.

Guo, R., Zhang, Q., Yu, X., Qi, Y., & Zhao, B. (2023). A deep spatio-temporal learning network for continuous citywide air quality forecast based on dense monitoring data. *Journal of Cleaner Production*, 414, 137568.

Guo, Y., Li, K., Zhao, B., Shen, J., Bloss, W. J., Azzi, M., & Zhang, Y. (2022). Evaluating the real changes of air quality due to clean air actions using a machine learning technique: Results from 12 Chinese mega-cities during 2013–2020. *Chemosphere*, 300, 134608.

Heo, S., Nam, K., Loy-Benitez, J., Li, Q., Lee, S., & Yoo, C. (2019). A deep reinforcement learning-based autonomous ventilation control system for smart indoor air quality management in a subway station. *Energy and Buildings*, 202, 109440.

Hu, K., Guo, X., Gong, X., Wang, X., Liang, J., & Li, D. (2022). Air quality prediction using spatio-temporal deep learning. *Atmospheric Pollution Research*, 13(10), 101543.

Idrees, Z., & Zheng, L. (2020). Low cost air pollution monitoring systems: A review of protocols and enabling technologies. *Journal of Industrial Information Integration*, 17, 100123.

Iskandaryan, D., Ramos, F., & Trilles, S. (2023). A set of deep learning algorithms for air quality prediction applications. *Software Impacts*, 17, 100562.

Jamei, M., Ali, M., Jun, C., Bateni, S. M., Karbasi, M., Farooque, A. A., & Yaseen, Z. M. (2023). Multi-step ahead hourly forecasting of air quality indices in Australia: Application of an optimal time-varying decomposition-based ensemble deep learning algorithm. *Atmospheric Pollution Research*, 14(6), 101752.

Jamei, M., Ali, M., Malik, A., Karbasi, M., Sharma, E., & Yaseen, Z. M. (2022). Air quality monitoring based on chemical and meteorological drivers: Application of a novel data filtering-based hybridized deep learning model. *Journal of Cleaner Production*, 374, 134011.

Kabir, M., Habiba, U., Iqbal, M. Z., Shafiq, M., Farooqi, Z. R., Shah, A., & Khan, W. (2023). Impacts of anthropogenic activities & climate change resulting from increasing concentration of carbon dioxide on environment in 21st Century; A Critical Review. In *IOP Conference Series: Earth and Environmental Science* (Vol. 1194, No. 1, p. 012010). IOP Publishing.

Kaginalkar, A., Kumar, S., Gargava, P., & Niyogi, D. (2021). Review of urban computing in air quality management as smart city service: An integrated IoT, AI, and cloud technology perspective. *Urban Climate*, 39, 100972.

Kang, G. K., Gao, J. Z., Chiao, S., Lu, S., & Xie, G. (2018). Air quality prediction: Big data and machine learning approaches. *International Journal of Environmental Science and Development*, 9(1), 8–16.

Kanmani, P., Selvaraj, P., & Burugari, V. K. (2022). An energy efficient approach of deep learning based soft sensor for air quality management. *Measurement: Sensors*, 24, 100460.

Kow, P. Y., Hsia, I. W., Chang, L. C., & Chang, F. J. (2022). Real-time image-based air quality estimation by deep learning neural networks. *Journal of Environmental Management*, 307, 114560.

Li, G., Tang, Y., & Yang, H. (2022). A new hybrid prediction model of air quality index based on secondary decomposition and improved kernel extreme learning machine. *Chemosphere*, 305, 135348.

Li, W., Ma, D., Fu, J., Qi, Y., Shi, H., & Ni, T. (2023). A quantitative exploration of the interactions and synergistic driving mechanisms between factors affecting regional air quality based on deep learning. *Atmospheric Environment*, 314, 120077.

Li, Y., Guo, J. E., Sun, S., Li, J., Wang, S., & Zhang, C. (2022). Air quality forecasting with artificial intelligence techniques: A scientometric and content analysis. *Environmental Modelling & Software*, 149, 105329.

Liao, H., Yuan, L., Wu, M., & Chen, H. (2023). Air quality prediction by integrating mechanism model and machine learning model. *Science of The Total Environment*, 899, 165646.

Liu, C. N., Lin, S. F., Awasthi, A., Tsai, C. J., Wu, Y. C., & Chen, C. F. (2014). Sampling and conditioning artifacts of PM2. 5 in filter-based samplers. *Atmospheric Environment*, 85, 48–53.

Livingston, S. J., Kanmani, S. D., Ebenezer, A. S., Sam, D., & Joshi, A. (2023). An ensembled method for air quality monitoring and control using machine learning. *Measurement: Sensors*, 30, 100914.

Lv, Y., Tian, H., Luo, L., Liu, S., Bai, X., Zhao, H., ... & Yang, J. (2023). Understanding and revealing the intrinsic impacts of the COVID-19 lockdown on air quality and public health in North China using machine learning. *Science of the Total Environment*, 857, 159339.

Ma, J., Cheng, J. C., Lin, C., Tan, Y., & Zhang, J. (2019). Improving air quality prediction accuracy at larger temporal resolutions using deep learning and transfer learning techniques. *Atmospheric Environment*, 214, 116885.

Majdi, A., Alrubaie, A. J., Al-Wardy, A. H., Baili, J., & Panchal, H. (2022). A novel method for indoor air quality control of smart homes using a machine learning model. *Advances in Engineering Software*, 173, 103253.

Maltare, N. N., & Vahora, S. (2023). Air quality index prediction using machine learning for Ahmedabad city. *Digital Chemical Engineering*, 7, 100093.

Masood, A., & Ahmad, K. (2021). A review on emerging artificial intelligence (AI) techniques for air pollution forecasting: Fundamentals, application and performance. *Journal of Cleaner Production*, 322, 129072.

Matthaios, V. N., Knibbs, L. D., Kramer, L. J., Crilley, L. R., & Bloss, W. J. (2024). Predicting real-time within-vehicle air pollution exposure with mass-balance and machine learning approaches using on-road and air quality data. *Atmospheric Environment*, 318, 120233.

Maung, T. Z., Bishop, J. E., Holt, E., Turner, A. M., & Pfrang, C. (2022). Indoor air pollution and the health of vulnerable groups: A systematic review focused on particulate matter (PM), volatile organic compounds (VOCs) and their effects on children and people with pre-existing lung disease. *International Journal of Environmental Research and Public Health*, 19(14), 8752.

Mehmood, K., Bao, Y., Cheng, W., Khan, M. A., Siddique, N., Abrar, M. M., ... & Naidu, R. (2022). Predicting the quality of air with machine learning approaches: Current research priorities and future perspectives. *Journal of Cleaner Production*, 379, 134656.

Miao, C., Peng, Z. R., Cui, A., He, X., Chen, F., Lu, K., ... & Chen, W. (2024). Quantifying and predicting air quality on different road types in urban environments using mobile monitoring and automated machine learning. *Atmospheric Pollution Research*, 15(3), 102015.

Middya, A. I., & Roy, S. (2022). Pollutant specific optimal deep learning and statistical model building for air quality forecasting. *Environmental Pollution*, 301, 118972.

Munir, S., Luo, Z., Dixon, T., Manla, G., Francis, D., Chen, H., & Liu, Y. (2022). The impact of smart traffic interventions on roadside air quality employing machine learning approaches. *Transportation Research Part D: Transport and Environmen*t, 110, 103408.

Prado-Rujas, I. I., García-Dopico, A., Serrano, E., Córdoba, M. L., & Pérez, M. S. (2024). A multivariable sensor-agnostic framework for spatio-temporal air quality forecasting based on deep learning. *Engineering Applications of Artificial Intelligence*, 127, 107271.

Rahman, M. M. (2023). Recommendations on the measurement techniques of atmospheric pollutants from in situ and satellite observations: A review. *Arabian Journal of Geosciences*, 16(5), 326.

Ravindiran, G., Hayder, G., Kanagarathinam, K., Alagumalai, A., & Sonne, C. (2023). Air quality prediction by machine learning models: A predictive study on the indian coastal city of Visakhapatnam. *Chemosphere*, 338, 139518.

Ravindiran, G., Rajamanickam, S., Kanagarathinam, K., Hayder, G., Janardhan, G., Arunkumar, P., ... & Muniasamy, S. K. (2023). Impact of air pollutants on climate change and prediction of air quality index using machine learning models. *Environmental Research*, 239, 117354.

Sarkar, N., Gupta, R., Keserwani, P. K., & Govil, M. C. (2022). Air Quality Index prediction using an effective hybrid deep learning model. *Environmental Pollution*, 315, 120404.

Shin, S., Baek, K., & So, H. (2023). Rapid monitoring of indoor air quality for efficient HVAC systems using fully convolutional network deep learning model. *Building and Environment*, 234, 110191.

Singh, N., Agarwal, R., Awasthi, A., Gupta, P. K., & Mittal, S. K. (2010). Characterisation of atmospheric aerosols for organic tarry matter and combustible matter during crop residue burning and non-crop residue burning months in the Northwestern region of India. *Atmospheric Environment*, 44(10), 1292–1300.

Subbiah, S. S., & Paramasivan, S. K. (2024). Prediction of air quality pollutants using artificial intelligence techniques: A review. In AIP Conference Proceedings (Vol. 2802, No. 1). AIP Publishing.

Tandon, A., Awasthi, A and. Pattnayak, K. C, 2023. Comparison of different Machine Learning methods on Precipitation dataset for Uttarakhand, 2023 2nd International Conference on Ambient Intelligence in Health Care (ICAIHC), Bhubaneswar, India, 2023 (pp. 1–6). doi: 10.1109/ICAIHC59020.2023.10431402

Tang, D., Zhan, Y., & Yang, F. (2024). A review of machine learning for modeling air quality: Overlooked but important issues. *Atmospheric Research*, 107261.

Wang, D., Wei, S., Luo, H., Yue, C., & Grunder, O. (2017). A novel hybrid model for air quality index forecasting based on two-phase decomposition technique and modified extreme learning machine. *Science of the Total Environment*, 580, 719–733.

Wang, J., Du, W., Lei, Y., Chen, Y., Wang, Z., Mao, K., ... & Pan, B. (2023). Quantifying the dynamic characteristics of indoor air pollution using real-time sensors: Current status and future implication. *Environment International*, 175, 107934.

Wang, S., Cai, W., Tao, Y., Sun, Q. C., Wong, P. P. Y., Huang, X., & Liu, Y. (2023). Unpacking the inter-and intra-urban differences of the association between health and exposure to heat and air quality in Australia using global and local machine learning models. *Science of The Total Environment*, 871, 162005.

Wang, S., McGibbon, J., & Zhang, Y. (2024). Predicting high-resolution air quality using machine learning: Integration of large eddy simulation and urban morphology data. *Environmental Pollution*, 344, 123371.

Wang, W., An, X., Li, Q., Geng, Y. A., Yu, H., & Zhou, X. (2022). Optimization research on air quality numerical model forecasting effects based on deep learning methods. *Atmospheric Research*, 271, 106082.

Wood, D. A. (2022). Local integrated air quality predictions from meteorology (2015 to 2020) with machine and deep learning assisted by data mining. *Sustainability Analytics and Modeling*, 2, 100002.

Wu, B., Wu, C., Ye, Y., Pei, C., Deng, T., Li, Y. J., … & Wu, D. (2024). Long-term hourly air quality data bridging of neighboring sites using automated machine learning: A case study in the Greater Bay area of China. *Atmospheric Environment*, 321, 120347.

Wu, C. L., Song, R. F., Zhu, X. H., Peng, Z. R., Fu, Q. Y., & Pan, J. (2023). A hybrid deep learning model for regional O3 and NO2 concentrations prediction based on spatio-temporal dependencies in air quality monitoring network. *Environmental Pollution*, 320, 121075.

Yadav, N., Sorek-Hamer, M., Von Pohle, M., Asanjan, A. A., Sahasrabhojanee, A., Suel, E., … & Ganguly, A. R. (2024). Using deep transfer learning and satellite imagery to estimate urban air quality in data-poor regions. *Environmental Pollution*, 342, 122914.

Yang, D., Wang, J., Yan, X., & Liu, H. (2022). Subway air quality modeling using improved deep learning framework. *Process Safety and Environmental Protection*, 163, 487–497.

Zhang, D., Wang, Q., Song, S., Chen, S., Li, M., Shen, L., … & Zheng, H. (2023). Machine learning approaches reveal highly heterogeneous air quality co-benefits of the energy transition. *Iscience*, 26(9), 1–15.

Zhou, S., Wang, W., Zhu, L., Qiao, Q., & Kang, Y. (2024). Deep-learning architecture for PM2. 5 concentration prediction: A review. *Environmental Science and Ecotechnology*, 21, 100400.

Zhou, Y., Chang, F. J., Chang, L. C., Kao, I. F., & Wang, Y. S. (2019). Explore a deep learning multi-output neural network for regional multi-step-ahead air quality forecasts. *Journal of Cleaner Production*, 209, 134–145.

2 Temporal Variations of Sulfur Dioxide Levels across India

A Biennial Assessment (2020–2021)

Bisma Nadeem, Navjot Hothi, Rahul Malik, Juhi Gupta, and Sarfraz Hussain

ABBREVIATIONS:

sulfur dioxide	**SO_2**
Central Pollution Control Board	**CPCB**
World Health Organization	**WHO**
parts per billion	**ppb**
photomultiplier tube	**PMT**
State Pollution Control Boards	**SPCB**
Pollution Control Committees	**PCC**
below detection limit	**BDL**
National Air Quality Monitoring Programme	**NAMP**
inter-state bus terminal	**ISBT**
Raipur Road	**RR**
clock tower	**CT**

2.1 INTRODUCTION

India's declining air quality has become a serious environmental and public health issue. The excessive release of pollutants, including sulfur dioxide (SO_2), is mostly caused by the burning of fossil fuels, industrial emissions, vehicle pollution, and agricultural practices (Agarwal et al., 2014; Singh et al., 2010). Burning coal, oil, and gas as well as industrial processes like smelting, producing electricity, and manufacturing are the main sources of SO_2 emissions. SO_2 is well-known for having harmful consequences on human health, including an increased risk of cardiovascular disease, respiratory ailments, and even early mortality. Additionally, it interacts with other

DOI: 10.1201/9781032683805-2

elements of the atmosphere to create secondary pollutants like sulphate aerosols, which are harmful to ecosystems, impair visibility, and alter climatic patterns. To develop an efficient air pollution management method, an understanding of the distribution and concentrations of SO_2 is required (Abramson et al., 1991). Some properties like morphology, temperature, structure, etc. not only affect the metals but also the gases (Ravikiran et al., 2019; Singh et al., 2020; Anand et al., 2017).

This chapter aims to examine and offer a thorough profile of ambient air quality at various locations in India concerning SO_2. The research is based on comprehensive atmospheric monitoring data gathered from several monitoring stations located strategically across the nation (Khalaf et al., 2022). The chapter describes the spatial and temporal patterns of SO_2 concentrations, locates pollution hotspots, and assesses how well SO_2 levels adhere to both national air quality standards established by the Central Pollution Control Board (CPCB) and international standards like those advised by the World Health Organization (WHO). To support decision-making based on facts, this research also strives to give policymakers, environmental organizations, and stakeholders useful insights for SO_2 reduction (Gulia et al., 2020). For enforcing focused mitigation strategies such as tighter emission limits, technology breakthroughs, and public awareness campaigns, the chapter identifies some of the main sources and variables impacting SO_2 pollution. This study is significant because it has the potential to guide efficient policy actions and methods for lowering SO_2 pollution, thereby, improving India's air quality and public health. We have identified the places that need immediate attention and subsequently require to distribute resources appropriately by knowing the changes in SO_2 concentrations. Additionally, this chapter can help advance our understanding of the dynamics of air pollution on a worldwide scale and serve as a guide for other nations dealing with comparable problems (LeFevre et al., 2015).

Section 2 describes the impact of SO_2 on the environment and methods to mitigate its effects. Section 3 provides the methods for collecting SO_2 data and its analysis. Section 4 elaborates the results obtained through analysis in Section 3. Finally, Section 5 provides a conclusion of the study.

2.2 IMPACT AND STUDY OF SO_2 ON AIR POLLUTION

SO_2 is commonly produced as a byproduct of burning coal or oil, and it has played a notable role in causing atmospheric corrosion in urban and industrial environments. It creates sulfuric acid in water and is only mildly water-soluble.

$$SO_2 + H_2O = H_2SO_3 \tag{2.1}$$

SO_2 has been linked to considerable nasal inflammation and irritation and is a prevalent component of urban air pollution. When exposed to SO_2 at work through inhalation, there has been some, if scant, evidence of clastogenic effects in people. Lymphocytes from employees who were exposed to 15.92 *ppm* SO_2 in a fertilizer industry in India showed sister chromatid exchanges and chromosomal abnormalities. Similar results were found in another study that involved workers at a sulfuric acid factory in China, where exposure quantities of SO_2 in the air ranged from 0.13 to 4.57 *ppm* (Pohl, 1998; Yadav & Kaushik, 1996). It should be highlighted that additional contaminants in these work situations might have influenced the outcomes (Lippmann et al., 2003). SO_2 irritates the skin, eyes, nose, throat, and mucous membranes of the lungs. When present in high concentrations, especially during intense physical activity, it can lead to respiratory system irritation and inflammation. Symptoms such as breathing difficulties, coughing, throat irritation, and discomfort while taking deep breaths may arise. Furthermore, elevated levels of SO_2 can impair lung function, worsen asthma attacks, and exacerbate existing heart conditions among vulnerable populations. This gas can also combine with other airborne substances, forming small particles which when inhaled, can have severe detrimental effects on health (Limaye et al., 2018).

2.2.1 Sampling Methods of Sulfur Dioxide Monitoring

2.2.1.1 Ultraviolet Fluorescence Technique

Ultraviolet (UV) fluorescence is utilized to assess ambient SO_2 levels, relying on the fact that SO_2 molecules absorb UV light at 214 *nm*. This absorption causes outer electrons to become excited and transition to a higher energy state (Linch et al., 2019). Subsequently, as the excited electrons return to their initial state, they emit photons with a wavelength of 390 nm. This principle forms the basis of SO_2 analyzers for evaluating ambient SO_2 concentrations.

In the process, the photomultiplier tube (PMT) is responsible for detecting the light emitted by the de-excited SO_2 electrons and converting it into an electrical signal. The analyzer's electronic system then calibrates and scales the PMT signal to generate a corresponding voltage response, which can be interpreted as measurements of SO_2 concentration in units like parts per billion (ppb). The fluorescence will increase in intensity when the SO_2 concentration in the air sample rises. This method is the most common procedure for monitoring SO_2 (Amann et al., 2013).

2.2.1.2 Metal Oxide-Based Sensor Technique

Another method for monitoring is based upon a metal oxide-based sensor. When SO_2 molecules interact with the metal oxide surface of a sensor, a reaction occurs, leading to the dissociation of SO_2 into charged ions. This process alters the resistance of the sensor's film. The change in resistance is measured as a signal, which is then converted into the corresponding gas concentration. However, it's worth noting that the energy consumption of these SO_2 sensors tends to be higher compared to other types of SO_2 monitors (Grubler et al., 2002).

2.2.1.3 Electrochemical Sensor Technique

The electrochemical mechanism relies on the diffusion of SO_2 gas into the appropriate sensor. This generates electrical signals whose magnitude is proportional to the concentration of SO_2. This is crucial for monitoring SO_2 levels in ambient air because it enables precise measurement of even low SO_x concentrations. Applications like ambient air monitoring favor SO_2 monitors based on electrochemistry out of all the mentioned SOx monitoring concepts. This is because they are less expensive compared to the alternatives and produce more precise SO_2 concentrations (Diamond et al., 2020).

2.2.2 Reduction of SO_2 Emissions

Numerous technological and non-technical solutions are available to reduce Sulfur emissions. Developing nations in Asia have only made tentative moves in this regard, while many developed nations throughout the world have been able to significantly reduce their sulfur emissions over the past few decades. The significant financial costs that these pollution control measures imposed on Asia's developing economies prevented the adoption of modern emission controls more quickly in the early 1990s. It was projected that annual SO_2 emissions in Asia could reach 80-110 Tg per yr by 2020 (Linch et al., 2019). However, new high-resolution projections covering the period from 1975 to 2000 suggest a lower estimate of 40–45 Tg yr per year for Asia's SO_2 emissions by 2020. This reduction is primarily attributed to a decrease in SO_2 emissions in China, which accounts for approximately two-thirds of the total Asian SO_2 emissions. The decline in China's emissions can be attributed to factors such as reduced industrial coal usage, a slower Chinese economy, and the closure of small and inefficient plants (Grubler et al., 2002). Consequently, the decrease in SO_2 emissions has resulted in a reduction in acid deposition, not only in China but also in Japan, leading to improved visibility and potential health benefits. The region's SO_2 emissions will alter depending on several conflicting factors (economic growth, pollution control regulations, etc.) throughout the ensuing decades. However, if current trends continue, sulfur emissions will be lower than any predictions made by the Intergovernmental Panel on Climate Change (IPCC) (Orellano et al., 2021).

2.2.3 Sulfur Deposition and Ambient SO_2 Concentrations

The World Health Organization (WHO) has recommended maintaining an annual average SO_2 concentration below 50 μg/m^3. Apart from its negative impact on human health and the environment, SO_2 serves as a precursor to other sulfur oxides, including SO_3, as well as various secondary air pollutants such as particulate matter, smog, and acid rain. High levels of SO_2 exposure can have long-term repercussions on the respiratory system, especially in young children, the elderly, and those who already have respiratory problems (Sharma et al., 2022). The effective detection and control of sulfur oxide exposure using SO_2 monitoring also has the added benefit of minimizing the development of particulate sulfur pollutants such as fine sulphate particles, acid rain, and smog. Real-time monitoring of SO_2 levels plays a crucial role in the formulation of an action plan to meet air quality criteria and the calculation of the air quality index to ensure public health. In contrast, the national standard in China permits a slightly higher limit of 60 μg/m^3 (WHO et al., 2006).

2.2.4 Sources of SO_2

SO_2 is a chemical compound, and it is produced by volcanoes, industrial processes, transportation, and coal fire-based power plants. Some 67% of SO_2 is naturally released from volcanoes into the atmosphere. Anthropogenic sources include the paper industry, excavation, and distribution of fossil fuels, smelting of metals like sulphide ore and lead, petroleum refining, and production of diesel and petrol vehicles (Morawska et al., 2018). Burning fuels like coal and wood for home heating contain sulfur and this can exaggerate emissions of SO_2. This is especially observed in places where there are few options for cleaner fuel or where traditional cooking methods are still widely used (Sharma et al., 2020). The creation of particulate matter, acid rain, and respiratory issues are just a few of the negative consequences that SO_2 emissions can have on the environment and people's health. Stricter rules on industrial operations are implemented along with the use of cleaner fuels and emission control technologies in efforts to limit SO_2 emissions (Kaur & Pandey, 2021).

2.3 METHODOLOGY

The main objective of this section is to establish the importance of monitoring ambient SO_2 levels and their effects on air quality and human health. The current state of the nation's air quality monitoring systems, including the tools and techniques utilized for SO_2 profiling is being studied (Kaur & Pandey, 2021). Different geographic regions are focused on SO_2-level monitoring. The sampling process, including the criteria used to choose the monitoring sites and the frequency of data collection, is also presented. In this study, the annual average data of SO_2 for 2 years from 2020–2021 is analyzed. This section also contains details on the sampling sites and sampling procedures (Verma et al., 2023). The study identifies the states and cities in India that have had the highest concentrations of SO_2 for the years 2020 and 2021. The study ultimately aims to identify SO_2 mitigations with the aid of data, thus enhancing atmospheric air quality (Mookherjee, 2022).

2.3.1 Study on India's Situation Regarding Ambient Air Quality of SO_2

Automatic and continuous ambient air quality monitoring stations are used to track ambient air quality around the nation. A total of 1257 monitoring stations make up the network as of right now. A total of 883 stations within 378 cities and towns in 28 states and 7 union territories are monitored manually (Gargava et al., 2000), while 374 stations with 190 cities and towns in 27 states and 4 union territories are continuously monitored (WHO et. al., 2006). Pollution Control Committees (PCC), State Pollution Control Boards (SPCB), and other respectable organizations participate in the monitoring of air pollution. To ensure that air quality data is accurate and consistent, CPCB collaborates with these organizations and provides technical and financial support. 2020 saw the operation of the CPCB's National Air Quality Monitoring Programme (NAMP) in India as a national ambient air quality monitoring project, based on the data that was available. This chapter looks at roadside, industrial, rural, urban, and background monitoring station types. The selection procedure is based on how far away an area is from areas that pollute, like industrial zones and metropolitan areas. The region's terrain and climate are also taken into account.

2.3.2 Mitigations of SO_2

Multiple steps may be taken to remediate the amount of SO_2 in Indian towns. This needs to be enforcement of strict pollution requirements for businesses, power plants, and cars to strengthen the current regulations. Regular monitoring and inspections are required to ensure compliance. Stricter guidelines for gasoline sulfur concentration can dramatically lower SO_2 emissions. Promoting the adoption of renewable energy sources like solar, wind, and hydropower is crucial. This approach helps reduce dependence on fossil fuel-powered plants, which are major contributors to SO_2 emissions. By shifting towards renewable energy, we can mitigate the release of SO_2 and its harmful effects on the environment and public health (Gupta et al., 2022). Cleaner fuels like LPG, CNG, and electric cars should be promoted in order to reduce SO_2 emissions and minimize their detrimental effects on the environment and human health. In order to enhance environmental surveillance, it is imperative to build a resilient system for monitoring air quality in Indian cities. This could facilitate prompt action by making it simpler to detect places with elevated SO_2 concentrations. As a result, we can reduce SO_2 emissions and maintain a stable SO_2 level in the upcoming years by using this knowledge.

2.4 RESULT AND DISCUSSIONS

The present assessment has been done based on annual average concentration of SO_2 from 2020–2021 across India. The cluster of cities were divided into five ranges. By using a group of cities, five ranges were created: Below Detection Limit (BDL), very low, low, moderate, and high. Graphs depicting SO_2 concentrations in various cities of India are plotted for analysis. The graphs represented places that need urgent attention and priorities where resources should be spent by evaluating the variability in SO_2 concentrations across different locations (Akan et al., 2022). The results can be used as a reference for other nations facing comparable problems, this research can also further our understanding of the dynamics of air pollution on a larger, global scale (Gordon, 1998).

Table 2.1 provides a categorization of yearly average concentrations of SO_2, expressed in $\mu g/m^3$. Nearly BDL range (0–5 $\mu g/m^3$) represents extremely low or negligible SO_2 concentrations, potentially near or below the measuring device's detection limit. Very low concentration range (5–10 $\mu g/m^3$) indicates an extremely low concentration of SO_2 in the air, pointing to an environment that is largely pure and

TABLE 2.1
Average Concentration of SO_2 Based on Five Different Categories

Annual average concentration of SO_2 ($\mu g/m^3$)	Category
0–5	Nearly BDL (Below Detection Limit)
5–10	Very low concentration
10–20	Low concentration
20–40	Moderate concentration
40–50	High concentration

TABLE 2.2
Annual Average Concentration Range of SO_2 in Different Cities across India for the years 2020 and 2021

Annual average concentration range of SO_2 in (μg/m³)	Number of cities in 2020	Number of cities in 2021
0–5	116	106
5–10	134	150
10–20	90	102
20–40	22	25
40–50	0	0

unpolluted. The low concentration range (10–20 μg/m³) shows that the amount of SO_2 present is modest yet discernible and there is probably no risk to health from the air quality. Moderate concentration range (20–40 μg/m³) denotes a moderate concentration of SO_2 in the atmosphere. This concentration may require attention even if it is still within allowable bounds, particularly in places where sensitive populations are present. The high concentration range (40–50 μg/m³) suggests that the amount of SO_2 in the air is very high. This concentration might be harmful to health, particularly for people who have respiratory issues, and it might raise questions about the quality of the air. By illustrating the possible effects of SO_2 concentrations on public health and air quality, these categories aid in the communication of pollution levels in a particular area.

Table 2.2 analyzes SO_2 concentration across India of 26 states and 5 union territories spanning over 336 cities in years 2020 and 2021 (WHO et al., 2021). It is observed that SO_2 concentration BDL is witnessed in 116 and 106 cities respectively in the years 2020 and 2021. Concentration from 5–10 μg/m³ which is characterized as very low, is seen in 134 cities for 2020 and 150 cities for 2021. Low concentrations of SO_2 are observed in 90 and 102 cities respectively for the year 2020 and 2021 respectively. Moderate concentrations between 20 and 40 μg/m³ are seen in a total of 22 cities in the year 2020 and 25 for the year 2021. (Gulia et al., 2015). There are no cities for concentration from 40–50 μg/m³ (Singh et al., 2023).

2.4.1 Analytical Graph Analysis of SO_2 Concentration for the Year 2020

Indian cities like Guntur, Nellore, Vijayawada, Bongaigaon, Bilaspur, Panaji, Damtal, Kala Amb, Nalagarh, Shimla, Jammu, Hassan, Tumkuru, Kannaur, Kozhikode, Suri, and many more have witnessed SO_2 concentrations in BDL. In Panaji and Umiam where population density is lower, it is found that the SO_2 concentration witnessed a minimal level of 2 μg/m³. Other densely populated areas, however, show more concentration in comparison to Panaji and Umiam. Cities like Anatapur, Vishakapatnam, Magherita, Sivasagar, Hajipur, Chandigarh, Amona, Cuncolin, Ponda, Usgao- Pale, Kaithal, Yamuna Nagar, Kochi, Maithar, Nagpur, Shilong, Jharsugunda, Dera Bassi,

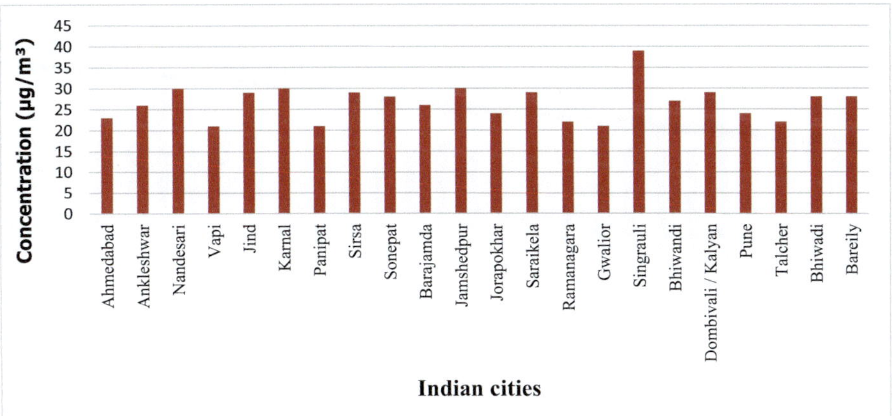

FIGURE 2.1 SO_2 concentration in 22 Indian cities in the year 2020 for a range limit between 20 to 40μg/m³.

Ludhiana, Raiganj, and many more have very low concentrations of SO_2 levels. Average concentration levels spikes are seen in Kota, Amona, Malhar, and Haldwani which are averagely populated areas. Monitoring at regular intervals is not required in the cities with low concentrations. Some of these cities are Amaravati, Darbhanga, Daman, Rajkot, Vatva, Bhiwani, Fatehabad, Mandikhera, Dhanbad, Sindri, Madikeri, Bhopal, Mandideep, Ratlam, Amravati, Chandrapur, Mumbai, Thane, Paradeep, Khanna, Trichny, Gajroula, Greater Noida, Meerut, Noida, Kashipur, Asansol, Haldia, Kharagpur, and Sankrali and many more.

The SO_2 annual average concentration for 2020 is shown in Figure 2.1, which spans 22 Indian cities and ranges from 20 μg/m³ to 40 μg/m³. The figure makes specific reference to the cities that were part of the study. It is imperative to conduct routine monitoring of SO_2 levels in these cities to preserve air quality and guarantee public health. The results show that no observed city surpassed the yearly average high SO_2 concentration limit. This is a good sign that the policies already in place are managing and controlling SO_2 emissions in these cities.

While no city exceeded the annual average concentration limit, a few cities came close to the concentration's top limit (20 μg/m³ to 40 μg/m³). Preventive actions need to be given top priority in these cities to keep the SO_2 levels from rising further. In cities where concentrations are getting close to this level, proactive measures are needed. The SO_2 concentration at 38 μg/m³ was seen in the city of Singrauli. This result indicates that preventive actions should be taken to preserve and improve air quality as the concentration limit is close to the threshold. Stricter industry rules, public awareness programs, and emission limits are a few examples of such actions that may be taken to reduce the possible health concerns linked to SO_2 exposure.

2.4.2 Analytical Graph Analysis of SO_2 Concentration for 2021

SO_2 concentration is up to 5 μg/m³ in 106 cities across 2021. In cities like Chittor, Eluru, Guntur, Kadapa, Nellore, Ongole, Vijayawada, Itanagar, Bongaigaon,

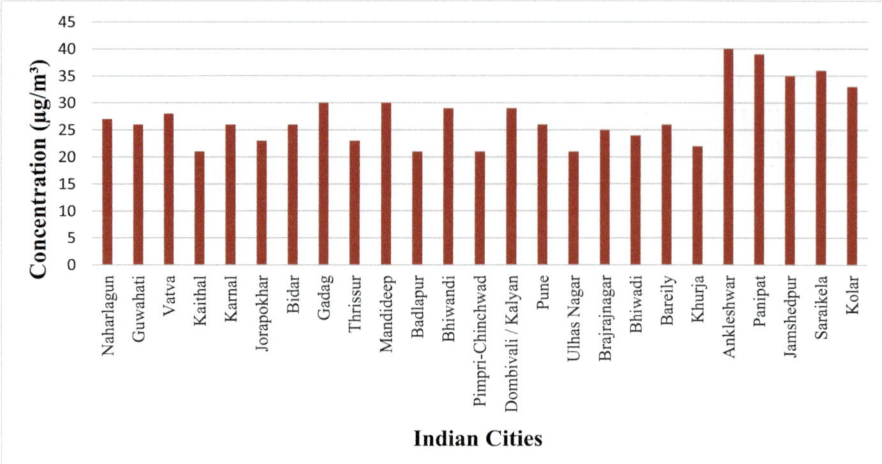

FIGURE 2.2 SO_2 concentration in 25 Indian cities in the year 2021 for a range limit between 20 and 40μg/m³.

Magherita, Hajipur, Tumkuru, Kannaur, Kozhikode, Suri, and many more the SO_2 concentration is within the BDL. Tribeni and Baddi have a low population density and it is found that the SO_2 concentration is minimal, which is 2 μg/m³. Other densely populated areas, however, show more concentration as compared to Tribeni and Baddi. No preventive measures are suggested in these cities because the concentration of SO_2 is below BDL. SO_2 concentration from 5 to 10 μg/m³ is observed across 150 cities in the year 2021 showcasing that concentration levels are very low. Cities like Anantapur, Vishakapatnam, Magherita, Sivasagar, Hajipur, Chandigarh, Amona, Cuncolin, Ponda, Usgao- Pale, Kaithal, Yamuna Nagar, Kochi, Maithar, Nagpur, Shilong, Jharsugunda, Dera Bassi, Ludhiana, Raiganj, and many more fall in this category. Regular monitoring is not required in these cities for maintaining the optimal concentration of $SO_{2.}$ SO_2 concentration from 10 to 20 μg/m³ is observed in 102 cities across India in 2021. Cities like Amaravati, Raipur, Jamnagar, Surat, Vapi, Sirsa, Jharia, Udupi, Bhopal, Indore, Maihar, Satna, Solapur, Kohira, Amritsar, Chennai, Madurai, Anpara, Greater Noida, Meerut, Noida, Haridwar, and many more fall in this category. Monitoring at regular intervals is not required in these cities. In a few cities like Gaya and Thane less concentration is observed (11 μg/m³) as compared to Rajkot, Ambarnath, and Kohima, where the annual average concentration of SO_2 is 20 μg/m³ in 2021.

Figure 2.2 gives a detailed picture of the average yearly concentration of SO_2 in 25 Indian cities for the year 2021, ranging from 20 to 40 μg/m³. Figure 2.2 provides comprehensive details about the cities that are being examined. To maintain the best possible air quality and to guarantee the health and welfare of the local populace, these cities must be routinely monitored. Importantly, the research shows that not even one of the examined cities had SO_2 concentrations above the set yearly average level. This successful result highlights how effective the current controls over SO_2 emissions are in limiting excessive pollution levels in these cities. The SO_2 concentration in the city of Ankleshwar is 40 µg/m³. While this number is within the range under investigation, its closeness to the standard limit indicates that precautionary steps should be taken to reduce any further rises in SO_2 levels. To reduce the health concerns connected to SO_2 exposure, these actions could include tighter industry rules, improved emission controls, and public awareness campaigns.

2.4.3 Ecological Sensitive Areas

Ecological cities, sometimes referred to as green cities or eco-cities, are places that place a high value on environmental responsibility and sustainability. These cities seek to reduce their environmental impact, enhance the standard of living for their citizens, and improve the health of the earth. The annual average standard concentration notified by CPCB for the ecologically sensitive area of SO_2 is 20 µg/m³. No ecologically sensitive city exceeded the national standards for SO_2 as observed from an annual average concentration of the air in 2020 and 2021 except Dehradun city in Uttarakhand (22 µg/m³) in 2020 and 2021 (Singh et al., 2023).

According to available data, in 2020, the NAMP administered by the CPCB, operated as a nationwide program dedicated to monitoring ambient air quality. The network consisted of 804 active stations, covering 344 cities and towns across 26 states and 5 union territories (Deep et al., 2019). As of the latest information in 2021, the NAMP in India has expanded its network to include 1257 monitoring stations. Among these, 883 stations covering 378 cities and towns in 28 states and 7 union territories are monitored manually. Additionally, 374 stations monitoring 190 cities and towns in 27 states and 4 union territories provide continuous monitoring. It's worth noting that the network still maintains 804 active stations that serve 336 cities and towns in 26 states and 5 union territories (Gupta & Dalei, 2019). Table 2.3 indicates the concentration of SO_2 across different ecologically sensitive cities of India from 2020 to 2021. Dehradun is the only city that reflects a higher SO_2 concentration than the national standard of SO_2 (20 µg/m³) which is 22 µg/m³. It's important to take preventive measures in the Dehradun city regarding the pollution from the SO_2 pollutant because excess concentration may lead formation of acid rain in this city and cause the bleaching of leaves. The graphical representation shown in Figure 2.3 reflects

TABLE 2.3
Concentration of SO_2 across Different Ecologically Sensitive Cities of India from 2020 to 2021

State	City	Concentration of SO_2 in µg/m³ for the year 2020	Concentration of SO_2 in µg/m³ for the year 2021
Rajasthan	Alwar	13	12
Uttarakhand	Dehradun	22	22
Uttar Pradesh	Agra	19	5
Uttar Pradesh	Firozabad	7	8
Uttar Pradesh	Mathura	11	12

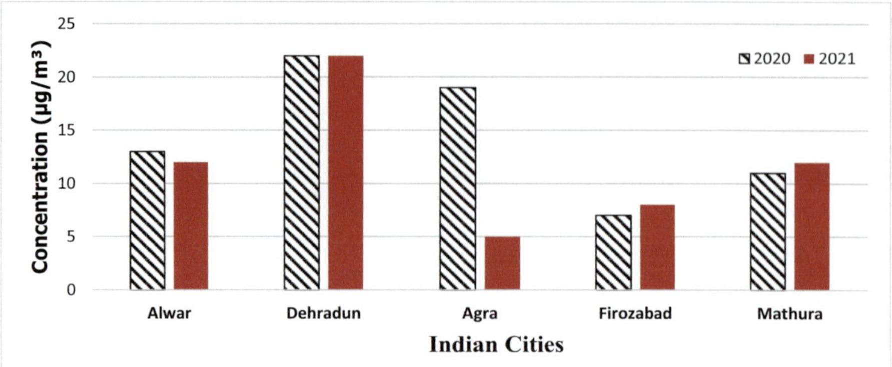

FIGURE 2.3 Graphical representation of comparison of SO_2 concentration in ecological cities notified by CPCB for 2020 and 2021.

the comparison of SO_2 concentration in ecological cities notified by CPCB. There are only five cities under ecological sensitive areas, i.e., Alwar, Dehradun, Agra, Firozabad, and Mathura. Dehradun highlights maximum concentration in both years.

Three heavily inhabited and heavily polluted major cities of Dehradun, Rishikesh, and Haridwar in Uttarakhand were chosen for the current investigation. Dehradun, the capital of Uttarakhand state, is situated in the western Himalayan foothills at 30.32° N and 78.03° E, or 700 AMSL. Dehradun is a well-known tourist destination that is used as a hub by millions of tourists every year on their way to the Himalayas. In addition, the area is semi-industrial. The city's air quality is declining as a result of increased anthropogenic activity like industry, large-scale construction, rapid population growth, deforestation, and careless development. To keep an eye on the air quality, three different areas in the city of Dehradun, the Inter-State Bus Terminal (ISBT), Raipur Road (RR), and Clock Tower (CT) – that may have significant pollution levels were selected. The CT site is the busiest area of the city due to its location in the business sector. Owing to its strategic location in Dehradun, the CT experiences steady traffic flow from most passing vehicles.

Table 2.3 analyzes the quantities of SO_2 in μg/m³ for several ecologically sensitive Indian cities in 2020 and 2021. The cities are mentioned with the states in which they are located, and the annual SO_2 concentrations are provided. In 2020, the SO_2 content in Alwar was 13 μg/m³ and it significantly dropped to 12 μg/m³ in 2021. This implies that Alwar's air quality was comparatively steady during the given time frame. In 2020 and 2021, Dehradun consistently recorded a SO_2 value of 22 μg/m³. The persistent level may suggest that future environmental management plans and continuous monitoring are required. In Agra, the concentration of SO_2 decreased significantly from 19 μg/m³ in 2020 to 5 μg/m³ in 2021. Emission limits, other environmental initiatives, or municipal interventions could be the reason for this considerable decline. In 2020, the SO_2 content in Firozabad was 7 μg/m³ in 2021 and it rose to 8 μg/m³. Even though the change is little, it emphasizes how crucial it is to keep an eye on things to find and deal with possible pollution sources. From 11 μg/m³ in 2020 to 12 μg/m³ in 2021, Mathura's SO_2 content increased slightly. This small change might call for more research into regional issues affecting air quality. Maintaining and improving air quality in environmentally sensitive areas requires ongoing monitoring and flexible environmental controls.

For environmental policymakers, Figure 3 provides useful information by highlighting locations that need attention. Targeted approaches can be developed based on the results to improve air quality in environmentally sensitive cities like Dehradun even further. The graph makes it possible to compare the air quality of various cities and determine which have improved or gotten worse over the given time frame.

2.5 CONCLUSIONS

This study provides an extensive understanding of the dynamics of air quality in a group of cities by analyzing SO_2 concentrations in India from 2020 to 2021. The CPCB has set a nationwide yearly average of 50 μg/m³ for air quality, and this adherence to the guidelines is a sign that air quality control in major urban centres is working well. By dividing these cities into five different ranges, our research provides a clearer understanding of the changes in SO_2 levels. Crucially, the analysis confirms that during the evaluated time, no city under observation had SO_2 levels above the federal guidelines. This observation is encouraging as it indicates a general adherence to the national benchmarks and highlights the efficacy of the current air quality management strategies. The study emphasizes how important it is to have ongoing surveillance and flexible environmental regulations to guarantee long-term improvements in air quality and safeguard the natural integrity of these important areas.

Notably, the study concentrates on five cities designated by the national government as ecologically sensitive namely Agra, Firozabad, Mathura, Alwar, and Dehradun. But there is one notable exception: Dehradun city of Uttarakhand, where the SO_2 concentrations were higher than those recommended for environmentally sensitive locations. Although the CPCB has established a yearly average threshold of 20 μg/m³ for certain locations, Dehradun reported that in 2020 and 2021, concentrations had reached 22 μg/m³. This draws attention to a troubling pattern of rising SO_2 levels in an environmentally sensitive area, which calls for focused measures and close monitoring.

ACKNOWLEDGMENT

This research is a culmination of hard work and contributions from several people in all sorts of ways possible. It is a pleasure to convey our gratitude to each one of them in our humble acknowledgment. Foremost, I would like to thank my Internal Guide Dr Juhi Gupta and my External Guide Shri Aditya Sharma (Scientist -E, DH, AQMN, CPCB) for their constant support and guidance and for allowing me to work under their guidance and making it a wonderful experience. I would also like to convey my gratitude to Shri Fasiur Rehman (Scientist-C, CPCB) for his constant guidance in every step of completing this report.

REFERENCES

Abramson, M., & Voigt, T. (1991). Ambient air pollution and respiratory disease. *Medical Journal of Australia*, *154*(8), 543–553.

Agarwal, R., Awasthi, A., Mital, S. K., Singh, N., & Gupta, P. K. (2014). Statistical model to study the effect of agriculture crop residue burning on healthy subjects. *Mapan, 29*, 57–65.

Akan, A. P., & Coccia, M. (2022). Changes of air pollution between countries because of lockdowns to face COVID-19 pandemic. *Applied Sciences*, *12*(24), 12806.

Amann, M., Klimont, Z., & Wagner, F. (2013). Regional and global emissions of air pollutants: Recent trends and future scenarios. *Annual Review of Environment and Resources*, *38*, 31–55.

Anand, G., Nippani, S. K., Kuchhal, P., Vasishth, A., & Sarah, P. (2017). Dielectric, ferroelectric and piezoelectric properties of Ca0. 5Sr0. 5Bi4Ti4O15 prepared by solid state technique. *Ferroelectrics*, *516*(1), 36–43.

Deep, A., Pandey, C. P., Nandan, H., Purohit, K. D., Singh, N., Singh, J., ... & Ojha, N. (2019). Evaluation of ambient air quality in Dehradun city during 2011–2014. *Journal of Earth System Science*, *128*, 1–14.

Diamond, M. S. (2020). *On the Role of Natural Laboratories and Natural Experiments in Elucidating Cloud-Aerosol-Climate Interactions: A Story of Ships, Smoke, and Shutdowns*. University of Washington.

Gargava, P., Sengupta, B., & Biswas, D. (2000). *Strategies for Prevention and Control of Air Pollution in India*. www.researchgate.net/profile/Udit-Varshney/publication/342923746_Comparative_Study_On_Particulate_Matter_Control_Techniques_A_Review/links/5f0dc7ad458515129998c90c/Comparative-Study-On-Particulate-Matter-Control-Techniques-A-Review.pdf

Gordon, G. E. (1988). Receptor models. *Environmental Science & Technology*, *22*(10), 1132–1142.

Grubler, A. (2002). Trends in global emissions: Carbon, sulfur, and nitrogen. *Encyclopedia of Global Environmental Change*, *3*, 35–53.

Gulia, S., Khanna, I., Shukla, K., & Khare, M. (2020). Ambient air pollutant monitoring and analysis protocol for low and middle income countries: An element of comprehensive urban air quality management framework. *Atmospheric Environment*, *222*, 117120.

Gulia, S., Nagendra, S. S., Khare, M., & Khanna, I. (2015). Urban air quality management: A review. *Atmospheric Pollution Research*, 6(2), 286–304.

Gupta, A., & Dalei, N. N. (Eds.). (2019). *Energy, Environment, and Globalization: Recent Trends, Opportunities and Challenges in India*. Springer.

Gupta, V., Bisht, L., Deep, A., & Gautam, S. (2022). Spatial distribution, pollution levels, and risk assessment of potentially toxic metals in road dust from major tourist city, Dehradun, Uttarakhand India. *Stochastic Environmental Research and Risk Assessment*, *36*(10), 3517–3533.

Jones, A. P. (1999). Indoor air quality and health. *Atmospheric Environment*, *33*(28), 4535–4564.

Kaur, R., & Pandey, P. (2021). Air pollution, climate change, and human health in Indian cities: A brief review. *Frontiers in Sustainable Cities*, *3*, 705131.

Khalaf, E. M., Mohammadi, M. J., Sulistiyani, S., Ramírez-Coronel, A. A., Kiani, F., Jalil, A. T., ... & Derikondi, M. (2022). Effects of sulfur dioxide inhalation on human health: A review. *Reviews on Environmental Health*. doi: 10.1515/reveh-2022-0237

LeFevre, G. H., Paus, K. H., Natarajan, P., Gulliver, J. S., Novak, P. J., & Hozalski, R. M. (2015). Review of dissolved pollutants in urban storm water and their removal and fate in bioretention cells. *Journal of Environmental Engineering*, *141*(1), 04014050.

Limaye, S. S., Mogul, R., Smith, D. J., Ansari, A. H., Słowik, G. P., & Vaishampayan, P. (2018). Venus' spectral signatures and the potential for life in the clouds. *Astrobiology*, *18*(9), 1181–1198.

Linch, A. L. (2019). *Evaluation Ambient Air Quality By Personnel Monitoring: Volume 1: Gases and Vapors*. CRC Press.

Lippmann, M., Frampton, M., Schwartz, J., Dockery, D., Schlesinger, R., Koutrakis, P., ... & Zelikoff, J. (2003). The US environmental protection agency particulate matter health effects research centers program: A midcourse report of status, progress, and plans. *Environmental Health Perspectives*, *111*(8), 1074–1092.

Mookherjee, P. (2022). India's Air Pollution Challenge: Translating Policies into Effective Action. www.orfonline.org/research/indias-air-pollution-challenge

Morawska, L., Thai, P. K., Liu, X., Asumadu-Sakyi, A., Ayoko, G., Bartonova, A., ... & Williams, R. (2018). Applications of low-cost sensing technologies for air quality monitoring and exposure assessment: How far have they gone?. *Environment International*, *116*, 286–299.

Orellano, P., Reynoso, J., & Quaranta, N. (2021). Short-term exposure to sulphur dioxide (SO2) and all-cause and respiratory mortality: A systematic review and meta-analysis. *Environment International*, *150*, 106434.

Pohl, H. R. (1998). Toxicological Profile for Sulfur Dioxide. https://stacks.cdc.gov/view/cdc/6358

Ravikiran, U., Sarah, P., Anand, G., & Zacharias, E. (2019). Influence of Na, Sm substitution on dielectric properties of SBT ceramics. *Ceramics International*, 45(15), 19242–19246.

Sharma, D., & Mauzerall, D. (2022). Analysis of air pollution data in India between 2015 and 2019. *Aerosol and Air Quality Research*, *22*(2), 210204.

Sharma, R., Kumar, R., Sharma, D. K., Son, L. H., Priyadarshini, I., Pham, B. T., ... & Rai, S. (2019). Inferring air pollution from air quality index by different geographical areas: Case study in India. *Air Quality, Atmosphere & Health*, *12*, 1347–1357.

Sharma, S., Zhang, M., Gao, J., Zhang, H., & Kota, S. H. (2020). Effect of restricted emissions during COVID-19 on air quality in India. *Science of the Total Environment*, *728*, 138878.

Singh, A., Deep, A., Pandey, C., & Nandan, H. (2023). Comparative study of gaseous pollutants for major cities in foothills of Garhwal Himalaya of Uttarakhand. *MAUSAM*, *74*(1), 57–72.

Singh, K., Thakur, A., Awasthi, A., & Kumar, A. (2020). Structural, morphological and temperature-dependent electrical properties of BN/NiO nanocomposites. *Journal of Materials Science: Materials in Electronics, 31*, 13158–13166.

Singh, N., Agarwal, R., Awasthi, A., Gupta, P. K., & Mittal, S. K. (2010). Characterization of atmospheric aerosols for organic tarry matter and combustible matter during crop residue burning and non-crop residue burning months in Northwestern region of India. *Atmospheric Environment, 44*(10), 1292–1300.

Verma, R. L., Gunawardhana, L., Kamyotra, J. S., Ambade, B., & Kurwadkar, S. (2023). Air quality trends in coastal industrial clusters of Tamil Nadu, India: A comparison with major Indian cities. *Environmental Advances*, *13*, 100412.

World Health Organization. (2006). *Air Quality Guidelines: Global Update 2005: Particulate Matter, Ozone, Nitrogen Dioxide, and Sulfur Dioxide*. World Health Organization.

World Health Organization. (2021). *WHO Global Air Quality Guidelines: Particulate Matter (PM2. 5 and PM10), Ozone, Nitrogen Dioxide, Sulfur Dioxide and Carbon Monoxide*. World Health Organization.

Yadav, J. S., & Kaushik, V. K. (1996). Effect of sulphur dioxide exposure on human chromosomes. *Mutation Research/Environmental Mzutagenesis and Related Subjects, 359*(1), 25–29.

3 The Effectiveness of Machine Learning Techniques in Enhancing Air Quality Prediction

Sushant Das and Pushp Raj Tiwari

3.1 INTRODUCTION

Ensuring air quality is a crucial concern for public health and livelihoods, and poses a challenge to economic development and society. Particulate Matter 2.5 ($PM_{2.5}$) is one of the primary contributors to pre-mature mortality as well as severe pollution events in both urban and rural areas. Briefly, $PM_{2.5}$ refers to fine particles suspended in the air, each with a diameter of 2.5 micrometers or smaller. These tiny particles can play a significant role in air quality and have far-reaching implications for public health (Pope et al., 2002). Ammonia (NH_3), a pollutant gas, can originate both from natural and anthropogenic sources with the main source being agricultural fields, landfill sites, live stocks, and sewage works (Sutton et al., 1998, Wilson et al., 2004) which contributes to the $PM_{2.5}$ indirectly by the formation of ammonium salts (e.g., ammonium sulfate and ammonium nitrate). Elevated levels of atmospheric ammonia can have detrimental effects on ecosystems. Ammonia deposition can lead to soil acidification and nutrient imbalances, affecting plant and microbial communities (Pearson and Stewart, 1993). In aquatic ecosystems, ammonia can contribute to eutrophication, leading to changes in water quality and biodiversity. Significant contribution of combustion-related sources to ammonia emissions has been reported by Chen et al., (2022). Therefore, it is important to predict ammonia levels accurately as it is an integral part of air quality and environmental management. It supports proactive measures to mitigate pollution, protect human health, and preserve ecosystems. Observation stations are set up to measure NH_3 to access air quality; however, the area coverage is limited to local areas. Recent advances in satellite-retrieved algorithms provide information about NH_3 content across the globe (Van Damme et al., 2018). However, this facilitates information about present to past concentrations of NH_3, but it is equally important to have information about the future to limit the adverse effects. Therefore, there is an increasing need for reliable and accurate air quality prediction models to forecast the accurate concentration of NH_3. Conventional models often rely on complex numerical simulations, but the abundance of information in climate model outputs offers a chance to enhance the accuracy of air quality

DOI: 10.1201/9781032683805-3

predictions. Briefly, the climate models do forecast by providing insights into future climate conditions based on current knowledge of the climate state variables. These models encompass intricate physical and chemical processes, each characterized by significant uncertainty which leads to substantial biases (Van Ulden and Oldenborgh, 2006; Knote et al., 2011, Das et al., 2020, Sokhi et al., 2021). These models require the creation of an updated regional to global emission source inventory (e.g., Ghosh et al., 2023), real-time meteorological data, and physical and chemical processes, including pollutant transport and diffusion. The models are highly complicated in terms of representing all the physical processes that are difficult to resolve and therefore use parameterization schemes that take a lot of computing resources. To improve the existing biases in the climate models, there has been a growing focus on intelligent data analysis approaches in further improving the prediction of air pollutant concentrations. The Machine Learning (ML) method demonstrates high precision in consistently predicting air pollutants across diverse environmental conditions. ML models can handle complex, non-linear relationships within air quality data more effectively (e.g., Ravidiran et al., 2023). By capturing intricate patterns and interactions among various pollutants and meteorological variables, ML models can provide more accurate predictions of air quality (e.g., Lee et al., 2020). This accuracy is attributed to the utilization of a growing volume of historical data available for research both from models and observations, enhancing the capability to generate more precise forecasts. Recent developments enable hybrid approaches to combine both climate models and ML techniques to improve prediction (Slater et al., 2023). In this chapter, we showcase a case study where we make use of a weather research forecasting model along with a ML technique to improve the prediction of NH_3 level over the United Kingdom (UK).

3.2 MODEL, METHODOLOGY, AND DATA USED

In this case study, we have used the National Centre for Atmospheric Research's (NCAR) Weather Research and Forecasting model (version 4.2.1) coupled with Chemistry (WRF-Chem) (Fast et al., 2006; Grell et al., 2005) and a supervised ML emulator. This is an atmospheric chemistry transport model widely applied to studies around the globe (Georgiou et al. 2018; Kumar et al., 2018). The model is made up of a staggered horizontal grid, terrain-following vertical coordinates, and completely compressible non-hydrostatic equations (Skamarock et al., 2019). Hourly European Centre for Medium-Range Weather Forecasts (ECMWF) reanalysis (ERA5) data is used for initializing the meteorology, boundary conditions, and nudging in the model (Hersbach et al., 2020). The model domain along with configuration is shown in Figure 3.1. The details of the model physics and chemistry are represented through various parameterized schemes are given in Table 3.1.

The time-varying boundary conditions for chemistry have been taken from the global 6-hourly Model for Ozone and Related Chemical Tracers (MOZART-4)/ Goddard Earth Observing System Model version 5. Carbon Bond Mechanism version Z (CBM-Z; Zaveri et al. 1999) which simulates tropospheric gas phase chemistry is a core module in the chemistry transport models (CTMs). In this study, we used the CBM-Z module to create a training dataset of about 1 million examples, with

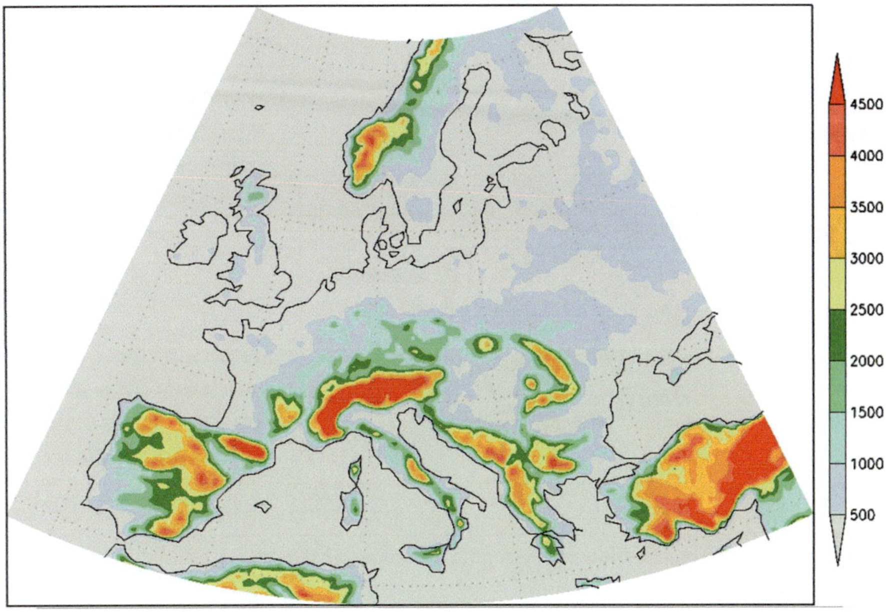

FIGURE 3.1 Model domain and topography (m) used in the present study.

TABLE 3.1
Model Configuration Used in the Present Work

Model	WRF 4.2.1
Number of domains	One
domain central point	50.57°N, 10.40°E
Horizontal grid size	5 km
Map projection	Lambert Conformal
Meteorological data	ECMWF-ERA5 (0.25° resolution)
Land surface data	IGBP MODIS Noah
Radiation scheme	RRTM long-wave radiation
Surface layer parameterization	Noah land surface scheme
PBL scheme	YSU
Microphysics scheme	MORRISON
Trace gas chemistry and aerosol scheme	MOZART trace gas chemistry with MOSAIC aerosol scheme

each example including input pollutant concentrations, meteorology conditions, and output changes in pollution concentrations after one hour of CBM-Z simulation time, which corresponds to minute time steps. Training examples are created by (i) using Latin hypercube sampling to generate pseudorandom values for each variable within the typical atmospheric ranges; (ii) initializing CBM-Z with these random values

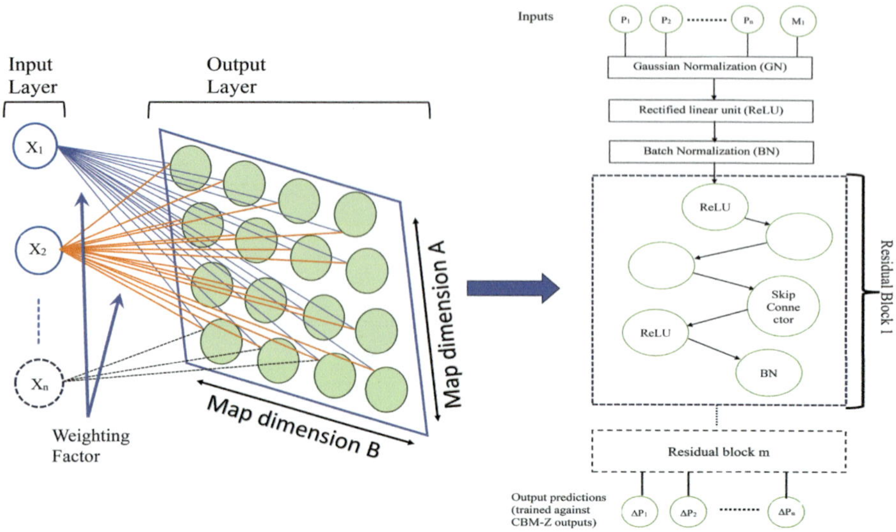

FIGURE 3.2 Schematic showing Machine Learning (ML) based emulator used in the present study.

and running it forward for 3hrs of simulation time (i.e., spin up) to allow the system to adjust to a state closer to what might occur in ambient air; and (iii) running the model forward for 24 hours of simulation time to simulate a diurnal cycle of air pollution, exogenously adjusting the solar zenith angle as the model runs, to represent the changing location of the sun. Further, we have trained our 21-layer neural network (Figure 3.2) by minimizing average mean-squared error (MSE) between CBM-Z and neural network-predicted concentration changes among all pollutants, using stochastic gradient descent.

We have tested our neural network-based ML approach for computational speed by comparing model run times between CBM-Z and ML for simulating chemistry in half a million independent grid cells during a simulation period of 24 hours. The number of grid cells approximately corresponds to one vertical layer of a CTM simulation over the UK at 0.10 × 0.10 horizontal resolution with 36 vertical layers. We have tested the performance of an ML-based emulator in predicting the changes in concentration of pollutant species over multiple time steps (i.e., 3–72 hours) and quantified ML-based performance in replicating CBM-Z predictions. Model performance has been evaluated against ground-based and satellite data sets obtained from the Department for Environment, Food and Rural Affairs (DEFRA) and Rutherford Appleton Laboratory (RAL). In this study, we identify the NH_3 hotspots over the UK to reduce emissions of ammonia, particularly from agricultural sources by implementing mitigation strategies. Further, ammonia (NH_3) is a gas that only stays in the atmosphere for a few hours once emitted. However, when ammonia mixes with other gases in the atmosphere, such as nitrogen oxides and sulfur dioxide, it can form particulate matter (PM) which can exist for several days and be transported large distances.

3.3 RESULT AND DISCUSSION

We first developed the ML-based emulator and analyzed the accuracy (in standalone mode) and computational time against WRF-Chem model generated NH_3 for April 2016. Further, we have replaced the chemical mechanism with ML and performed a simulation to see whether the ML-based emulator can reproduce the behavior of a chemical mechanism and its impact on computational time. Table 3.2 shows the computational time gain achieved with the ML emulator compared to WRF-Chem simulation. It is noticed that the ML emulator roughly halves the computational time or, in other words, achieves a twofold time gain. The decrease in the computational time using various ML techniques is well demonstrated by Sarker (2021). We have computed the Root Mean Square Error (RMSE) for variables (e.g., wind components at 10 m (U10, V10), temperature at 2 m (T2), Sea Level Pressure (SLP)). The detailed performances are shown in Table 3.3. Compared with the ground-based observation, the RMSE of 2 m temperature (T2), wind at 10 m (U10, V10), and SLP of the ML-based approach is reduced by 30%, 29%, and 27% respectively. Earlier, Xu et al., (2020) also found that RMSE significantly reduced using gradient boosting decision tree algorithm compared to standard WRF simulation in predicting wind speed. Similar arguments were given by Wang et al (2021) where they found that the average RMSE of SLP decreased by more than 50% in random forest-based adjusting method compared to the numerical weather forecast model.

Further we have evaluated model performance (i.e., WRF-Chem and ML emulator) against RAL space-based satellite NH_3 observations (Figure 3.3). It is noticed that though the ML-based emulator achieves a significant computational time gain, it is unable to capture the spatial pattern of NH_3 in certain regions compared

TABLE 3.2
Computational Time Gain with ML Compared to WRF-Chem

Experiments	Simulation time (hr)	Time gain
WRF-Chem	8	~1.9
ML	4.3	

TABLE 3.3
Root Mean Square Error (RMSE) for ML and WRF-Chem Model

Variables	WRF-Chem	ML
T2	1.92	1.36
U10	3.64	2.76
V10	3.37	2.19
SLP	2.81	2.04

to WRF-Chem during the month of April. The underestimation in ML is mostly in western and southern UK, whereas it predicts well over the Midlands of the UK compared to WRF-Chem, where most agricultural activities happen and produce huge NH_3. Moreover, it is also worth mentioning here that though WRF-Chem can capture the spatial extent of NH_3 over the region, it still lacks in capturing the correct amount and takes significant computational time to simulate NH_3. The possible reason for the ML-based emulator failure in capturing the spatial amount of NH_3, especially over the west and southern UK, might be due to the limited number of neural net layers and the absence of physical process representation and their feedback as compared to WRF-Chem.

In addition, we have also analyzed the seasonality of NH_3 over UK as seen in Figure 3.4. The plot provides the NH_3 estimates as the percentage change in emissions

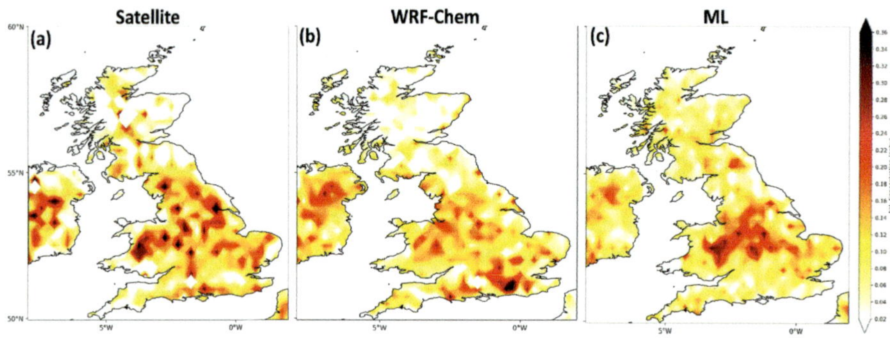

FIGURE 3.3 Observed and WRF-Chem, Machine Learning (ML) simulated NH_3 for the period April 2016.

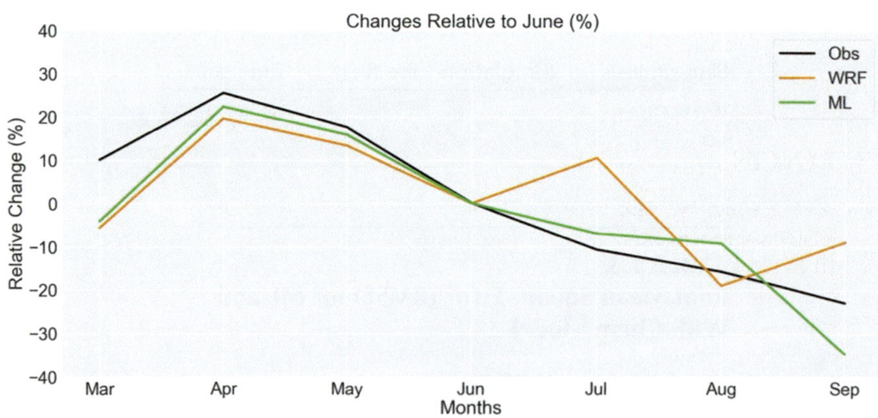

FIGURE 3.4 NH_3 emissions seasonality from March to September over the UK. Lines show the percentage change in emissions in each month relative to those in June for observed (black), WRF-Chem (orange) and Machine Learning (ML) (green).

in each month relative to those in June. It can be noticed that the ML-based emulator is able to capture the changes with some biases but at the same time are close to observation compared to WRF-Chem.

3.4 CONCLUSION

Air Quality Forecasting Systems (AQFS) have an integrated chemistry module and simulations through them are computationally very expensive. In this study, we tested and tried to demonstrate the concept of replacing the CTM module from AQFS with a novel Machine Learning (ML) approach to achieve orders of magnitude speed up in chemistry simulations which mostly through traditional methods are slow and tend to suffer from numerical instability. We have used WRF-Chem and an ML-based emulator for this work to simulate the NH_3 over the UK for April 2016. The key findings of this study are summarized below:

- The ML-based emulator shows a twofold computational time gain compared to the conventional WRF-Chem modeling framework.
- Spatial and temporal variation of NH_3 is reasonably simulated by WRF-Chem while ML captures NH_3 close to observation, especially in the Midlands where most of the agricultural activity takes place.
- The ML-based emulator shows better closeness as compared to WRF-Chem in capturing the percentage change in NH3 emissions in each month relative to those in June.

One of the primary limitations of the study is that the ML technique used lacks an explicit physical understanding of the processes. They learn patterns from the training data but may not capture underlying physical mechanisms accurately. Therefore, a physics-driven ML technique is required to enhance the quality of NH_3 simulation. In a future study, we will be implementing a greater number of layers, providing information about wind, temperature, etc. from radar/satellite data, and will investigate avenues on how we can represent the processes using image processing approaches to strengthen our ML-based emulator.

REFERENCES

Chen, Z. L., Song, W., Hu, C. C., et al. (2022): Significant contributions of combustion-related sources to ammonia emissions. *Nature Communication* 13, 7710. https://doi.org/10.1038/s41467-022-35381-4

Das, S., Giorgi, F., Giuliani, G., Dey, S., and Coppola, E. (2020): Near-future anthropogenic aerosol emission scenarios and their direct radiative effects on the present-day characteristics of the Indian summer monsoon. *Journal of Geophysical Research: Atmospheres*, 125, e2019JD031414. https://doi.org/10.1029/2019JD031414

Fast, J. D., Gustafson Jr., W. I., Easter, R. C., Zaveri, R. A., Barnard, J. C., Chapman, E. G., Grell, G. A., and Peckham, S. E. (2006): Evolution of ozone, particulates, and aerosol direct radiative forcing in the vicinity of Houston using a fully coupled meteorology-chemistry-aerosol model. *Journal of Geophysical Research: Atmospheres*, 111, 1–29. https://doi.org/10.1029/2005JD006721.

Georgiou, G. K., Christoudias, T., Proestos, Y., Kushta, J., Hadjinicolaou, P., and Lelieveld, J. (2018): Air quality modelling in the summer over the eastern Mediterranean using WRF-Chem: Chemistry and aerosol mechanism intercomparison. *Atmospheric Chemistry and Physics*, 18, 1555–1571. https://doi.org/10.5194/acp-18-1555-2018.

Ghosh, S., Dey, S., Das, S., Riemer, N., Giuliani, G., Ganguly, D., Venkataraman, C., Giorgi, F., Tripathi, S. N., Ramachandran, S., Rajesh, T. A., Gadhavi, H., and Srivastava, A. K. (2023): Towards an improved representation of carbonaceous aerosols over the Indian monsoon region in a regional climate model: RegCM. *Geoscience Model Development*, 16, 1–15. https://doi.org/10.5194/gmd-16-1-2023.

Grell, G. A., Peckham, S. E., Schmitz, R., McKeen, S. A., Frost, G., Skamarock, W. C., and Eder, B. (2005): Fully coupled "online" chemistry within the WRF model. *Atmospheric Environment*, 39, 6957–6975. https://doi.org/10.1016/j.atmosenv.2005.04.027.

Hersbach, H., Bell, B., Berrisford, P., et al. (2020): The ERA5 global reanalysis. *Quarterly Journal of the Royal Meteorological Society*, 146, 1999–2049. https://doi.org/10.1002/qj.3803.

Knote, C., Brunner, D., Vogel, H., Allan, J., Asmi, A., Äijälä, M., ... & Vogel, B. (2011). Towards an online-coupled chemistry-climate model: evaluation of trace gases and aerosols in COSMO-ART. Geoscientific Model Development, 4(4), 1077–1102.

Kumar, M., Parmar, K. S., Kumar, D. B., Mhawish, A., Broday, D. M., Mall, R. K., and Banerjee, T. (2018): Long-term aerosol climatology over Indo-Gangetic Plain: Trend, prediction and potential source fields. *Atmospheric Environment*, 180, 37–50. https://doi.org/10.1016/j.atmosenv.2018.02.027.

Lee, M., Lin, L., Chen, C. Y., et al. (2020): Forecasting air quality in Taiwan by using machine learning. *Science Reports,* 10, 4153. https://doi.org/10.1038/s41598-020-61151-7

Pearson, J., & Stewart, G. R. (1993): The deposition of atmospheric ammonia and its effects on plants. New *Phytologist*, 125(2), 283–305.

Pope, A., Burnett, R. T., Thun, M. J., Calle, E. E., Krewski, D., Thurston, G. D. (2002): Cardiopulmonary mortality, and long-term exposure to fine particulate air pollution. *Lung Cancer*, 287(9), 1132–1141.

Ravindiran, G., Hayder, G., Kanagarathinam, K., Alagumalai, A., & Sonne, C. (2023): Air quality prediction by machine learning models: A predictive study on the Indian coastal city of Visakhapatnam. Chemosphere, 338, 139518.

Sarker, I. H. (2021). Deep learning: A comprehensive overview on techniques, taxonomy, applications and research directions. *SN Computer Science,* 2, 420. https://doi.org/10.1007/s42979-021-00815-1

Skamarock, W. C., Snyder, C., Klemp, J. B., & Park, S.-H. (2019): Vertical resolution requirements in atmospheric simulation. *Monthly Weather Review*, 147(7), 2641–2656.

Slater, L. J., Arnal, L., Boucher, M.-A., Chang, A. Y. -Y., Moulds, S., Murphy, C., Nearing, G., Shalev, G., Shen, C., Speight, L., Villarini, G., Wilby, R. L., Wood, A., and Zappa, M. (2023): Hybrid forecasting: blending climate predictions with AI models. *Hydrology and Earth System Sciences*, 27, 1865–1889, https://doi.org/10.5194/hess-27-1865.

Sokhi, R. S., Tiwari P. R., and de Medeiros J. S. N. (2021): Changes in extreme events over Asia for present and future climate conditions based on a modelling analysis of atmospheric circulation anomalies. *Theoretical and Applied Climatology*. https://doi.org/10.1007/s00704-021-03742-6.

Sutton, M. A., Milford, C., Dragosits, U., Place, C. J., Singles, R. J., Smith, R. I., Pitcairn, C. E. R., Fowler, D., Hill, J., ApSimon, H. M., Ross, C., Hill, R., Jarvis, S. C., Pain, B. F., Phillips, V. C., Harrison, R., Moss, D., Webb, J., Espenhahn, S. E., Lee, D. S., Hornung,

M., Ullyett, J., Bull, K. R., Emmett, B. A., Lowe, J., and Wyers, G. P. (1998): Dispersion, deposition and impacts of atmospheric ammonia: Quantifying local budgets and spatial variability. *Environmental Pollution,* 102, 349–361.

Van Damme, M., Clarisse, L., Whitburn, S., et al. (2018): Industrial and agricultural ammonia point sources exposed. *Nature*, 564, 99–103. https://doi.org/10.1038/s41586-018-0747-1

Van Ulden, A. P., & Van Oldenborgh, G. J. (2006): Large-scale atmospheric circulation biases and changes in global climate model simulations and their importance for climate change in Central Europe. Atmospheric Chemistry and Physics, 6(4), 863–881.

Wang, A., Xu, L., Li, Y., Xing, J., Chen, X., Liu, K., et al. (2021): Random-forest based adjusting method for wind forecast of WRF model. *Computer Geoscience* , 155, 104842. http://dx.doi.org/10.1016/j.cageo.2021.104842.

Wilson, L. J., Bacon, P. J., Bull, J., Dragosits, U., Blackall, T. D., Dunn, T. E., Hamer, K. C., Sutton, M. A., Wanless, S. (2004): Modelling the spatial distribution of ammonia emissions from seabirds in the UK. *Environmental Pollution*, 131, 173–185.

Xu, W., Ning L., and Luo, Y. (2020): Wind speed forecast based on post-processing of numerical weather predictions using a gradient boosting decision tree algorithm. *Atmosphere*, 11, 738. https://doi.org/10.3390/atmos11070738.

Zaveri, R. A., and Peters, L. K. (1999): A new lumped structure photochemical mechanism for large-scale applications. *Journal of Geophysical Research: Atmospheres,* 104, 30387–30415.

4 Enhancing Environmental Resilience

Precision in Air Quality Monitoring through AI-Driven Real-Time Systems

Ankit Mahule, Kaushik Roy, Ankush D. Sawarkar, and Sagar Lachure

4.1 INTRODUCTION

The condition of the air we inhale significantly influences the health of our communities, the state of our environment, and our overall quality of life. (Panneerselvam et al., 2023). The air we inhale affects our immediate health, our long-term susceptibility to various diseases, and the state of our planet. There are closed and open burning reasons that deteriorate the air quality, which poses short and long-term direct effects on the respiratory system (Agarwal et al., 2014; Singh et al., 2010). Recognizing the significance of this, monitoring air quality has become imperative in our modern world (Oh & Kim, 2020). Traditionally, this monitoring relied on periodic data collection through fixed sensors and manual sampling (Asha et al., 2022; Masood & Ahmad, 2021). In the present rapidly evolving society, traditional air quality monitoring methods prove inadequate in delivering the prompt and exhaustive data necessary for informed decision-making and safeguarding the health of public (Heo et al., 2019).

In this context, technological progress has unveiled novel possibilities for real-time air quality monitoring systems (Cheng et al., 2014; Fadhil et al., 2023; Shah et al., 2020). These advancements, frequently propelled by artificial intelligence (AI), have the potential to revolutionize how we assess and address air quality concerns (Dhanalakshmi et al., 2020; Zivelonghi & Giuseppi, 2024). This chapter investigates the amalgamation of AI-driven systems, weather model data, and case studies tailored to specific regions. It aims to present a comprehensive insight into how these technologies can elevate air quality monitoring, offering valuable perspectives for policymakers, researchers, and practitioners dedicated to building a sustainable future.

 DOI: 10.1201/9781032683805-4

4.1.1 Background and Importance of Real-Time Air Quality Monitoring

Escalating environmental challenges, including swift urbanization, industrialization, and the intricate consequences of climate change, have intensified the need for accurate and real-time air quality information (Enigella & Shahnasser, 2018; Kim et al., 2014; Yadav et al., 2023). These challenges mandate a shift from traditional monitoring approaches to real-time solutions. Real-time air quality monitoring introduces a revolutionary method, delivering continuous and high-frequency data streams to effectively monitor changes in air quality in real-time (Moharana et al., 2020; Shah et al., 2020; Sharma et al., 2020). This capability is not only advantageous but is increasingly becoming an imperative necessity. It enables swift responses to pollution, diminishes health risks, and guides the development of effective environmental policies for the betterment of global communities (Kim et al., 2014; Sharma et al., 2020).

In today's interconnected world, where information and data play a pivotal role in decision-making, real-time air quality monitoring is poised to be a cornerstone in ensuring the health of the public and the sustainability of the environment (Saini et al., 2020a; Sharma et al., 2020). The ability to access immediate data on air quality empowers governments, communities, and individuals to make informed choices, implement swift interventions, and collaborate in safeguarding the well-being of our planet and the health of its inhabitants (Hable-Khandekar & Srinath, 2017). It is, therefore, essential to explore the potential of this technology and understand its significance in addressing the intricate challenges of our time.

4.1.2 Challenges with Conventional Monitoring Approaches

Traditional methods of monitoring air quality have their role in tracking long-term patterns and establishing fundamental data, but they prove inadequate when confronted with the dynamic challenges of today (Moursi et al., 2021). First, their intermittent approach to data collection is ill-suited for detecting and responding to rapid shifts in air quality. For instance, abrupt increases in pollution levels resulting from industrial accidents, wildfires, or fleeting weather patterns may go unnoticed for hours or even days using traditional monitoring techniques. Second, these systems often tend to be concentrated in urban areas, where industrial activities and vehicular emissions present significant air quality concerns. This urban-centric focus leaves rural and underserved communities with inadequate coverage, impeding their capacity to address local air quality issues.

Lastly, traditional methods, while valuable in some respects, entail a significant amount of manual labor and are susceptible to technical issues that can undermine their reliability over time (Asha et al., 2022; Kaginalkar et al., 2021; D. Zhang & Woo, 2020). In light of these challenges, it becomes evident that a shift towards real-time air quality monitoring, bolstered by modern technology and data-driven solutions, is essential to address the pressing needs of our evolving environmental landscape and ensure a more comprehensive and responsive approach to safeguarding air quality.

4.1.3 Aims and Structure of the Chapter

The central aim of this chapter is to explore the transformative impact of AI on reshaping the field of air quality monitoring. It delves into the applications of AI algorithms, specifically Machine Learning (ML) and data analytics, for processing data from diverse sources, including satellite observations, sensor networks, and weather models. Emphasizing the crucial integration of weather model data into real-time assessments, the chapter elucidates how this fusion enhances the accuracy of pollution forecasting. To illustrate these concepts, a series of region-specific case studies are presented, offering tangible examples of the implementation of AI-based monitoring systems and highlighting their adaptability and scalability across various contexts.

In summary, this chapter serves as a comprehensive guide to understanding how AI is revolutionizing air quality monitoring. It illuminates the pivotal role of AI in protecting community health, facilitating data-driven policy interventions, and advancing our comprehension of the intricate relationship between environmental factors and empirical observations. The exploration of AI's potential in air quality monitoring aligns with the broader global endeavor toward a sustainable and healthier future.

4.2 ARTIFICIAL INTELLIGENCE REVOLUTIONIZING AIR QUALITY MONITORING

In the ever-evolving landscape of technological innovation, AI stands out as a cornerstone, revolutionizing how we perceive and address complex challenges. Within this transformative wave, AI technologies have emerged as powerful tools, fundamentally reshaping the realm of air quality monitoring. Harnessing the capabilities of ML and data analytics, AI processes extensive and diverse datasets with unparalleled speed and accuracy (Ali et al., 2015; Fadhil et al., 2023). This capability is instrumental in enabling real-time air quality assessment, allowing AI-driven systems to identify pollution patterns, forecast pollution events, and provide valuable insights into air quality dynamics (Alexandrova & Ahmadinia, 2018; Fadhil et al., 2023; Karagulian et al., 2019; J. Wang et al., 2023). Unlike traditional methods constrained by human oversight and periodic data collection, AI empowers policymakers, urban planners, and local health officials with tools for informed decision-making and effective interventions to safeguard community health.

In the face of rapid urbanization and increased industrial activity, AI-driven air quality monitoring not only addresses immediate concerns but also lays the groundwork for long-term environmental sustainability. Integrating AI into air quality monitoring represents a significant leap forward, responding to the demands of our changing world (Akselsen et al., 2019; Zhu et al., 2023). This evolution is driven by technological advancements that align with contemporary challenges. The section unfolds a revolutionary transformation in air quality monitoring through the infusion of AI algorithms. It delves into the applications of AI, encompassing ML and data analytics, elucidating their pivotal role in reshaping the evaluation and oversight of air quality (Akselsen et al., 2019; Y. Li et al., 2022). The narrative aims to illuminate the innovative solutions AI offers, demonstrating how it empowers decision-makers to navigate complex environmental concerns (Kaginalkar et al., 2021). This exploration underscores AI's potential as a cornerstone in the pursuit of a cleaner and healthier world, prioritizing public health and the sustainability of the environment (Akselsen et al., 2019; Zhu et al., 2023).

4.2.1 Demystifying AI Algorithms for Air Quality Monitoring

ML, a subset of AI, empowers air quality monitoring systems to collect data and make predictions or decisions autonomously, eliminating the need for explicit programming. In the context of air quality, ML models undergo training using historical and real-time data to identify patterns and relationships (Tandon et al., 2023). These models excel in classifying pollutants, forecasting pollution levels, and detecting anomalies, thereby enhancing our understanding of air quality dynamics.

Consider the classification of pollutants as an example. While traditional methods rely on predefined thresholds to categorize pollutant levels as "good" or "bad," ML leverages insights from patterns in historical data, considering factors like meteorological conditions, traffic patterns, and industrial activities (Sharma et al., 2020; Völgyesi et al., 2008; D. Zhang & Woo, 2020). This dynamic approach enables a more nuanced and precise evaluation of air quality. The application of ML in air quality monitoring not only improves the accuracy of assessments but also allows for a more adaptable and responsive approach to the complexities of air quality management in our ever-changing world. As we explore further the domain of AI-driven solutions, the potential for innovative insights and more effective strategies to protect our environment and health continues to grow.

4.2.2 Data Analytics

Data analytics in air quality monitoring entails the methodical examination of data to extract practical insights. By employing statistical techniques, data preprocessing, and visualization, analytics play a key role in unveiling concealed trends, correlations, and anomalies(Shah et al., 2020). This process aids in making well-informed decisions, optimizing pollution control strategies, and enriching our overall comprehension of air quality dynamics. Data analytics has the capacity to unveil intricate relationships within air quality data. For instance, it can divulge how specific meteorological conditions, like temperature inversions, influence pollutant levels (Borrajo & Cao, 2020; Sree Devi & Rahamathulla, 2020). Equipped with these insights, authorities can issue precise advisories to mitigate health risks during adverse weather conditions.

The utilization of data analytics in air quality monitoring is crucial, not only for improving decision-making but also for advancing our comprehension of the intricate interactions occurring in our environment. This knowledge empowers us to develop more targeted and effective strategies for safeguarding public health and maintaining the quality of our air, particularly when facing challenges posed by varying weather conditions and industrial activities (Roy et al., 2024). As we harness the potential of data analytics in this field, we are better equipped to tackle air quality issues and mitigate their impacts on communities and the environment.

4.2.3 Data Sources for AI-Powered Monitoring

4.2.3.1 Satellite Observations

Artificial intelligence harnesses satellite observations to monitor extensive air quality patterns. Satellites, equipped with specialized sensors, capture data pertaining to atmospheric composition, aerosols, and pollutant concentrations. AI processes this satellite data, offering a global perspective on air quality that facilitates the monitoring of long-distance pollutant dispersion and the identification of pollution sources (Engel-Cox et al., 2004; Gupta et al., 2006; Kaginalkar et al., 2021; Shelestov et al., 2018). The integration of satellite data holds immense value in comprehending transboundary air pollution. For instance, AI can trace the movement of pollutants across international borders, thereby aiding diplomatic efforts to mitigate cross-border pollution.

4.2.3.2 Sensor Networks

Ground-level air quality monitoring relies on networks of sensors collecting real-time data at specific locations. These networks incorporate both stationary and mobile sensors designed to measure various parameters, including particulate matter, gases, and meteorological conditions. AI plays a vital role in orchestrating and analyzing diverse sensor data, enabling continuous and localized monitoring (Alexandrova & Ahmadinia, 2018; Molka-Danielsen et al., 2017). The real-time sensor data offers a detailed perspective on air quality at street level. AI algorithms can adeptly process this data, generating pollution heatmaps that precisely identify pollution hotspots. Subsequently, urban planners can leverage this valuable information to develop urban environments that are cleaner and healthier.

4.2.3.3 Weather Models

Weather models simulate environmental variables such as meteorology, atmospheric circulation, and pollutant dispersion. AI integrates the outputs of these weather models with real-time monitoring data to enhance the accuracy of pollution forecasting (Balogun et al., 2021; Nam et al., 2020). This integration offers valuable insights into the complex interplay between meteorology and air quality, aiding in the establishment of early warning systems for pollution events.

The pivotal role of AI in this integration stems from its capability to analyze extensive datasets generated by weather models and combine them with real-time monitoring data. Through this process, AI can predict pollution events with exceptional accuracy, enabling authorities to proactively implement measures to safeguard community health. The seamless integration of these data sources represents a significant advancement in our ability to understand and respond effectively to air quality challenges.

4.2.4 Real-Time Air Quality Monitoring System

Creating a sophisticated real-time air quality monitoring system driven by AI is a multifaceted process that requires integrating various components and interactions. The following presents an illustrative example of constructing a flowchart to depict such a system, inspired by existing works (Ali et al., 2015; Fadhil et al., 2023). It's important to note that the provided representation offers a high-level overview and can be further detailed to correspond with the specific architecture and components of a particular system. Figure 4.1 showcases a simplified representation of the system's structure.

- **Sensors:** These encompass a variety of air quality sensors responsible for data collection.
- **Data Collector:** This component gathers data from the sensors.
- **Database:** The data collected is stored within a database, serving the purpose of maintaining historical records and facilitating the training of AI models.

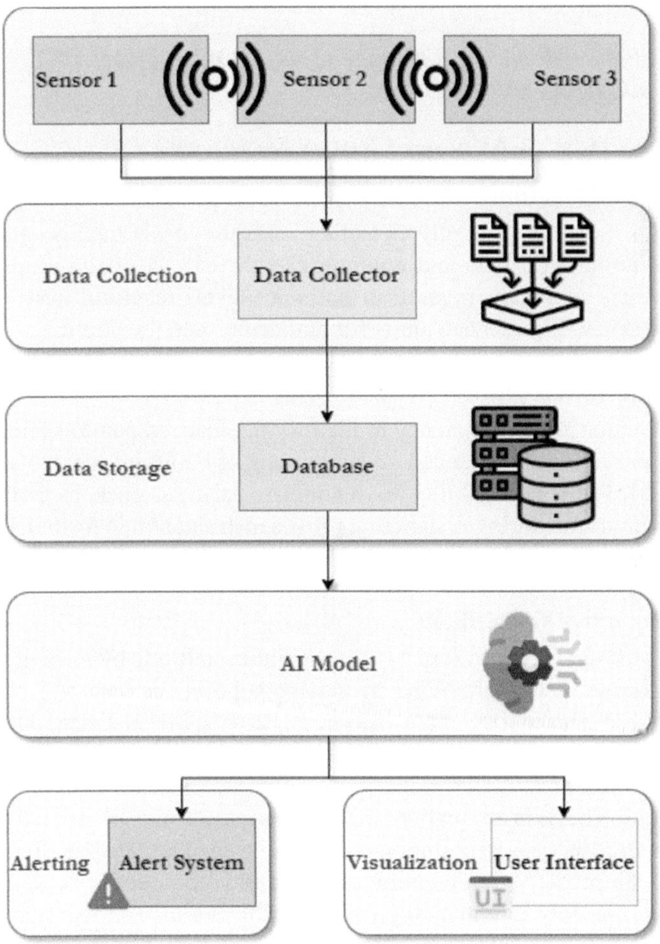

FIGURE 4.1 Dynamic air quality monitoring infrastructure: a visual overview.

- **AI Model:** This model processes data, generates predictions, and provides insights.
- **User Interface:** The user interface presented here acts as a platform through which users can effectively visualize essential information regarding air quality.
- **Alert System:** This system functions by generating notifications based on the analysis conducted by the AI model, which are subsequently transmitted to users.

Establishing a real-time air quality monitoring system of this nature is essential for providing timely and precise information about air quality, a key element in protecting public health and the environment. It forms a critical part of our modern environmental management strategies, providing actionable data for decision-makers, urban planners, and the public. The seamless integration of these components is central to the success of such a system.

4.3 BENEFITS OF AI IN IDENTIFYING POLLUTION PATTERNS AND PREDICTING EVENTS

4.3.1 Advantages of AI in Air Quality Monitoring

4.3.1.1 Swift Detection

AI algorithms excel at promptly detecting changes in air quality, enabling swift responses to pollution events and reducing health risks (Motlagh et al., 2023). For example, if there is a sudden surge in pollutant levels resulting from an industrial accident, AI can swiftly activate alerts for authorities and the public.

4.3.1.2 Pattern Recognition

ML models demonstrate proficiency in identifying nuanced patterns and correlations within data, providing an enhanced understanding of pollution sources and dynamics (Su et al., 2021). AI's capability to recognize patterns, such as pollution peaks associated with traffic during rush hours, proves instrumental in formulating effective traffic management strategies.

4.3.1.3 Predictive Capabilities

AI possesses the capability to forecast future pollution levels by leveraging historical data and real-time inputs, providing invaluable support for proactive pollution control and public health management (Mo et al., 2019). For instance, AI systems can anticipate upcoming air quality conditions, allowing vulnerable individuals to take precautionary measures.

The utilization of AI in air quality monitoring yields a multitude of advantages, ranging from swift responses to pollution events and a deeper understanding of pollution dynamics to the proactive anticipation of air quality conditions. These benefits play a pivotal role not only in enhancing public health but also in facilitating informed decision-making and the implementation of effective pollution control strategies. The

TABLE 4.1
Key Benefits of AI in Air Quality Monitoring

Benefit	Description	Reference
Rapid detection	AI algorithms facilitate the swift identification of shifts in air quality, allowing for timely responses.	(Aguénounon et al., 2020)
Pattern recognition	ML models are adept at discerning subtle patterns and associations in air quality data.	(Hilpert et al., 2019)
Predictive capabilities	AI demonstrates the ability to anticipate future pollution levels, assisting in proactive pollution management and safeguarding public health.	(Bae et al., 2021)
Actionable insights for decision-makers	AI generates practical insights for various stakeholders, including those involved in public health, urban planning, and policy development.	(Wilson & van der Velden, 2022)
Integration of weather model data	The amalgamation of weather model data enhances forecasting precision and offers a comprehensive perspective on air quality dynamics.	(Lee et al., 2020)

proactive nature of AI-driven monitoring aligns seamlessly with the global pursuit of healthier and more sustainable environments.

Table 4.1 summarizes the key benefits of AI in air quality monitoring. It highlights how AI enables rapid detection of changes in air quality, enhances pattern recognition for a deeper understanding of pollution sources, and provides predictive capabilities for proactive pollution control and public health management.

4.3.2 Generation of Practical Guidance for Decision-Makers

4.3.2.1 Public Health

Health authorities can receive instantaneous alerts and recommendations in real-time to protect vulnerable populations during episodes of poor air quality, thereby mitigating the health impacts of pollution (Saini et al., 2020a; L. Wang et al., 2016). For instance, during a wildfire event, AI can offer tailored guidance to asthma patients on minimizing their exposure to smoke.

4.3.2.2 Urban Planning

City planners can harness AI-generated data to develop more sustainable urban environments, optimizing traffic flow and reducing emissions (Gulia et al., 2015). AI can suggest adjustments to traffic signal timings based on real-time air quality data, alleviating congestion and pollution.

4.3.2.3 Policy Development

Policymakers gain valuable evidence-based data for crafting effective air quality regulations and emission reduction strategies. AI plays a pivotal role in simulating the potential impacts of different policy scenarios, providing policymakers with insightful information for making well-informed decisions. The revolutionary impact of AI algorithms, including ML and data analytics, on air quality monitoring is evident through their ability to swiftly detect changes, recognize patterns, and make predictions (Engel-Cox & Hoff, 2005). Drawing on data from satellite observations, sensor networks, and weather models, AI provides practical insights that empower decision-makers to mitigate the impact of pollution and safeguard public health.

4.3.2.4 Integration of Weather Model Data

Weather models, which simulate various environmental variables and atmospheric conditions, play a crucial role in enhancing the accuracy and comprehensiveness of air quality assessments. This section delves deeply into weather models and their integration into real-time air quality monitoring systems (Engel-Cox & Hoff, 2005). The integration of weather model data signifies a substantial advancement in our ability to comprehend and address air quality challenges. It provides a comprehensive perspective on how environmental conditions impact air quality, facilitating more informed decision-making and the formulation of effective pollution control strategies.

4.3.3 An Overview of Weather models and Their Role in Air Quality Assessment

Weather models represent complex mathematical depictions of Earth's climate system (Juda-rezler, 2010). They simulate interactions among various atmospheric components, including composition, meteorological parameters, and mechanisms for transporting pollutants (Fiore et al., 2012). Although weather models are primarily designed for long-term climate predictions, they provide valuable insights into short-term air quality dynamics. In the realm of air quality assessment, weather models serve a dual purpose (Tendelilin, 2010). First, they furnish essential information about atmospheric conditions and pollutant dispersion over extensive spatial and temporal scales. This includes factors such as wind patterns, temperature inversions, and regional climate phenomena, all exerting significant influences on air quality. Second, when integrated into real-time monitoring systems, weather model data enhances our understanding of pollution sources, dispersion patterns, and prevailing trends.

The integration of weather model data significantly enhances our capacity to comprehend and manage air quality issues. It provides a holistic view of how environmental conditions impact air quality, facilitating more effective decision-making and strategies for pollution control. By uniting the capabilities of weather models with real-time monitoring, we are better equipped to address the evolving challenges of air quality management. Weather models play a vital role in advancing our understanding of air quality dynamics and providing accurate forecasts.

Figure 4.2 illustrates how weather model data seamlessly integrates into air quality monitoring systems. Weather models offer essential information about atmospheric conditions, pollutant dispersion, and regional climate phenomena. Combining this data with real-time monitoring provides a more comprehensive understanding of air quality, including its sources, trends, and forecasting. The integration of weather

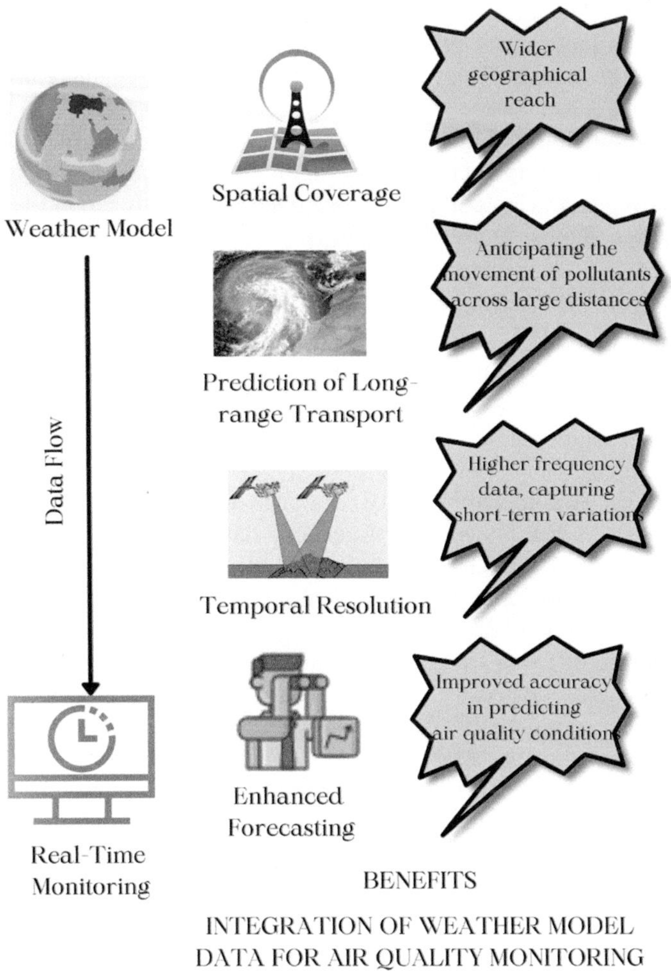

FIGURE 4.2 Synergizing weather models and real-time monitoring for enhanced air quality insight.

model data is instrumental for assessing air quality globally, tracking transboundary pollution, and improving the accuracy of both short-term and long-term forecasts. This approach empowers decision-makers, researchers, and policymakers with valuable insights to enhance air quality management and safeguard public health.

- **Weather Models:** Placed at the apex of the diagram, the "weather models" component represents sophisticated mathematical simulations of Earth's climate system. These models encapsulate a diverse range of atmospheric variables, incorporating meteorological conditions, pollutant dispersion mechanisms, and mechanisms for long-range transport.

- **Data Flow:** The arrow linking "weather models" with "real-time monitoring" symbolizes the transfer of data from weather models to real-time air quality monitoring systems. This data flow is essential for augmenting air quality evaluations.
- **Real-Time Monitoring:** Located at the bottom, "real-time monitoring" represents the comprehensive and ongoing monitoring of air quality across various regions. This component integrates data from ground-level sensors, satellites, and other monitoring devices.

4.3.3.1 Benefits of Weather Model Data Integration

1. **Spatial Coverage:** Adjacent to "weather models," the "spatial coverage" benefit denotes the broader geographical reach achieved by incorporating weather model data. This integration extends air quality monitoring to regions lacking traditional monitoring infrastructure, including remote or underserved areas.
2. **Temporal Resolution:** Positioned alongside "spatial coverage," "temporal resolution" reflects the higher-frequency data provided by weather models. This enhanced temporal resolution enables real-time monitoring systems to track short-term variations and changes in air quality, ensuring timely responses to pollution events.
3. **Prediction of Long-range Transport:** The "prediction of long-range transport" benefit, located on the right, underscores the capacity to predict the movement of pollutants over substantial distances. Weather models are proficient in replicating the long-range transport of pollutants, facilitating the assessment of transboundary air pollution.
4. **Enhanced forecasting:** The fourth benefit, "enhanced forecasting," underscores the improved accuracy in predicting air quality conditions. By integrating weather model outputs with real-time monitoring data, these systems offer more precise and longer-term air quality forecasts, thereby enhancing public health protection.

Figure 4.2 operates as a visual representation of the synergy between weather model data and real-time air quality monitoring. It showcases how this integration amplifies the capabilities of these systems in safeguarding community health and the environment.

4.3.3.2 Exploring Weather Models in Air Quality Assessment

To comprehend the intricate dynamics of air quality, an exploration into the realm of weather models becomes imperative. Weather models, sophisticated mathematical representations of the Earth's climate system, play a pivotal role in evaluating the impact of climate change on various domains, including air quality. There are two primary categories of weather models: Global Weather Models (GCMs) and Regional Weather Models (RCMs).

- **Global Weather Models (GCMs):** Global Weather Models, designed to emulate climate patterns across the entire planet, offer a broad perspective on atmospheric dynamics (Pierce et al., 2009). Integration of GCMs with chemistry models enhances the accuracy of air pollutant movement and transformation simulations. This integration allows for a thorough evaluation of the interplay between climate change and air quality, elucidating how changes in climate patterns impact the dispersion of air pollutants.
- **Regional Weather Models (RCMs):** In contrast, Regional Weather Models focus on simulating climate phenomena within more confined areas, such as specific countries or continents (Rummukainen, 2010). These models provide a localized understanding of weather patterns, contributing valuable insights into the interaction between regional climatic conditions and air quality dynamics.

The incorporation of weather models into air quality monitoring is instrumental in unravelling the complex interplay between climate and air quality. By elucidating how climatic variations influence the dispersion and behavior of air pollutants, these models contribute crucial knowledge for devising effective strategies to mitigate air pollution impacts amid a changing climate.

4.3.3.3 The Importance of Weather Models in Assessing Air Quality

Weather models are essential in assessing air quality through various mechanisms:

- **Identification of Pollution Sources:** Weather models facilitate the simulation of the transport and dispersion of air pollutants originating from various sources, aiding in the identification of pollution origins. This data is instrumental in determining the sources of air pollution exacerbating issues within specific geographical areas.
- **Anticipation of Future Air Quality:** The use of weather models allows for the anticipation of forthcoming air quality conditions by considering various factors, including climate change, population growth, and economic development. This information can be harnessed for the development and implementation of strategies to manage air quality effectively.
- **Addressing Data Gaps:** Weather models possess the capability to overcome data deficiencies in air quality monitoring, particularly in regions with limited or no air quality monitoring infrastructure. Utilizing weather model data allows for accurate and reliable air quality assessments, addressing a critical need in regions where monitoring is lacking.

The significance of weather models in evaluating air quality extends beyond mere assessment; it serves as a pivotal tool in the endeavor to comprehend and manage the complex relationship between climate change and air quality. By providing valuable insights and bridging data gaps, weather models contribute to more informed decision-making and the development of effective strategies for preserving air quality in an ever-changing world.

4.3.3.4 Weather Models in Air Quality Assessment: Applications

Weather models find practical applications in various facets of air quality assessment. Several specific instances of their utilization for evaluating air quality include:

- **Assessing Climate Change Effects on Ozone Levels:** Weather models are employed to evaluate the influence of climate change on ozone levels. Ozone is a significant atmospheric pollutant capable of causing respiratory and health-related issues. Weather models have indicated that climate change is expected to lead to increased ozone levels in various regions worldwide.
- **Evaluating Climate Change Effects on Particulate Matter Levels:** Weather models are employed to evaluate the impact of climate change on particulate matter concentrations. Particulate matter encompasses solid particles and liquid droplets that can remain suspended in the atmosphere. Weather models have revealed that climate change is likely to result in elevated levels of particulate matter in specific regions while concurrently causing a reduction in particulate matter levels in other regions.
- **Analyzing the Impact of Air Pollution on Climate Change:** Weather models are essential for evaluating the influence of air pollution on climate change. Airborne pollutants, such as black carbon and ozone, possess the ability to absorb solar radiation, leading to an increase in atmospheric temperatures. Weather models have demonstrated that air pollution can contribute to climate change, emphasizing its significant influence on the Arctic region.

The application of weather models in these contexts not only enhances our understanding of the intricate interplay between climate change and air quality but also provides essential insights for formulating effective strategies to mitigate the consequences of these changes on human health and the environment.

4.3.4 The Complementary Role of Weather Model Data in Real-Time Monitoring

The integration of weather model data into real-time air quality monitoring systems offers several compelling advantages:

1. **Spatial Coverage:** Weather models provide information over extensive geographical areas, effectively bridging gaps in real-time monitoring networks. This ensures that even remote or underserved regions are encompassed in comprehensive air quality assessments. Whether it is a rural area or a region lacking monitoring infrastructure, weather models extend the reach of air quality assessment.
2. **Temporal Resolution:** While real-time monitoring systems provide high-frequency data, weather models contribute insights into long-term trends and seasonal patterns. Combining these perspectives offers a more comprehensive view of air quality dynamics. The inclusion of weather model data enables

a nuanced understanding of how air quality evolves over different seasons and years.

3. **Prediction of Long-Range Pollutant Transport:** Weather models excel at simulating the long-range transport of pollutants. By integrating weather model outputs with real-time monitoring data, it becomes possible to predict the movement of pollutants across regional and even international borders. This is invaluable for assessing transboundary air pollution and elucidating the sources of pollutants affecting a particular region.
4. **Enhanced Forecasting:** The incorporation of weather model data significantly enhances the accuracy of both short-term and long-term air quality forecasts. This is critically important for issuing timely warnings and implementing preventive measures during pollution episodes. With weather model data, forecasting models can provide more precise predictions of air quality conditions, ultimately leading to improved public health outcomes.

The synergy between weather model data and real-time monitoring is pivotal in strengthening our capacity to assess and respond to air quality challenges effectively. It bridges gaps in spatial and temporal coverage, enhances predictive capabilities, and ultimately contributes to safeguarding public health and the environment.

4.3.5 Exemplifying Enhanced Forecasting Accuracy Through Weather Model Integration

To demonstrate the tangible benefits of integrating weather model data, consider the following examples:

1. **Wildfire Smoke:** During a wildfire event, weather models can predict the direction and dispersion of smoke plumes. Integrating this data with real-time monitoring empowers communities to proactively prepare for and respond to hazardous air quality conditions. Evacuation plans, advisories on respiratory protection, and emergency response strategies can be tailored based on these predictions.
2. **Seasonal Trends:** Weather models capture seasonal variations in air quality, such as increased particulate matter during winter months due to heating-related emissions. When coupled with real-time sensor data, these models assist in proactive air quality management. Cities can implement measures like restricting wood burning during winter or promoting alternative heating methods to mitigate seasonal air quality challenges.
3. **Urban Heat Island Effect:** Weather models can simulate the urban heat island effect, where urban areas experience higher temperatures than their surrounding rural areas. This effect has the potential to intensify air pollution. By incorporating these models into monitoring systems, cities can develop strategies to mitigate these effects. For instance, urban planners can design open spaces and employ cool roofs to counteract the heat island effect, ultimately enhancing air quality.

The integration of weather model data with real-time monitoring systems is an exemplary case of harnessing scientific insights for practical benefits. It equips decision-makers with the tools and knowledge needed to prepare for, mitigate, and respond to air quality challenges effectively, safeguarding public health and the well-being of communities.

4.3.6 A Comprehensive Insight into Pollution Sources, Dispersion, and Trends

The integration of weather model data fosters a comprehensive understanding of air quality dynamics, equipping researchers, policymakers, urban planners, and local health officials to achieve several crucial objectives:

1. **Identification of Pollution Sources:** By considering regional climate phenomena and pollutant transport mechanisms, weather model data assists in precisely identifying pollution sources. This foundational understanding is crucial for devising effective pollution control strategies and focusing emissions reductions where they are most urgently required.
2. **Tracking Pollutant Dispersion:** Weather models play a significant role in monitoring the dispersion of pollutants over extensive spatial scales. This information is invaluable for assessing how pollutants travel and disperse, potentially affecting neighboring regions or even neighboring countries.
3. **Assessment of Long-Term Trends:** The long-term perspective offered by weather modelling provides insights into trends in air quality over extended periods. This historical viewpoint assists in recognizing evolving patterns and assessing the effectiveness of past interventions aimed at improving air quality.
4. **Evidence-Based Decision-Making:** With a comprehensive perspective on air quality, decision-makers gain access to evidence-based data. This information guides the formulation of effective air quality regulations, strategies for emission reduction, and measures to safeguard public health.

In summary, the incorporation of weather model data significantly enhances real-time air quality monitoring. It offers crucial spatial and temporal context, improves forecasting accuracy, and promotes a deeper understanding of pollution sources and trends. This integration serves as a potent tool in the continuous effort to protect air quality and the health of our communities.

4.4 REGION-SPECIFIC CASE STUDIES

In this section, we provide a series of case studies specific to various regions to serve as practical illustrations of the implementation and efficacy of AI-driven air quality monitoring systems in diverse contexts. Each case study provides insights into the geographical location, pollution characteristics, contextual factors, and the impact of AI-powered monitoring systems. These cases illustrate the flexibility and scalability of such systems while offering insights into the challenges encountered and valuable lessons learned. They are instrumental in showcasing the real-world implications of AI-based air quality monitoring, highlighting the adaptability of these systems across different settings, and emphasizing the importance of addressing region-specific nuances for effective air quality management.

4.4.1 Case Study 1: Managing Urban Air Quality in Beijing, China

Geographical Location: Beijing, China (Fang et al., 2009).

Pollution Profile: Beijing faces heightened levels of particulate matter ($PM_{2.5}$) and volatile organic compounds (VOCs), primarily originating from industrial activities and vehicular emissions.

Contextual Factors: The city contends with rapid urbanization, high population density, and meteorological conditions that exacerbate pollution events.

Adaptability and Scalability: AI-driven air quality monitoring systems were strategically deployed throughout the city, integrating data from both fixed and mobile sensors. These systems showcased exceptional adaptability by continuously adjusting to shifting pollution patterns and delivering real-time alerts. This adaptability enabled precise implementation of emission controls where and when they were most critical.

Insights Gained: The AI-powered system provided invaluable insights, aiding in the identification of pollution hotspots and facilitating targeted interventions. Furthermore, it significantly enhanced forecasting accuracy during smog episodes, resulting in more effective public health advisories and a reduction in health risks.

***Challenges and Lessons Learned*:** The implementation of AI-powered monitoring in Beijing brought forth several challenges. Data privacy concerns, particularly regarding the collection and utilization of personal health data, emerged as a prominent issue. Additionally, ensuring the continuous maintenance of sensors posed logistical challenges. Nevertheless, this case study underscores the potential of AI to be a cornerstone in combatting air pollution in densely populated urban areas. The adaptability and responsiveness of AI systems to dynamic pollution patterns hold the promise of saving lives and enhancing overall air quality in rapidly growing megacities.

This case study illuminates the practical application of AI in addressing air quality issues, not only as an adaptable tool but also as an avenue for overcoming the unique challenges posed by large, urban environments.

4.4.2 Case Study 2: Air Quality Monitoring in the Amazon Rainforest, Brazil

Geographical Location: The Amazon Rainforest, Brazil (Reddington et al., 2015).

Pollution Profile: Seasonal forest fires lead to elevated levels of particulate matter (PM_{10}) and carbon monoxide (CO).

Contextual Factors: *The* region presents unique challenges with its remote and rugged terrain, limited infrastructure, and the paramount importance of preserving the rainforest's rich biodiversity.

Adaptability and Scalability: In the heart of the expansive Amazon rainforest, strategic placements of AI-powered air quality monitoring systems were meticulously designed to encompass a diverse range of ecosystems. Augmented by satellite data, these systems offered a comprehensive perspective on air quality. Their adaptability emerged as a key asset in addressing the dynamic and often unpredictable patterns of forest fires. The AI-driven systems played a crucial role in the early detection of fires, enabling swift responses and providing vital support for the preservation of this ecologically significant treasure.

Insights Gained: The Amazon Rainforest case study stands as a paradigm of how AI technology facilitated timely fire detection, proving instrumental in protecting the region's rich biodiversity. The utilization of AI-driven forecasts empowered the prediction of areas prone to fires, facilitating proactive fire management and conservation efforts.

Challenges and Lessons Learned: A central challenge in this scenario revolved around maintenance in remote and challenging environments. Logistical hurdles related to sensor maintenance and data retrieval were ingeniously addressed through innovative solutions, including the deployment of autonomous drones. Overall, this case study underscores the pivotal role of AI technology in monitoring and conserving fragile ecosystems like the Amazon Rainforest. Its adaptability and capacity to provide early warnings are invaluable for safeguarding biodiversity and mitigating the environmental impacts of forest fires.

This case serves as a testament to the utility of AI in preserving the unique and irreplaceable ecosystems of the Amazon Rainforest, demonstrating that technology can play a vital role in safeguarding our planet's ecological diversity.

4.4.3 Case Study 3: Monitoring Industrial Emissions in Rotterdam, Netherlands

Geographical Location: Rotterdam, Netherlands (Super et al., 2017).

Pollution Profile: The city contends with emissions from a dense concentration of industrial facilities, leading to elevated levels of sulfur dioxide (SO_2) and nitrogen oxides (NOx).

Contextual Factors: Rotterdam boasts a major industrial port and faces the critical necessity of adhering to stringent air quality standards.

Adaptability and Scalability: In the industrial epicenter of Rotterdam, AI-driven monitoring systems were strategically deployed in proximity to crucial industrial sites. These systems continuously gathered data from various sources, including emissions monitoring equipment and environmental sensors. Their adaptability was manifest in their capacity to accommodate changes in industrial processes and furnish real-time emissions data.

Insights Gained: The implementation of AI-based monitoring enabled the identification of non-compliant facilities, simplifying regulatory enforcement. Furthermore, industries themselves reaped benefits from the system by gaining real-time insights into their emissions. This knowledge empowered them to optimize emission control measures, contributing to both regulatory compliance and improved air quality.

Challenges and Lessons Learned: A significant challenge in this case pertained to data integration from various industrial sources, often employing different standards and protocols. The effective functioning of the AI-powered system necessitated standardization efforts and the development of data-sharing agreements. The Rotterdam case underscores the pivotal role of AI in enhancing industrial compliance and air quality in heavily industrialized regions.

This case illuminates the practical application of AI in the industrial heart of Rotterdam, where it serves not only as a tool for regulatory enforcement but also as a means for industries to voluntarily contribute to cleaner air and sustainable operations.

4.5 FUTURE DIRECTIONS AND IMPLICATIONS

- **Advancements on the Horizon:** The relentless evolution of AI-powered air quality monitoring systems beckons promising advancements. The near future holds the potential for even more precise and widespread deployment of these systems. Enhanced data analytics, ML algorithms, and data fusion capabilities may offer deeper insights into air quality dynamics, ultimately leading to better public health protection.
- **Implications for Diverse Sectors:** The impact of AI-driven air quality monitoring resonates across various sectors. Public health stands to benefit from timely alerts and tailored recommendations during poor air quality episodes. Urban planning can harness data-driven insights to create cleaner and more sustainable cities. Policymakers can craft well-informed regulations, and industries can optimize emissions control measures. Additionally, education and awareness initiatives can thrive on data generated by these systems, fostering a broader understanding of air quality issues.
- **Ethical Considerations:** The ethical dimension of AI in air quality monitoring requires vigilant attention. Data privacy concerns, consent, and transparency is paramount. Striking a balance between harnessing data for the greater good and safeguarding individual privacy necessitates meticulous ethical frameworks.
- **Global Pursuit of Healthier Environments:** The quest for healthier environments is a global endeavor. As AI technologies transcend boundaries, international collaboration in developing and deploying AI-powered air quality monitoring systems gains significance. The shared goal is to protect public health, preserve the environment, and pave the way for a sustainable future.

In the unfolding narrative of air quality monitoring, AI serves as a potent ally, enabling informed decision-making, raising awareness, and advancing the collective pursuit of cleaner, healthier air.

4.5.1 Exploring Future Frontiers: Advancements in AI-Powered Air Quality Monitoring

The future trajectory of AI-powered air quality monitoring presents substantial potential, marked by various key directions for advancement:

- **Enhanced Spatial Resolution:** Refinement of AI algorithms can lead to higher spatial resolution, enabling precise identification of pollution sources and urban hotspots.
- **Integration with IoT:** The Internet of Things (IoT) can further enhance AI monitoring by connecting numerous sensors, vehicles, and devices. This network facilitates real-time data collection from a wide array of sources.
- **Cross-Domain Integration:** AI-driven systems have the potential to integrate data from various domains, including transportation, weather, and health records. This comprehensive approach offers a deeper understanding of how air quality impacts various aspects of society.
- **Predictive Analytics:** Advancements in ML will enable more accurate and long-term air quality forecasts. These forecasts are instrumental in proactive decision-making and risk mitigation.

4.5.1.1 Implications for Public Health, Policy Interventions, and Environmental Research

These advancements have far-reaching implications. For public health, they mean more timely alerts and tailored recommendations, reducing the health impact of poor air quality. Policymakers gain access to more accurate data for crafting effective regulations and emissions reduction strategies (Masood & Ahmad, 2021; Y. Zhang et al., 2012). Environmental research benefits from a wealth of data that fosters a deeper understanding of air quality dynamics. In essence, AI-powered air quality monitoring systems are poised to be indispensable tools for creating cleaner and healthier environments (Asha et al., 2022; V. O. K. Li et al., 2021).

4.5.2 Implications of AI-Powered Air Quality Monitoring

The implications of AI-powered air quality monitoring are vast and multifaceted:

- **Public Health:** The availability of timely and precise air quality data empowers individuals to take precautionary measures during pollution events, thereby reducing health risks. Healthcare systems can proactively allocate resources to address potential surges in medical needs.
- **Policy Interventions:** Policymakers can formulate evidence-based regulations and incentives to mitigate pollution sources. AI-driven data equips governments with the means to enforce compliance and assess the effectiveness of intervention strategies.
- **Environmental Research:** Researchers have access to a wealth of real-time data, enabling the study of air quality's impact on ecosystems, biodiversity, and climate change. This data contributes to the development of sustainable practices for environmental conservation.

- **Economic Benefits:** Improved air quality yields a range of economic benefits, including increased labor productivity, reduced healthcare costs, and potential economic gains through the adoption of cleaner technologies.

4.5.2.1 Ethical Considerations and Data Privacy in AI-Driven Monitoring

While the benefits are evident, ethical considerations regarding data privacy in AI-driven monitoring systems must not be overlooked. Striking a balance between data collection for the common good and protecting individual privacy is of paramount importance. Clear guidelines and regulations are necessary to ensure the responsible and ethical use of data in the pursuit of healthier environments.

4.5.2.2 Ethical Considerations in AI-Powered Air Quality Monitoring

While AI offers significant benefits, ethical consi
derations are paramount:

- **Data Privacy:** The collection of personal data from IoT devices and sensors necessitates stringent privacy measures. Implementing anonymization and secure data handling protocols is essential to protect individuals' information.
- **Transparency:** AI algorithms must be transparent and interpretable. Users and decision-makers should have a clear understanding of how these systems arrive at their conclusions to ensure trust and accountability.
- **Bias Mitigation:** Efforts to mitigate bias in AI algorithms are critical to prevent discrimination and ensure that monitoring systems are fair and equitable.
- **Community Engagement:** Engaging with local communities affected by air quality issues is essential. Transparency in data sharing and involving communities in decision-making processes fosters trust and inclusivity.

4.5.3 The Global Transition Towards Sustainable and Healthier Environments

As AI-powered air quality monitoring systems become more widespread, they play a pivotal role in the global transition towards sustainable and healthier environments. These systems empower individuals, governments, and organizations to make informed decisions, reduce pollution, and safeguard public health (Pulimeno et al., 2020; Saini et al., 2020b, 2020a). By addressing ethical concerns and fostering community engagement, the path to cleaner air becomes not only technologically advanced but also ethically sound and inclusive.

4.5.3.1 AI-Driven Air Quality Monitoring in the Context of Global Sustainability

The adoption of AI-driven air quality monitoring aligns with the global transition toward sustainability:

- **International Collaboration:** Nations can foster international cooperation by sharing data and best practices, allowing for a coordinated response to transboundary air pollution and climate change. This collective effort facilitates the achievement of common environmental goals.

- **Climate Resilience:** AI's predictive capabilities in forecasting extreme weather events and air quality changes enhance climate resilience efforts, particularly safeguarding vulnerable populations from the adverse effects of climate-related challenges.
- **Urban Planning:** Cities worldwide can harness AI-generated data to reimagine urban environments, giving precedence to green spaces and endorsing cleaner transportation methods. This data-driven approach empowers urban planners to make informed decisions for more sustainable and healthier urban landscapes.
- **Environmental Stewardship:** AI supports a paradigm shift towards environmental stewardship, encouraging industries and individuals to assume responsibility for minimizing their environmental impact. This shift entails a shared commitment to protecting the environment, reducing pollution, and embracing sustainable practices.

4.6 DISCUSSION

In the exploration of real-time air quality monitoring systems driven by AI, we have unraveled fundamental concepts, practical implementations, and the synergistic integration of weather model data. The deployment of AI-driven monitoring systems has been vividly portrayed through region-specific case studies, shedding light on both their execution and advantages. These case studies offer not only blueprints for establishing comprehensive air quality assessment networks but also profound insights into the complex interplay between contextual factors and research-driven findings.

- **Transformative Impact of AI:** One key finding lies in the transformative impact of AI on air quality monitoring. ML and data analytics algorithms, central to AI, process extensive data in real time, enabling the identification of pollution patterns and the prediction of pollution events. This real-time capability proves essential for safeguarding public health and formulating effective environmental policies.
- **Vital Role of Weather Model Integration:** Another significant takeaway is the indispensable role of integrating weather model data. The fusion of real-time monitoring data with weather models enhances forecasting accuracy, provides insights into pollution sources and dispersion, and offers a broader perspective on long-term trends. This integration emerges as a pivotal component in effective air quality management.

4.6.1 Summary of Key Findings and Takeaways

The exploration reveals key findings and takeaways:

- **Practical Guidance:** Region-specific case studies illustrate the practical implementation of AI-based monitoring systems. Emphasizing adaptability and scalability, these cases offer invaluable guidance for stakeholders considering AI-powered air quality monitoring.

- **Future Research Directions:** Identifying vast potential, future research should focus on enhancing spatial resolution, integrating IoT devices, and refining predictive analytics. Collaboration, ethical considerations, and community engagement are critical for a healthier and more sustainable environment.

4.7 CONCLUSION AND FUTURE SCOPE

This chapter contributes significantly to air quality monitoring and environmental science, offering practical guidance through case studies and outlining future research directions. It highlights the transformative potential of AI and the crucial role of weather model integration. As the world grapples with environmental challenges, the fusion of AI, real-time data streams, and weather models emerges as a promising pathway. This chapter serves as a compass for researchers, practitioners, and policymakers, envisioning cleaner air and improved public health for generations to come. Future research endeavors should focus on harnessing the full potential of AI, addressing challenges, and ensuring ethical considerations are embedded in the evolving landscape of real-time air quality assessment.

REFERENCES

Agarwal, R., Awasthi, A., Mital, S. K., Singh, N., & Gupta, P. K. (2014). Statistical model to study the effect of agriculture crop residue burning on healthy subjects. *Mapan*, 29, 57–65.

Aguénounon, E., Smith, J. T., Al-Taher, M., Diana, M., Intes, X., & Gioux, S. (2020). Real-time, wide-field and high-quality single snapshot imaging of optical properties with profile correction using deep learning. *Biomedical Optics Express*, *11*(10), 5701–5716. https://doi.org/10.1364/BOE.397681

Akselsen, S., Aurdal, P. E., Bach, K., & ... Nguyen, H. T. (2019). On the need for explanations, visualisations and measurements in data-driven air quality monitoring and forecasting. *... on Evaluation and ...*, *September*. www.researchgate.net/profile/Weiqing-Zhang-10/publication/337444299_On_the_need_for_explanations_visualisations_and_measurements_in_data-driven_air_quality_monitoring_and_forecasting/links/5dd7db59299bf10c3a274cab/On-the-need-for-explanations-visu

Alexandrova, E., & Ahmadinia, A. (2018). Real-time intelligent air quality evaluation on a resource-constrained embedded platform. *2018 IEEE 4th International Conference on Big Data Security on Cloud (BigDataSecurity), IEEE International Conference on High Performance and Smart Computing, (HPSC) and IEEE International Conference on Intelligent Data and Security (IDS)*, 165–170. https://doi.org/10.1109/BDS/HPSC/IDS18.2018.00045

Ali, H., Soe, J. K., & Weller, S. R. (2015). A real-time ambient air quality monitoring wireless sensor network for schools in smart cities. *2015 IEEE First International Smart Cities Conference (ISC2)*, 1–6. https://doi.org/10.1109/ISC2.2015.7366163

Asha, P., Natrayan, L., Geetha, B. T., Beulah, J. R., Sumathy, R., Varalakshmi, G., & Neelakandan, S. (2022). IoT enabled environmental toxicology for air pollution monitoring using AI techniques. *Environmental Research*, *205*, 112574. https://doi.org/10.1016/J.ENVRES.2021.112574

Bae, W. D., Kim, S., Park, C.-S., Alkobaisi, S., Lee, J., Seo, W., Park, J. S., Park, S., Lee, S., & Lee, J. W. (2021). Performance improvement of ML techniques predicting the association of exacerbation of peak expiratory flow ratio with short term exposure level to indoor air quality using adult asthmatics clustered data. *PLOS ONE*, *16*(1), 1–16. https://doi.org/10.1371/journal.pone.0244233

Balogun, A.-L., Tella, A., Baloo, L., & Adebisi, N. (2021). A review of the inter-correlation of climate change, air pollution and urban sustainability using novel ML algorithms and spatial information science. *Urban Climate*, *40*, 100989. https://doi.org/https://doi.org/10.1016/j.uclim.2021.100989

Borrajo, L., & Cao, R. (2020). Big-but-biased data analytics for air quality. *Electronics (Switzerland)*, *9*(9), 1–11. https://doi.org/10.3390/electronics9091551

Cheng, Y., Li, X., Li, Z., Jiang, S., Li, Y., Jia, J., & Jiang, X. (2014). AirCloud: A Cloud-Based Air-Quality Monitoring System for Everyone. *Proceedings of the 12th ACM Conference on Embedded Network Sensor Systems*, 251–265. https://doi.org/10.1145/2668332.2668346

Dhanalakshmi, S., Poongothai, M., & Sharma, K. (2020). IoT based indoor air quality and smart energy management for HVAC system. *Procedia Computer Science*, *171*, 1800–1809. https://doi.org/10.1016/j.procs.2020.04.193

Engel-Cox, J. A., & Hoff, R. M. (2005). Science–policy data compact: Use of environmental monitoring data for air quality policy. *Environmental Science & Policy*, *8*(2), 115–131. https://doi.org/https://doi.org/10.1016/j.envsci.2004.12.012

Engel-Cox, J. A., Holloman, C. H., Coutant, B. W., & Hoff, R. M. (2004). Qualitative and quantitative evaluation of MODIS satellite sensor data for regional and urban scale air quality. *Atmospheric Environment*, *38*(16), 2495–2509. https://doi.org/https://doi.org/10.1016/j.atmosenv.2004.01.039

Enigella, S. R., & Shahnasser, H. (2018). Real time air quality monitoring. *2018 10th International Conference on Knowledge and Smart Technology (KST)*, 182–185. https://doi.org/10.1109/KST.2018.8426102

Fadhil, M. J., Gharghan, S. K., & Saeed, T. R. (2023). Air pollution forecasting based on wireless communications: Review. *Environmental Monitoring and Assessment*, *195*(10), 1152. https://doi.org/10.1007/S10661-023-11756-Y

Fang, M., Chan, C. K., & Yao, X. (2009). Managing air quality in a rapidly developing nation: China. *Atmospheric Environment*, *43*(1), 79–86. https://doi.org/https://doi.org/10.1016/j.atmosenv.2008.09.064

Fiore, A. M., Naik, V., Spracklen, D. V, Steiner, A., Unger, N., Prather, M., Bergmann, D., Cameron-Smith, P. J., Cionni, I., Collins, W. J., Dalsøren, S., Eyring, V., Folberth, G. A., Ginoux, P., Horowitz, L. W., Josse, B., Lamarque, J.-F., MacKenzie, I. A., Nagashima, T., … Zeng, G. (2012). Global air quality and climate. *Chemical Society Reviews*, *41*(19), 6663–6683. https://doi.org/10.1039/C2CS35095E

Gulia, S., Shiva Nagendra, S. M., Khare, M., & Khanna, I. (2015). Urban air quality management: A review. *Atmospheric Pollution Research*, *6*(2), 286–304. https://doi.org/https://doi.org/10.5094/APR.2015.033

Gupta, P., Christopher, S. A., Wang, J., Gehrig, R., Lee, Y., & Kumar, N. (2006). Satellite remote sensing of particulate matter and air quality assessment over global cities. *Atmospheric Environment*, *40*(30), 5880–5892. https://doi.org/https://doi.org/10.1016/j.atmosenv.2006.03.016

Hable-Khandekar, V., & Srinath, P. (2017). ML techniques for air quality forecasting and study on real-time air quality monitoring. *2017 International Conference on Computing,*

Communication, Control and Automation (ICCUBEA), 1–6. https://doi.org/10.1109/ICCUBEA.2017.8463746

Heo, S. K., Nam, K. J., Loy-Benitez, J., Li, Q., Lee, S. C., & Yoo, C. K. (2019). A deep reinforcement learning-based autonomous ventilation control system for smart indoor air quality management in a subway station. *Energy and Buildings*, *202*, 109440. https://doi.org/10.1016/J.ENBUILD.2019.109440

Hilpert, M., Johnson, M., Kioumourtzoglou, M.-A., Domingo-Relloso, A., Peters, A., Adria-Mora, B., Hernández, D., Ross, J., & Chillrud, S. N. (2019). A new approach for inferring traffic-related air pollution: Use of radar-calibrated crowd-sourced traffic data. *Environment International*, *127*, 142–159. https://doi.org/https://doi.org/10.1016/j.envint.2019.03.026

Juda-Rezler, K. (2010). New challenges in air quality and weather modeling. Archives of Environmental Protection, *36*(I), 3–28.

Kaginalkar, A., Kumar, S., Gargava, P., & Niyogi, D. (2021). Review of urban computing in air quality management as smart city service: An integrated IoT, AI, and cloud technology perspective. *Urban Climate*, *39*, 100972. https://doi.org/10.1016/J.UCLIM.2021.100972

Kang, G. K., Gao, J. Z., Chiao, S., Lu, S., & Xie, G. (2018). Air quality prediction: Big data and ML approaches. *International Journal of Environmental Science and Development*, *9*(1), 8–16. https://doi.org/10.18178/ijesd.2018.9.1.1066

Karagulian, F., Barbiere, M., Kotsev, A., Spinelle, L., Gerboles, M., Lagler, F., Redon, N., Crunaire, S., & Borowiak, A. (2019). Review of the performance of low-cost sensors for air quality monitoring. *Atmosphere, 10*(9), 506. https://doi.org/10.3390/atmos10090506

Kim, J.-Y., Chu, C.-H., & Shin, S.-M. (2014). ISSAQ: An integrated sensing systems for real-time indoor air quality monitoring. *IEEE Sensors Journal*, *14*(12), 4230–4244. https://doi.org/10.1109/JSEN.2014.2359832

Lee, M., Lin, L., Chen, C.-Y., Tsao, Y., Yao, T.-H., Fei, M.-H., & Fang, S.-H. (2020). Forecasting air quality in Taiwan by using ML. *Scientific Reports*, *10*(1), 4153. https://doi.org/10.1038/s41598-020-61151-7

Li, V. O. K., Lam, J. C. K., Han, Y., & Chow, K. (2021). A big data and artificial intelligence framework for smart and personalized air pollution monitoring and health management in Hong Kong. *Environmental Science & Policy*, *124*, 441–450. https://doi.org/10.1016/J.ENVSCI.2021.06.011

Li, Y., Guo, J. e., Sun, S., Li, J., Wang, S., & Zhang, C. (2022). Air quality forecasting with artificial intelligence techniques: A scientometric and content analysis. *Environmental Modelling & Software*, *149*, 105329. https://doi.org/10.1016/J.ENVSOFT.2022.105329

Masood, A., & Ahmad, K. (2021). A review on emerging artificial intelligence (AI) techniques for air pollution forecasting: Fundamentals, application and performance. *Journal of Cleaner Production*, *322*, 129072. https://doi.org/10.1016/J.JCLEPRO.2021.129072

Mo, X., Zhang, L., Li, H., & Qu, Z. (2019). A novel air quality early-warning system based on artificial intelligence. *International Journal of Environmental Research and Public Health*, *16*(19), 3505. https://doi.org/10.3390/ijerph16193505

Moharana, B. K., Anand, P., Kumar, S., & Kodali, P. (2020). Development of an IoT-based real-time air quality monitoring device. *2020 International Conference on Communication and Signal Processing (ICCSP)*, 191–194. https://doi.org/10.1109/ICCSP48568.2020.9182330

Molka-Danielsen, J., Engelseth, P., Olešnaníková, V., Šarafín, P., & Žalman, R. (2017). Big data analytics for air quality monitoring at a logistics shipping base via autonomous wireless sensor network technologies. *2017 5th International Conference on Enterprise Systems (ES)*, 38–45. https://doi.org/10.1109/ES.2017.14

Motlagh, N. H., Kortoçi, P., Su, X., Lovén, L., Hoel, H. K., Haugsvær, S. B., Srivastava, V., Gulbrandsen, C. F., Nurmi, P., & Tarkoma, S. (2023). Unmanned aerial vehicles for air pollution monitoring: A survey. *IEEE Internet of Things Journal*, *10*(24), 21687–21704. https://doi.org/10.1109/JIOT.2023.3290508

Moursi, A. S., El-Fishawy, N., Djahel, S., & Shouman, M. A. (2021). An IoT enabled system for enhanced air quality monitoring and prediction on the edge. *Complex and Intelligent Systems*, *7*(6), 2923–2947. https://doi.org/10.1007/s40747-021-00476-w

Nam, K. J., Heo, S. K., Li, Q., Loy-Benitez, J., Kim, M. J., Park, D. S., & Yoo, C. K. (2020). A proactive energy-efficient optimal ventilation system using artificial intelligent techniques under outdoor air quality conditions. *Applied Energy*, *266*, 114893. https://doi.org/10.1016/J.APENERGY.2020.114893

Oh, H. J., & Kim, J. (2020). Monitoring air quality and estimation of personal exposure to particulate matter using an indoor model and artificial neural network. *Sustainability (Switzerland)*, *12*(9), 13–18. https://doi.org/10.3390/su12093794

Panneerselvam, B., Ravichandran, N., Dumka, U. C., Thomas, M., Charoenlerkthawin, W., & Bidorn, B. (2023). A novel approach for the prediction and analysis of daily concentrations of particulate matter using ML. *Science of the Total Environment*, *897*, 166178. https://doi.org/10.1016/j.scitotenv.2023.166178

Pierce, D. W., Barnett, T. P., Santer, B. D., & Gleckler, P. J. (2009). Selecting global weather models for regional climate change studies. *Proceedings of the National Academy of Sciences of the United States of America*, *106*(21), 8441–8446. https://doi.org/10.1073/pnas.0900094106

Pulimeno, M., Piscitelli, P., Colazzo, S., Colao, A., & Miani, A. (2020). Indoor air quality at school and students' performance: Recommendations of the UNESCO chair on health education and sustainable development & the Italian society of environmental medicine (SIMA). *Health Promotion Perspectives*, *10*(3), 169–174. https://doi.org/10.34172/HPP.2020.29

Reddington, C. L., Butt, E. W., Ridley, D. A., Artaxo, P., Morgan, W. T., Coe, H., & Spracklen, D. V. (2015). Air quality and human health improvements from reductions in deforestation-related fire in Brazil. *Nature Geoscience*, *8*(10), 768–771. https://doi.org/10.1038/ngeo2535

Roy, A., Mandal, M., Das, S., Kumar, M., Popek, R., Awasthi, A., ... & Sarkar, A. (2024). Non-exhaust particulate pollution in Asian countries: A comprehensive review of sources, composition, and health effects. *Environmental Engineering Research*, 29(3), 230384.

Rummukainen, M. (2010). State-of-the-art with regional weather models. *WIREs Climate Change*, *1*(1), 82–96. https://doi.org/https://doi.org/10.1002/wcc.8

Saini, J., Dutta, M., & Marques, G. (2020a). A comprehensive review on indoor air quality monitoring systems for enhanced public health. *Sustainable Environment Research*, *30*(1), 1–12. https://doi.org/10.1186/S42834-020-0047-Y

Saini, J., Dutta, M., & Marques, G. (2020b). A comprehensive review on indoor air quality monitoring systems for enhanced public health. *Sustainable Environment Research*, *30*(1), 6. https://doi.org/10.1186/s42834-020-0047-y

Shah, S. K., Tariq, Z., Lee, J., & Lee, Y. (2020). Real-Time ML for Air Quality and Environmental Noise Detection. *2020 IEEE International Conference on Big Data (Big Data)*, 3506–3515. https://doi.org/10.1109/BigData50022.2020.9377939

Sharma, E., Deo, R. C., Prasad, R., & Parisi, A. V. (2020). A hybrid air quality early-warning framework: An hourly forecasting model with online sequential extreme learning machines and empirical mode decomposition algorithms. *Science of The Total Environment*, *709*, 135934. https://doi.org/10.1016/J.SCITOTENV.2019.135934

Shelestov, A., Kolotii, A., Lavreniuk, M., Medyanovskyi, K., Vasiliev, V., Bulanaya, T., & Gomilko, I. (2018). Air Quality Monitoring in Urban Areas Using in-Situ and Satellite Data Within Era-Planet Project. *IGARSS 2018 - 2018 IEEE International Geoscience and Remote Sensing Symposium*, 1668–1671. https://doi.org/10.1109/IGARSS.2018.8518368

Singh, N., Agarwal, R., Awasthi, A., Gupta, P. K., & Mittal, S. K. (2010). Characterization of atmospheric aerosols for organic tarry matter and combustible matter during crop residue burning and non-crop residue burning months in Northwestern region of India. *Atmospheric Environment, 44*(10), 1292–1300.

Sree Devi, M., & Rahamathulla, V. (2020). Air quality through IoT and big data analytics. In S. Borah, V. Emilia Balas, & Z. Polkowski (Eds.), *Advances in Data Science and Management* (pp. 181–187). Springer Singapore.

Su, X., Liu, X., Motlagh, N. H., Cao, J., Su, P., Pellikka, P., Liu, Y., Petäjä, T., Kulmala, M., Hui, P., & Tarkoma, S. (2021). Intelligent and scalable air quality monitoring with 5G edge. *IEEE Internet Computing*, *25*(2), 35–44. https://doi.org/10.1109/MIC.2021.3059189

Super, I., van der Gon, H. A. C., van der Molen, M. K., Sterk, H. A. M., Hensen, A., & Peters, W. (2017). A multi-model approach to monitor emissions of CO2 and CO from an urban–industrial complex. *Atmospheric Chemistry and Physics*, *17*(21), 13297–13316. https://doi.org/10.5194/acp-17-13297-2017

Tandon, A., Awasthi, A., & Pattnayak, K. C. 2023. Comparison of different machine learning methods on precipitation dataset for Uttarakhand. *2023 2nd International Conference on Ambient Intelligence in Health Care (ICAIHC), Bhubaneswar, India, 2023*, 1–6. doi: 10.1109/ICAIHC59020.2023.10431402

Tendelilin. (2010). No 主観的健康感を中心とした在宅高齢者における 健康関連指標に関する共分散構造分析. *Energies*, *6*(1), 7. http://journals.sagepub.com/doi/10.1177/1120700020921110%0Ahttps://doi.org/10.1016/j.reuma.2018.06.001%0Ahttps://doi.org/10.1016/j.arth.2018.03.044%0Ahttps://reader.elsevier.com/reader/sd/pii/S1063458420300078?token=C039B8B13922A2079230DC9AF11A333E295FCD8

Völgyesi, P., Nádas, A., Koutsoukos, X., & Lédeczi, Á. (2008). Air Quality Monitoring with SensorMap. *Proceedings - 2008 International Conference on Information Processing in Sensor Networks, IPSN 2008*, 529–530. https://doi.org/10.1109/IPSN.2008.50

Wang, J., Yoo, C. K., & Liu, H. (2023). An overview of artificial intelligence in subway indoor air quality prediction and control. *Process Safety and Environmental Protection*, *178*, 652–662. https://doi.org/10.1016/j.psep.2023.08.055

Wang, L., Zhong, B., Vardoulakis, S., Zhang, F., Pilot, E., Li, Y., Yang, L., Wang, W., & Krafft, T. (2016). Air quality strategies on public health and health equity in Europe—A systematic review. *International Journal of Environmental Research and Public Health*, *13*(12), 1196. https://doi.org/10.3390/ijerph13121196

Wilson, C., & van der Velden, M. (2022). Sustainable AI: An integrated model to guide public sector decision-making. *Technology in Society*, *68*, 101926. https://doi.org/https://doi.org/10.1016/j.techsoc.2022.101926

Yadav, N., Rajendra, K., Awasthi, A., Singh, C., & Bhushan, B. (2023). Systematic exploration of heat wave impact on mortality and urban heat island: A review from 2000 to 2022. *Urban Climate, 51*, 101622.

Zhang, D., & Woo, S. S. (2020). Real time localized air quality monitoring and prediction through mobile and fixed IoT sensing network. *IEEE Access*, *8*, 89584–89594. https://doi.org/10.1109/ACCESS.2020.2993547

Zhang, Y., Bocquet, M., Mallet, V., Seigneur, C., & Baklanov, A. (2012). Real-time air quality forecasting, part I: History, techniques, and current status. *Atmospheric Environment*, *60*, 632–655. https://doi.org/https://doi.org/10.1016/j.atmosenv.2012.06.031

Zhu, Y., Al-Ahmed, S. A., Shakir, M. Z., & Olszewska, J. I. (2023). LSTM-based IoT-enabled CO2 steady-state forecasting for indoor air quality monitoring. *Electronics (Switzerland)*, *12*(1). https://doi.org/10.3390/electronics12010107

Zivelonghi, A., & Giuseppi, A. (2024). Smart healthy schools: An IoT-enabled concept for multi-room dynamic air quality control. *Internet of Things and Cyber-Physical Systems*, *4*, 24–31. https://doi.org/10.1016/J.IOTCPS.2023.05.005

5 Forecasting Air Pollution with Artificial Intelligence

Recent Advancements at Global Scale and Future Perspectives

Prem Rajak, Satadal Adhikary, Suchandra Bhattacharya, and Abhratanu Ganguly

5.1 INTRODUCTION

Air pollution is continuously increasing in many developing countries due to rapid urbanization and industrialization (Yadav et al., 2023; Wang et al., 2018). The increase in global population has led to a sharp rise in the usage of automobiles that add to green house and other harmful gases daily in the environment. Moreover, burning of waste products, coal, wooden materials, agriculture crop residue burning, etc. further deteriorates air quality, posing serious threats to human health (Agarwal et. al, 2014; Verma et al., 2016). Not only outdoor pollution, indoor pollution is also increasing due to cooking, dust, allergic pollen grains, use of pesticides, etc. Air pollutants are reported to fuel a wide spectrum of diseases. Primarily the contaminants in air target the respiratory system leading to development of asthma, chronic obstructive pulmonary disease (COPD), and pulmonary fibrosis (Jiang et al., 2016). Air pollutants can also impart detrimental impacts on other systems including nervous system, endocrine system, reproductive functions, and others. Air pollutants and other environmental contaminants have been reported to be associated with carcinogenesis, neurodegeneration, oxidative injuries, and vision impairment (Wong et al., 2014). Hence it is essential to minimize air pollution around the world to combat these ailments and improve the quality of human well beings.

To overcome the environmental burden of air pollution, monitoring different air pollutants for better air quality management is essential. AI could be more effective in precise monitoring of various air pollutants (Yildirim & Bayramoglu, 2006). Different tools can also be employed to monitor other environmental factors like temperature; rainfall, density of vegetation etc. that are linked to air pollutants. Several machine

DOI: 10.1201/9781032683805-5

learning (ML) tools and sensors are being tested globally for their efficacy in sensing and managing numerous air pollutants of various chemical natures.

AI can be employed by a local body of environmentalists to make better decisions in a short time so that pollution risks for humans can be minimized. Notably, AI assists in fast processing of multiple environmental parameters so that they can precisely forecast the pollution status in a specified area for which previous data on different variables are available (Fernandez & Vico, 2013). Therefore AI senses the baseline of air pollutant levels and accurately identifies pollution hot spots over a large geographical area. AI also integrates data on air pollutants with the data on different variables including meteorological data to predict visibility, haze, and smoke for better air quality management.

With the advent of improved processors, more reliable sensors, and fast data analyzing tools, AI has shown its promising application in monitoring ambient air pollution in a short amount of time. The system can also be used friendly and easy to operate. Researchers have devised multiple models that might have implications for weather forecasting and air quality measurement and to measure carbon monoxide, sulfur dioxide, and suspended particulate matters in industrial environments. Studies have claimed that various environmental pollutants are detrimental to health. Therefore it is indispensable to explore tools for detection and monitoring of these pollutants. In studies different AI algorithms like Support Vector Regression (SVR), LASSO, Artificial Neural Network (ANN), random forest (RF), and xGBoost have been compared and found to be efficient for their precision in forecasting air pollution (Bozdag et al., 2020).

AI tools have better fault tolerance and produce more accurate data than conventional approaches for forecasting air pollution status. AI efficiently measures concentration of harmful gases and particles in air that help detect the pollution status of any area.

5.2 COMMON AI TECHNIQUES

AI has recently experienced substantial momentum in various fields of science, like medicine, environment surveillance, and contaminant monitoring. Significant numbers of studies are now focusing on the implication of AI in this area (Guo et al., 2022). Below, we present some of the most recent advancements in AI to monitor air pollution.

5.2.1 ANN Models

ANNs are archetypal examples of specially designed promising AI techniques which follow the organization and machinery of a human brain to some extent and processes and simulates non-linear information by applying algorithms that imitate the functioning of a biological brain (Sapra et al., 2015; Chatterjee & Pandya, 2016; Allegrini & Olivieri, 2019). ANNs consist of a number of non-linear processing units (Chatterjee & Pandya, 2016). These units collaborate collectively to address a diverse range of assignments, which include approximation of functions, classification of patterns, clustering, forecasting, optimization, content retrieval, and control

(Anand et al., 2017; Dorohoi et al., 2017; Hasanuzzaman & Rahim, 2019; Ravikiran et al., 2019).

An ANN consists of three essential elements: neurons or nodes, which serve as the core processing units; the network architecture, which defines the connections between these nodes; and the training algorithm, which is used to determine the network parameters (Cordova et al., 2021). The conventional approach to ANN modeling comprises three distinct stages. In first phase, one must choose an effective topology for the ANN, determine the ideal layer and neuron numbers, select an appropriate learning algorithm, and establish the input–output attributes (Subramanyam, 2007). The second phase is training which deals with expected and real outcomes. The validation phase evaluates the effectiveness of an ANN by assessing its capacity to generalize and accurately predict unknown inputs. Once the ANN model has been successfully validated, it can be considered appropriate for predicting unfamiliar variables. The multilayer perceptron (MLP) is a frequently employed pattern in ANN for forecasting various air pollutants (Agirre-Basurko et al., 2006). Pollutants could be harmful to living organisms. Hence, ANN mediated forecasting of air pollutants might be helpful in minimizing detrimental impacts on health.

The MLP comprises three distinct layers: input, hidden, and output layers. Furthermore, the models contain multiple hidden levels inside their structure, depending on the complexity and features of the specific problem. This enables them to fully understand the patterns of data with respect to input output variables. The selection of the number of hidden layer neurons is a critical component of MLP set-up and significantly affects the ultimate output of the network (da Silva et al., 2017). The MLP propagates information in a unidirectional manner inside the network. Introducing the input data to the neural network is the major role of input layer. The aforementioned inputs are forwarded to the hidden layer with different magnitudes. The hidden layer encompasses intricacy of data and imparts non-linear attributes. Inputs can be subsequently transmitted to output layer to generate final result (Masood & Ahmad, 2021).

In a study a comparison of performance was made between ANN and Multiple Linear Regression (MLR) models to calculate PM_{10} with respect to weather variables. Upon conducting a comparison between the two models, nonlinear ANN method had superior performance in predicting PM_{10} (Chaloulakou et al., 2003). At the same time, it was also reported that, when compared to alternative methodologies such as the Weather Research and Forecasting model and the Community Multiscale Air Quality model, ANN provides a greater suitability for the modeling and forecasting of air quality indices due to their utilization of statistical modeling approaches. These models have been extensively utilized for the purpose of predicting environmental data, specifically the levels of particulate matter, across several countries (Li et al., 2011; Li et al., 2016).

5.2.2 Deep Neural Networks (DNN)

DNNs represent a more advanced form of artificial neural networks, using deep architecture and sophisticated algorithms for training to achieve heightened degrees of data processing (Abiodun et al., 2018). From a structural point of view a neural network that consists of more than three layers is considered a DNN (Sahiner et al.,

2019; Wardah et al., 2019). The term "deep" in DNN refers to the organizational framework it possess, which contains one input, hidden, and output layers. Due to their increased structural depth, these networks are capable of creating more concise representations of the input-output relationship. One distinguishing characteristic between DNNs and standard ANNs is the "feature learning," defined as the capability of DNNs to perform automatic feature extraction from unprocessed inputs. Additionally, DNNs are also capable of refining their outcomes by incorporating additional data and employing more expansive network architectures (Choi, 2018; Ganapathy et al., 2018). The deep learning framework is well-suited for addressing challenges related to air pollution prediction, as well as handling complex issues such as nonlinear relationships, cyclical patterns, seasonal variations, and sequential dependencies among pollutant data (Drewil & Al-Bahadili, 2022). Several environmental contaminations pose adverse impacts on organisms. Notably, DNN could be helpful in monitoring of such contaminants.

The training of DNNs is a very intricate process that requires powerful graphics processing units (GPUs), sophisticated algorithms, and large amounts of data (Yasaka et al., 2018). Conventional Back-Propagation-Algorithm commonly serves as a primary training technique of DNNs. Nonetheless, recent deep networks trained using such approach frequently demonstrate unsatisfactory results. To tackle this difficulty, researchers have devised and applied sophisticated algorithms like Random Forest, Support Vector Machine (SVM), Greedy Layer-wise, and Dropout within the realm of deep learning (DL), with the goal of alleviating these worries. Recently, there has been a surge in the creation of DL tools, like Long Short-Term Memory networks (LSTMs), Recurrent Neural Networks (RNNs), Autoencoders, and Convolutional Neural Networks (CNNs). These architectures demonstrate a wide range of applications and have achieved remarkable test performances. The Autoencoders, LSTM, and CNN have become prominent deep neural architectures for air pollution prediction due to their exceptional performance (Kim et al., 2018; Zhang et al., 2020).

The main feature of this program is to ascertain the correlation between influential variables and target parameters utilizing a sophisticated network design. The observed patterns of associations in the original data are influenced by several frequency factors. Improper utilization of the deep network might result in the entangled multi-frequency data obscuring correlations, which in turn leads to inadequate recording of correlation information. The wavelet transform's ability to separate complex multi-frequency data and the deep learning methods' aptitude for identifying correlations have led to the development of a new hybrid model termed wavelet transform and transformer-like (WTformer) (Xu et al., 2023). The methodology utilized in this work is based on the application of wavelets, which are mathematical functions that can be transformed into a signal.

5.2.3 SVM Model

Another widely utilized machine learning algorithm is SVM, which has made significant advancements since its introduction (Kaur et al., 2018; Chervonenkis, 2013). SVMs are guided learning models that may be used for tasks such as regression as well as classification. Support Vector Regression (SVR) algorithms are frequently denoted as SVM regression algorithms (Maltare & Vahora, 2023). SVM refers to a collection of supervised learning techniques utilized for the analysis of data in the context of classification, density estimation, numerical prediction, and pattern recognition tasks (Roy et al., 2015). From an algorithmic point of view, the SVM aims to identify an N-dimensional hyperplane that maximizes the margin between data points belonging to different specified classes or labels (Thanki & Borra, 2019). The linear function required to accurately represent the information is commonly known as the hyperplane. The support vectors refer to the collection of data points which lie in the closest vicinity to the hyperplane. The best separating hyperplane, which lies at the midpoint of the margin, is designed to achieve the greatest distance of separation for the data points being used for training (Morio & Balesdent, 2015). The support vector regression model utilizes kernel functions to map data that is linearly separable into a higher-dimensional space (Dun et al., 2020). SVM utilizes kernel functions, such as polynomial, Radial Basis Function (RBF), and Fisher & Bayesian. These functions are crucial in bridging linear to non-linear relationships.

The efficacy of SVM has been extensively corroborated by numerous studies. An investigation carried out in India between 2015 and 2021 (Maltare & Vahora, 2023) assessed the effectiveness of machine learning models, namely SVM with various kernel functions, LSTM, and SARIMA with adjustable hyperparameter to create a more accurate air pollution prediction model. The SVM model utilizing the RBF kernel demonstrated greater performance in predicting the air quality index (AQI) when compared to other models. The presented approach holds potential for future investigation in predicting AQI across diverse geographic regions. In another contemporary research aimed at predicting the occurrence of asthma by using environmental factors, lifestyle choices, dietary patterns, and other relevant variables, SVM have depicted a notable level of accuracy rate of 93.19% (Noh & Park, 2020). Primary and secondary air pollutants like harmful gases, pesticides, heavy metals, etc., are hazardous to health. Thus, SVM have potential to precisely monitor and forecast the primary and secondary air pollutants in atmosphere.

5.2.4 Fuzzy Logic Models

Fuzzy logic is a type of multi-valued logic that allows the truth values of variables to range between 0 and 1, spanning real numbers. The concept of partial truth is applied to manage scenarios in which the truth value can vary between being entirely true and entirely untrue (Novak et al., 1999). In contrast, within the context of Boolean logic, variables are restricted to having truth values that are only represented by the integers 0 and 1. The methodology is derived from the concept of fuzzy sets, allowing the model to identify patterns in data by including adjectives such as low, high, little, medium, major, minor, and so on.

A conventional fuzzy logic modeling system generally includes three basic stages, such as fuzzification, inference, and defuzzification. Initially, input numerical values undergo a process known as fuzzification, wherein they are converted into membership grades. Furthermore, the grades run through a series of IF-THEN rules, resulting in the generation of a fuzzy output (inference). Finally, the fuzzy output has the potential to undergo a subsequent transformation into either a quantitative or qualitative yield, a process known as defuzzification (Erdik, 2009; Lamaazi & Benamar, 2018)

The mathematical expression of a hypothetical fuzzy set, Set-A is denoted as: $A = \{(x, \mu A(x)) | x \in X\}$. A fuzzy set refers to a category of entities that has a range of membership grades. A set of this nature is distinguished by a function that provides a membership grade, ranging from zero to one, to each object within the set (Zadeh, 1965). The aforementioned sets are a collection of components made up of a membership function called $\mu A(x)$ and a universe of discourse (X) of categories with a particular feature. An element x that is a member of a fuzzy Set-A possesses a partial membership, typically $\in [0,1]$. The extent to which the component x is owed to A is estimated by the membership function $\mu A(x)$ (Z. Li et al., 2006). In the last few years, there have been advancements in the field of indoor air quality (IAQ) inspection utilizing fuzzy logic in conjunction with reasoning methodologies. These studies have yielded diverse approaches to address various types of air pollutants. The utilization of a fuzzy logic controller is employed for the computation of the comfort and air quality index. This is achieved by integrating various air pollution data, including CO, NO_2, PMs, and temperature. The purpose of this index is to furnish users with information regarding acceptable levels of air quality in relation to the toxicity of different pollutants. In addition to this, the application of a fuzzy logic controller is employed to ascertain the appropriate output action, which manifests as the turning on or turning off of an air conditioner, the adjustment of a fresh air dumper to name a few (Dionova et al., 2020) (Figure 5.1).

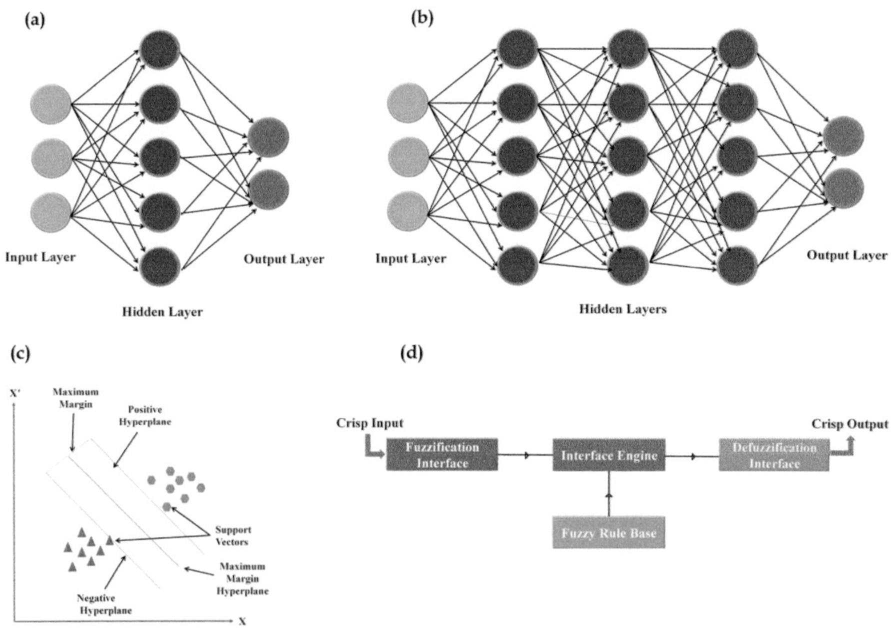

FIGURE 5.1 Principle of AI-based machine learning approaches used for air pollution monitoring: (a) ANN; (b) DNN; (c) SVM; (d) fuzzy logic.

5.3 AI IN AIR POLLUTION FORECASTING

5.3.1 Fuzzy Logic Models

Fuzzy logic is a well-accepted tool to forecast air pollution in various localities. It provides multiple benefits than the conventional AI tools. It exhibits superior tolerance to faults, high efficiency in generating non-linear functions of any complexity, and seamless integration with other data mining approaches (Abiodun et al., 2018). Cheng et al. (2011) utilized a unique fuzzy time series model that employed a two-phase linguistic partition strategy to forecast ozone levels. The setup was enhanced by using the cumulative probability distribution approach (CPDA), which divided entire range of possibilities into seven fuzzy sets. Subsequently, additional divisions were made using the uniform discretion method (UDM). This model predicted ozone level with better efficacy compared to the classical fuzzy model implying logical relationship. In another study carbon monoxide was measured using a neuro-fuzzy system (Jain and Khare, 2010). In this study several parameters like cloud cover, atmospheric pressure, number of vehicles, atmospheric temperature, humidity, etc., were considered. The Index of Agreement for CO detection was found to be 0.88–0.93.

In another investigation, a novel fuzzy algorithm such as "Sigma operator" was used to monitor air quality based on historical data, and impacts of air pollutants based on the toxicity results (Carbajal et al., 2012). Findings of the study have suggested precision air quality predictions compared to other models. A neuro-fuzzy model was also used to predict the sulfur dioxide and total suspended particulate matter. The system was integrated with meteorological parameters like precipitation to accurately detect the air pollution status. It was found that the model predicted the sulfur dioxide and total suspended particulate matters with a prediction accuracy of 75–90% and 69–80% respectively (Yildirim and Bayramoglu, 2006). Pollutants could impair the health of organisms. Hence search for suitable approaches for accurate monitoring of environmental contaminants are essential. Fuzzy-logic based models have been comparatively studied with the stochastic based models to predict ozone levels. Results of the study have revealed that fuzzy-logic based approaches have better prediction accuracy compared to the stochastic based methods (Al-Shammari, 2013).

The fuzzy inference ensemble (FIE) has been examined for overall air quality forecasting in the city of Athens. Results have suggested that FIE can effectively forecast total air quality with enhanced accuracy and greater values for correlation coefficient (Bougoudis et al., 2016). An adaptive neuro-fuzzy model was formulated to predict particulate matter concentrations. This system was integrated with various probability density functions. Results of the study suggested that this integrated model provides better predictive accuracy for particulate matters (Song et al., 2015).

A hybrid model known as grey wolf optimizer was used to perform prediction of sulfur dioxide and particulate matters. For this study, hourly data on these pollutants were recorded throughout the year and split into the datasets for the training of the system. Several statistical matrices such as RMSE, MAE, determination coefficient, etc., were implied to examine performance of model. The results of this analysis have shown that enriched data could be helpful in enhancing the precision of the prediction (Xu et al., 2017). Another hybrid model combining a fuzzy time series model and uncertainty analysis have efficiently forecasted the concentrations of nitrogen dioxide and particulate matter (Wang et al., 2018). The model had three distinct levels of analysis: deterministic forecasting, uncertainty analysis, and assessment. The hybrid model was deemed superior to the separate techniques for forecasting.

The air quality forecasting was performed using the neuro-fuzzy inference system with the application of the forward selection technique. The system utilized parameters such as air pollution concentrations, humidity, temperature, visibility, and wind speed as inputs. The results indicate that including the forward selection strategy has improved the performance of the adaptive neuro-fuzzy inference tool (Ghasemi & Amanollahi, 2019). A sensor utilizing an adaptive neural fuzzy inference tool was created for the purpose of monitoring benzene levels in an Internet of Things (IoT) context. Nanotechnology is also a burgeoning technology in detecting pollutants. It may also be helpful in construction of various sensors to detect air pollutants. The monitoring of air pollutants was conducted using hourly records of various pollutants such as benzene, PMs, SO_2, CO, NO_2, O_3, etc. These records were utilized as inputs to ensure accurate monitoring of air pollutants (Behal & Singh, 2020). Thus, fuzzy models show great potential in effectively predicting various air contaminants.

5.3.2 SVM for Air Pollutant Measurement

SVMs have been examined in conjunction with the machine learning models to predict various air pollutants like SO_2, CO, PM, and O_3. They are effective in recognizing non-linear input and output datasets that are complex to analyze mathematically (Awad & Khanna, 2015). Models like Back Propagation Neural Network (BPNN), Genetic algorithm optimized BPNN (GABPNN) and ensemble tool SVM-GABPNN have been developed and compared to predict daily ozone (Feng et al., 2011). Several meteorological parameters like temperature, humidity, wind speed, and UV radiation were used as inputs in this approach. It was suggested that the SVM-based system was better than the individual approaches for accurate prediction of air pollutants. In another study, partial least squares (PLS) were integrated with an SVM-based model to accurately predict the carbon monoxide concentrations (Yeganeh et al., 2012). This model performed better and had short computation time. Moreover, the results were more accurate and acceptable.

SVM approach was also examined to study the criteria pollutants in the city. This was used to analyze the relation between the primary and secondary pollutants affecting any region's air quality. Significant similarities between the observed and predicted data were observed that indicate the effectiveness of the SVM approach in forecasting air pollution (Nieto et al., 2013). In another study, an MLP, a vector aggressive moving average (VARMA), a linear stochastic autoregressive moving average (ARMA), and an RBF-SVM model were examined for 1–7 months for predicting particulate matter. Though all the results were acceptable, the SVM model was more accurate in forecasting future air pollution events (Nieto et al., 2018).

Another study employed principal component analysis (PCA) integrated with ANNs and SVM to forecast O_3 level. This study indicates that data of meteorological attributes & levels of other pollutants were useful in predicting ozone concentrations.

The results were acceptable and found that both ANN and SVM models produced good results that were similar to that of experimental data (Luna et al., 2014). Support vector regression (SVR) models were compared with ANN and SVM models for forecasting CO concentrations. Results obtained for SVR models showed high efficacy for predicting CO concentrations (Moazami et al., 2016).

Hybrid models of ANN and SVM were examined for their efficacy in forecasting air pollution. Results have demonstrated that the hybrid models were more effective in prediction of air pollution compared to individual models (Wang et al. 2015). Another hybrid approach referred to as hybrid-Garch model, employing SVM and ARIMA technology, was examined to analyze particulate matters. Different variables like meteorological and air quality were considered as inputs. Air pollutants like SO_2, CO, NO_2, and O_3 were considered environmental factors, whereas humidity, temperature, and pressure were considered air quality parameters. The hybrid model outperformed all the individual approaches (Wang et al., 2017).

5.3.3 Implications of ANN in Air Pollution Monitoring

In recent times, the utilization of ANNs has offered many opportunities to forecast air pollution. These techniques led to a significant change in perspective and have garnered the interest of scientists around the globe. Several studies have demonstrated that ANN could effectively predicted both gaseous and particle pollutants with satisfactory precision across a range of time intervals.

The research conducted by Mlakar et al. (1994) represented one of the initial attempts of ANN to forecast air pollution. The researchers have utilized a three-layered feed-forward non-linear multilayer perceptron (MLP) model. SO_2 concentrations were predicted at the thermal power plant at Sostanj employing several inputs such as temperature change, wind direction, relative humidity, SO_2 concentrations, etc. The findings indicated that the network could accurately predict SO_2 concentrations at the corresponding sites with a 30-minute lead time. In another study, Arena et al. (1995) documented applying the MLP model to forecast SO_2 concentrations in an industrial region with a lead time of 6 hours. Findings of this study have demonstrated a high level of similarity between the predictions generated by the ANN. The model demonstrated a notable level of accuracy and its predictive capabilities were deemed reasonable across various climatic circumstances. ANN is effective in monitoring atmospheric contaminants. Sohn et al. (1999) employed a neural network model to predict concentrations of various major pollutants, specifically NO_2, O_3, total hydrocarbons, SO_2, CH_4, and NO. The ANN utilizing the backpropagation algorithm was optimized for a configuration consisting of 7 output nodes, 30 hidden nodes, and 16 input nodes. The model incorporated several characteristics, including time and the concentration of contaminants, as input variables. ANN showed satisfactory accuracy predictions with in a specific range. Additionally, the accuracy of these predictions can be improved by including additional input factors related to meteorological events. Slini et al. (2003) conducted a study wherein they designed ANN to forecast the hourly amounts of ozone, NO_2, and CO. The model inputs consisted of pollutant concentration levels measured on an hourly basis, as well as geological and climatic characteristics. Results of this analysis have revealed that the ANN model based on NO_2 exhibited a very weak correlation coefficient (R) of 0.45 as well as relatively greater standard deviation (SD) value of 0.614. In a separate study, a Principal Component Analysis (PCA)-Multilayer Perceptron model has been utilized to predict O_3 concentrations. As input variables, the model incorporated hourly measurements of NO_2, O_3, wind direction, and humidity. Principal Component Analysis was utilized as a preprocessing technique for the environmental variables. The suggested approach effectively simulated O_3 concentrations with enhanced precision (Karatzas and Kaltsatos, 2007). Furthermore, Kandya et al. (2013) enhanced predictive efficacy of the ANN by developing 34 multilayer feed forward neural networks (MLFFNNs). The topology of the ANN has been modified by adjusting hidden layer node numbers and modifying quantity of training epochs. According to the study, the model M-21 demonstrated superior predicting capabilities when the hidden nodes were set at 105, accompanied by input parameters of 17 as well as an epoch value of 2000. Mishra and Goyal (2015) have assessed two models for their accuracy in air pollutant measurement. Authors compared PCA-based ANN with the multiple linear regression (MLR). Here PCA-ANN performed superior in estimating NO_2 concentrations. A three-layer Multilayer Perceptron (MLP) model has been suggested by Elangasinghe et al. (2014) to forecast NO_2 concentrations. The Lavernburg Marquardt Backpropagation (BP) algorithm was adopted to train the model. Several environmental parameters like temperature, sun light, wind direction, humidity, etc., were considered as input data. Results of the study indicated that the MLP model integrated with meteorological data display better prediction efficiency. In their study, Noori et al. (2010) conducted a comparison between adaptive neuro-fuzzy inference system (ANFIS) and ANN to assess their respective performances in forecasting daily concentrations of CO over a 1-day interval period. The researchers utilized the forward selection (FS) and gamma test (GT) for quick optimization of input data and for reduction in computational time. The ANN model was trained with daily data on NOX, THC, PM_{10}, SO_2, O_3 levels and environmental parameters like direction of wind, temperature, humidity, etc. The utilization of Monte-Carlo simulation-based uncertainty analysis evaluated the model performances. Results of the study demonstrated that employing ANN could provide more accurate data compared to ANFIS. Therefore, different models of ANN could be beneficial in measuring concentrations of various environmental pollutions.

5.3.4 DNN-Based Measurement and Forecasting of Air Pollutants

DNN models have been examined for their potential to monitor various air pollutants. They are also found effective in measuring the amount of air pollutants. Freeman et al. (2018) integrated RNN and LSTM approaches to forecast ozone levels in a heterogeneous metropolitan environment. The dataset used for the RNN model consisted of hourly data of different environmental and air pollutant characteristics. According to the study, the model demonstrated favorable predictability in estimating 8-hour mean O_3 concentrations. But, instances of overfitting have been identified in the model. The MAE values obtained throughout training and examining phases were 0.41 and 0.37, respectively. Furthermore, the RNN have superior predictive capabilities in comparison to standard feed-forward neural networks (FFNN). Additionally, RNNs have demonstrated their effectiveness in addressing acute air pollution outbreaks and can potentially be extended to similar scenarios.

In a study, Wang and Song (2018) developed an ensemble methodology for predicting air pollution. This approach involved utilizing past air pollution and meteorological measurements. The researchers advocated the utilization of three hidden layers containing a LSTM network. The data was divided into categories depending on several variables, such as weather patterns, using a fuzzy c-means clustering technique. The findings indicated that the ensemble model's short-term (few hours) and long-term (more than one day) predictions exhibited significantly higher accuracy compared to the individual models. A notable investigation by Zhou et al. (2019) explored the application of LSTM with deep learning algorithms for predicting NOx, PM_{10}, and $PM_{2.5}$.

In their study, Li et al. (2017) presented novel approach for modeling the extended spatiotemporal distribution of air pollutants. This approach involved the utilization of various models, namely stacked LSTM (LSTME), time delay neural network (TDNN), autoregressive moving average (ARMA), spatiotemporal deep learning (STDL), regular LSTM and support vector regression (SVR). Around 20,196 observations were considered as input data derived in two years from an independent test set. These observations were acquired from 12 air quality monitoring stations and represented hourly $PM_{2.5}$ values. An autocorrelation analysis was conducted to assess the association between $PM_{2.5}$ values at various time intervals for each site. It is worth noting that the authors achieved the maximum performance (MAPE 11.93%) by incorporating auxiliary inputs into their model. In conclusion, the LSTME exhibited superior performance in comparison to alternative machine learning methodologies, as evidenced by MAPE, MAE, and RMSE values. In a recent study, Soh et al. (2018) introduced a DL model called the spatial-temporal DNN (ST-DNN). This model was integrated with ANN and convolutional neural networks (CNN) and LSTM. The primary objective of this model was to accurately forecast the levels of $PM_{2.5}$ across a 48-hour time frame. The DL model was trained on a restricted dataset comprising air pollution and meteorological attributes from the immediate and preceding hour. The findings indicate that the ST-DNN with CNN module consistently demonstrates high performance over extended periods of time. (Figure 5.2) (Table 5.1).

5.4 CASE STUDIES EMPLOYING AI FOR AIR QUALITY MONITORING

In a very recent study, the 3D neural network was employed in precise forecasting of air pollutants. Deep learning networks were integrated with earth-specific priors and found effective in resolving complex patterns of data. The system was trained using 39 years of data to produce more reliable deterministic data. The authors also suggested that the system can track the environmental variables more effectively (Bi et al., 2023). In another recent investigation, AI models like ANN and ANFIS were tested to evaluate their efficacy in forecasting moisture in the environment. Geographical and periodicity data collected over the year 2011–2021 was utilized

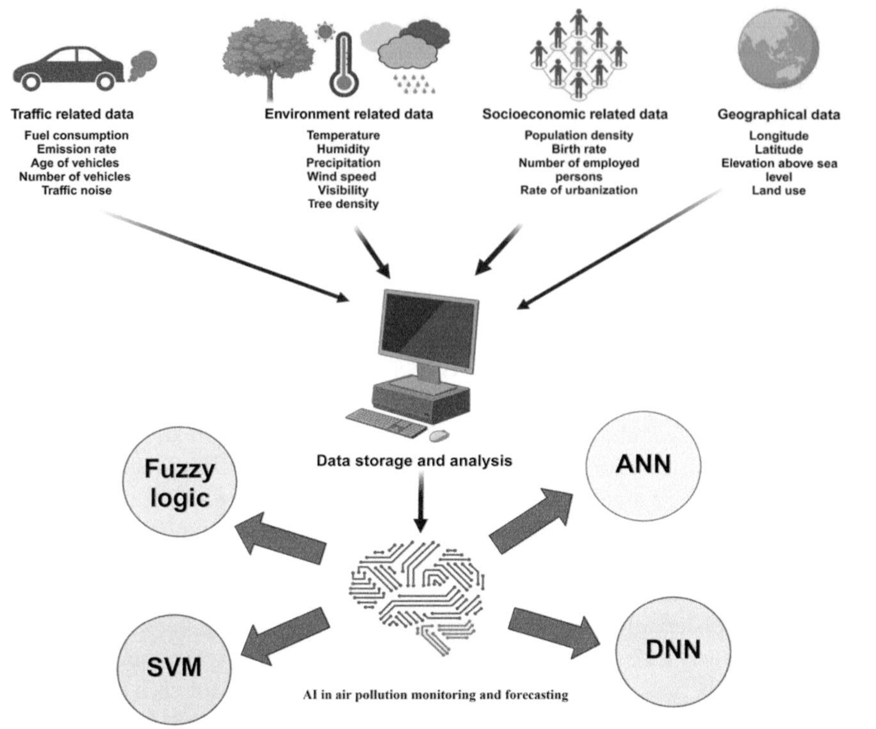

FIGURE 5.2 Various models of AI implicated in air pollution forecasting and air pollutant measurements (created using BioRender.com).

TABLE 5.1
Different AI Models and Their Application in Air Pollutant Measurement

AI model	Air pollutant measured	Input parameters	Reference
Neuro-fuzzy system	CO	Cloud cover, atmospheric pressure, number of vehicles, atmospheric temperature, humidity	Jain & Khare, 2010
Neuro-fuzzy model	Sulfur dioxide and total suspended particulate matter	Precipitation level	Yildirim & Bayramoglu 2006
Forward selection technique with neuro-fuzzy inference system	PM_{10}, O_3, CO, SO_2, NO_2,	Humidity, temperature, visibility, wind speed	Ghasemi & Amanollahi 2019

TABLE 5.1 (Continued)
Different AI Models and Their Application in Air Pollutant Measurement

AI model	Air pollutant measured	Input parameters	Reference
Support vector machines	O_3, CO, SO_2,	Temperature, humidity, wind speed, and UV radiation	Awad & Khanna, 2015
Hybrid-Garch model, employing SVM and ARIMA technology	NO_2, O_3, CO, SO_2,	Humidity, temperature, and pressure	Wang et al., 2017
ANN model	NO_2, O_3, Total hydrocarbons, SO_2, CH_4, and NO	Time and the concentration of contaminants	Sohn et al., 1999
ANN model	O_3, NO_2, and CO	Geological and climatic characteristics	Slini et al. (2003)
Principal component analysis-multilayer perceptron model	O_3	Wind direction, ambient temperature, NO_2, O_3, and humidity	Karatzas & Kaltsatos, 2007
Multilayer perceptron (MLP) model	NO_2	Temperature, sun light, wind direction, humidity	Elangasinghe et al., 2014
Adaptive neuro-fuzzy inference system (ANFIS)	NOX, THC, PM_{10}, SO_2, O_3	Direction of wind, temperature, humidity,	Noori et al., 2010

to enhance forecasting accuracy. Results suggested that ANFIS outperforms ANN through all testing stations. Interestingly, ANFIS scored 0.995 and ANN scored 0.935 as Nash–Sutcliffe efficiency coefficients over all testing stations (Abebe & Endalie, 2023).

Authors of a recent study have examined the efficacy of a deep learning model in forecasting environmental attributes. Bi-LSTM recurrent neural network models have been constructed in the TensorFlow deep learning framework and the previous five years of data were integrated to forecast environmental variables for 24–72 hours. Both models were trained with five years of historical data. In this study, one trained model depicted temperature forecast accuracy with plus minus 2°C with RMSE of 1.45°C. Another model predicted temperature and relative humidity with RMSE of 2.26°C and 14% respectively (Zenkner & Navarro-Martinez, 2023). In a study, a feed-forward neural network and resilient propagation learning algorithm were employed to forecast site-specific humidity for six localities of São Paulo state, Brazil. Several environmental parameters like specific humidity, air temperature, the vertical component of the wind, perceptible water, and moisture divergence flux have been considered in the study to train neural networks for better prediction. Results of the investigation suggested that ANN models were superior compared to data obtained through linear regression model (Ramirez et al., 2005).

In the southwest of Iran, AI models like WRF numerical model, ANN, and SVM model were evaluated to forecast air temperature. Analysis of different schemas of the WRF model revealed that the MYJLG schema outperforms all other schemas of the model. In addition, the partial mutual information (PMI) algorithm of the ANN and SVM model reflected the most important variables such as relative humidity, current rain status, and temperature of 12 hours ago. Finally, the comparative analysis of peak discharge and runoff height values has suggested that the SVM model is superior to ANN and WRF models and provides more accurate prediction regarding air temperature (Shirali et al., 2020). Anandharajan et al. (2016) employed an AI approach to perform environmental monitoring. Authors considered available data to train machine learning tools. The system was based on linear regression to accurately predict environmental attributes of the next day. Results showed more than 90% accuracy. Therefore, several studies equivocally suggest the promising implications of AI and various models in air quality monitoring.

5.5 LIMITATIONS AND FUTURE PERSPECTIVES

Despite the extensive literature describing the AI-based air pollution forecasting, there remains a need for a comprehensive study on the integration of ANN and DNN in this field. Furthermore, optimization methods and model design enhancements are required to enhance their efficacy. AI models are specific equipment that might be costly and hence cannot be easily accessible. Moreover, operating various tools and software of AI require expertise and hence cannot be used by general population. Therefore, trained manpower is required to operate these models along with precise knowledge on multiple meteorological factors like temperature, pressure, humidity, wind directions, etc. AI models are usually trained based on data that are gathered previously at different time frames. Hence, data storage is also essential for successful implementation of AI models in air quality monitoring. Moreover, data stored in cloud platforms may be vulnerable to cyber manipulation that can interfere with the accurate prediction of air pollutant levels by AI tools. In future research, it is essential to develop DL models for air quality image recognition. These models should be designed to yield favorable outcomes across various natural lighting conditions (such as daylight, twilight, and darkness) as well as various meteorological parameters (including rain, clear weather, and snow). Moreover, AI-based technologies need simplification so that it can be accessible to general population of the globe. The high cost of AI models could also be an issue that prevents its global implementation. Therefore, studies are required in development of cost effective tools, accessories, and software so that AI enabled air pollutant monitoring and air pollution forecasting can be performed at low cost-effective and reasonable way.

5.6 CONCLUSION

AI-enabled approaches are extensively regarded as an influential technology for air pollution forecasting. These approaches generally have greater fault tolerance, enhanced accuracy, and improved specificity. These methods have promising scopes to monitor environmental variables that substantially influence air pollution levels. The utilization of DNN has demonstrated promising capabilities in revolutionizing atmospheric modeling, enabling enhanced precision in long-term and abrupt heavy pollution prediction. In the foreseeable future, these methodologies will be employed in a diverse array of intricate non-linear prediction challenges. Nevertheless, despite the multiple benefits, certain issues exist that need to be addressed. One of the primary issues encountered while utilizing AI for air pollution forecasting is selection of the most suitable AI technique as several AI models show varied degrees of accuracy. Studies have reported that AI approaches like ANN, SVM, fuzzy logic, and DNN have greater precision in monitoring the air pollutant concentrations and ensuing forecasting. These techniques require further development for application on a large scale. Moreover, AI tools and equipment may be costly and require expert knowledge in this field.

REFERENCES

Abebe, W. T., & Endalie, D. 2023. "Artificial intelligence models for prediction of monthly rainfall without climatic data for meteorological stations in Ethiopia". *Journal of Big Data* 10 (1): 2. https://doi.org/10.1186/s40537-022-00683-3

Abiodun, O. I., Jantan, A. B., Omolara, A. E., Dada, K. V., Mohamed, N., & Arshad, H. 2018. "State-of-the-art in artificial neural network applications: A survey". *Heliyon* 4, 1–41. https://doi.org/10.1016/j.heliyon.2018.e00938.

Abiodun, O. I., Jantan, A., Omolara, A. E., Dada, K. V., Mohamed, N. A., & Arshad, H. 2018. "State-of-the-art in artificial neural network applications: A survey". *Heliyon* 4 (11): e00938. https://doi.org/10.1016/j.heliyon.2018.e00938

Agarwal, R., Awasthi, A., Mital, S. K., Singh, N., & Gupta, P. K. 2014. "Statistical model to study the effect of agriculture crop residue burning on healthy subjects". *Mapan* 29: 57–65.

Agirre-Basurko, E., Ibarra-Berastegi, G., & Madariaga, I. 2006. "Regression and multilayer perceptron-based models to forecast hourly O_3 and NO_2 levels in the Bilbao area". *Environmental Modeling & Software* 21 (4): 430–446. https://doi.org/10.1016/j.envsoft.2004.07.008

Allegrini, F., & Olivieri, A. C. 2019. "Chemometrics and statistics: Neural networks". In P. Worsfold, C. Poole, A. Townshend, & M. Miró (Eds.), *Encyclopedia of Analytical Science (Third Edition)* (pp. 487–499). Elsevier.

Al-Shammari, E. T. 2013. "Public warning systems for forecasting ambient ozone pollution in Kuwait". *Environmental Systems Research* 2: 1–9. https://doi.org/10.1186/2193-2697-2-2.

Anand, G., Nippani, S. K., Kuchhal, P., Vasishth, A., & Sarah, P. 2017. "Dielectric, ferroelectric and piezoelectric properties of Ca0. 5Sr0. 5Bi4Ti4O15 prepared by solid state technique". *Ferroelectrics* 516 (1): 36–43. https://doi.org/10.1080/00150193.2017.1362277

Anandharajan, T. R. V., Hariharan, G. A., Vignajeth, K. K., & Jijendiran, R. 2016. "Weather monitoring using artificial intelligence". *In 2016 2nd International Conference on Computational Intelligence and Networks (CINE),* 106–111.

Arena, P., Fortuna, L., Gallo, A., Nunnari, G. & Xibilia, M. G. 1995. "Air pollutionestimation via neural networks". *IFAC Proceedings* 28 (10): 787–792.https://doi.org/10.1016/S1474-6670(17)51616-X.

Awad, M., Khanna, R. 2015. "Deep neural networks". In: *Efficient Learning Machines* (pp. 127–147). Apress, Berkeley. https://doi.org/10.1007/978-1-4302-5990-9_4.

Behal, V., & Singh, R. 2020. "Personalised healthcare model for monitoring and prediction of airpollution: Machine learning approach". *Journal of Experimental & Theoretical Artificial Intelligence* 33: 425–449. https://doi.org/10.1080/0952813X.2020.1744197.

Bi, K., Xie, L., Zhang, H., Chen, X., Gu, X., & Tian, Q. 2023. "Accurate medium-range global weather forecasting with 3D neural networks". *Nature* 619 (7970): 533–538. https://doi.org/10.1038/s41586-023-06185-3

Bougoudis, I., Demertzis, K., & Iliadis, L. S. 2016. "HISYCOL a hybrid computational intelligence system for combined machine learning: The case of air pollution modeling in Athens". *Neural Computing and Applications* 27: 1191–1206. https://doi.org/10.1007/s00521-015-1927-7.

Bozdağ, A., Dokuz, Y., & Gökçek. Ö. B. 2020. "Spatial prediction of PM10 concentration using machine learning algorithms in Ankara, Turkey". *Environmental Pollution* 263 (Pt A): 114635. https://doi.org/10.1016/j.envpol.2020.114635.

Carbajal-Hernández, J. J., Sánchez-Fernández, L. P., Carrasco-Ochoa, J. A., & Martínez-Trinidad, J. F. 2012. "Assessment and prediction of air quality using fuzzy logic and autoregressive models". *Atmospheric Environment* 60: 37–50. https://doi.org/10.1016/j.atmosenv.2012.06.004.

Chaloulakou, A., Grivas, G., & Spyrellis, N. 2003. "Neural network and multiple regression models for PM10 prediction in Athens: A comparative assessment". *Journal of the Air & Waste Management Association* 53 (10): 1183–1190. https://doi.org/10.1080/10473289.2003.10466276

Chatterjee, S. P., & Pandya, A. S. 2016. "Chapter 12—Artificial neural networks in drug transport modeling and simulation–II". In M. Puri, Y. Pathak, V. K. Sutariya, S. Tipparaju, & W. Moreno (Eds.), *Artificial Neural Network for Drug Design, Delivery and Disposition* (pp. 243–261). Academic Press. https://doi.org/10.1016/B978-0-12-801559-9.00012-0

Cheng, C., Huang, S., & Teoh, H. J. 2011. "Predicting daily ozone concentration maxima using fuzzy time series based on a two-stage linguistic partition method". *Computers & Mathematics with Applications* 62*:* 2016–2028. https://doi.org/10.1016/j.camwa.2011.06.044.

Chervonenkis, A. Y. 2013. "Early History of Support Vector Machines". In Bernhard Schölkopf, Zhiyuan Luo, & Vladimir Vovk (Eds.), *Empirical Inference* (pp. 13–20). Springer, Berlin, Heidelberg. https://doi.org/10.1007/978-3-642-41136-6_3

Choi, H. 2018. "Deep learning in nuclear medicine and molecular imaging: Current perspectives and future directions". *Nuclear Medicine and Molecular Imaging* 52 (2): 109–118. https://doi.org/10.1007/s13139-017-0504-7

Cordova, C. H., Portocarrero, M. N. L., Salas, R., Torres, R., Rodrigues, P. C., & López-Gonzales, J. L. 2021. "Air quality assessment and pollution forecasting using artificial neural networks in Metropolitan Lima-Peru". *Scientific Reports* 11: 24232. https://doi.org/10.1038/s41598-021-03650-9

da Silva, I. N., Hernane Spatti, D., Andrade Flauzino, R., Liboni, L. H. B., & dos Reis Alves, S. F. 2017. "Artificial Neural Network Architectures and Training Processes". In *Artificial*

Neural Networks (pp. 21–28). Springer, Cham. https://doi.org/10.1007/978-3-319-43162-8_2

Dionova, B. W., Mohammed, M. N., Al-Zubaidi, S., & Yusuf, E. 2020. "Environment indoor air quality assessment using fuzzy inference system". *ICT Express* 6 (3): 185–194. https://doi.org/10.1016/j.icte.2020.05.007

Dorohoi, D. O., Barzic, A. I., & Aflori, M. (Eds.), 2017. *Electromagnetic Radiation in Analysis and Design of Organic Materials: Electronic and Biotechnology Applications* (1st ed.). CRC Press, Boca Raton.

Drewil, G. I., & Al-Bahadili, R. J. 2022. "Air pollution prediction using LSTM deep learning and metaheuristics algorithms". *Measurement Sensors* 24: 100546. https://doi.org/10.1016/j.measen.2022.100546

Dun, M., Xu, Z., Chen, Y., & Wu, L. 2020. "Short-term air quality prediction based on fractional grey linear regression and support vector machine". *Mathematical Problems in Engineering* 2020: e8914501. https://doi.org/10.1155/2020/8914501

Elangasinghe, M. A., Singhal, N., Dirks, K. N., & Salmond, J. A. 2014. "Development of an ANN–based air pollution forecasting system with explicit knowledge throughsensitivity analysis". *Atmospheric Pollution Research* 5 (4): 696–708. https://doi.org/10.5094/APR.2014.079.

Erdik, T. 2009. "Fuzzy logic approach to conventional rubble mound structures design." *Expert Systems with Applications* 36 (3, Part 1): 4162–4170. https://doi.org/10.1016/j.eswa.2008.06.012

Feng, Y., Zhang, W., Sun, D., & Zhang, L. 2011. "Ozone concentration forecast method based on genetic algorithm optimized back propagation neural networks and support vector machine data classification". *Atmospheric Environment* 45: 1979–1985. https://doi.org/10.1016/j.atmosenv.2011.01.022.

Fernandez, J. D., & Vico. F. 2013. "AI methods in algorithmic composition: A comprehensive survey". *Journal of Artificial Intelligence Research* 48: 513–582. https://doi.org/10.1613/jair.3908.

Freeman, B. S., Taylor, G. W., Gharabaghi, B., & Thé, J. (2018). "Forecasting air quality time series using deep learning". *Journal of the Air & Waste Management Association* 68: 866–886. https://doi.org/10.1080/10962247.2018.1459956.

Ganapathy, N., Swaminathan, R., & Deserno, T. M. 2018. "Deep learning on 1-D biosignals: A taxonomy-based survey". *Yearbook of Medical Informatics* 27 (1): 98–109. https://doi.org/10.1055/s-0038-1667083

Ghasemi, A., & Amanollahi, J. 2019. "Integration of ANFIS model and forward selection method for air quality forecasting". *Air Quality, Atmosphere & Health* 12: 59–72. https://doi.org/10.1007/s11869-018-0630-0.

Guo, Q., Ren, M., Wu, S., Sun, Y., Wang, J., Wang, Q., Ma, Y., Song, X., & Chen, Y. 2022. "Applications of artificial intelligence in the field of air pollution: A bibliometric analysis". *Frontiers in Public Health* 10: 933665. https://doi.org/10.3389/fpubh.2022.933665

Hasanuzzaman, M., & Rahim, N. A. 2019. *Energy for Sustainable Development: Demand, Supply, Conversion and Management*. Academic Press. https://doi.org/10.1016/C2017-0-01639-7

Jain, S., & Khare, M. 2010. "Adaptive neuro-fuzzy modeling for prediction of ambient CO concentration at urban intersections and roadways". *Air Quality, Atmosphere & Health* 3: 203–212. https://doi.org/10.1007/s11869-010-0073-8.

Jiang, X., Mei, X., & Feng, D. 2016. Air pollution and chronic airway diseases: what should people know and do? *Journal of Thoracic Disease* 8: E31–40 . https://doi.org/10.3978/j.issn.2072-1439.2015.11.50

Kandya, A., Sm, S. N., & Tiwari, V. 2013. "Forecasting the tropospheric ozone using artificial neural network modelling approach: A case study of Megacity Madras, India". *Journal of Civil and Environmental Engineering* 2013: 1–5. https://doi.org/10.4172/2165-784X.S1-006

Karatzas, K. D., & Kaltsatos, S. 2007. "Air pollution modelling with the aid of computational intelligence methods in Thessaloniki, Greece". *Simulation Modelling Practice and Theory* 15: 1310–1319. https://doi.org/10.1016/j.simpat.2007.09.005.

Kaur, P., Chaudhary, P., Bijalwan, A., & Awasthi, A. (2018). "Network traffic classification using multiclass classifier". In *Advances in Computing and Data Sciences: Second International Conference, ICACDS 2018, Dehradun, India, April 20-21, 2018, Revised Selected Papers, Part I 2* (pp. 208–217). Springer Singapore.

Kim, K., Kim, D.-K., Noh, J., & Kim, M. 2018. "Stable forecasting of environmental time series via long short term memory recurrent neural network". *IEEE Access 6*: 75216–75228. https://doi.org/10.1109/ACCESS.2018.2884827

Lamaazi, H., & Benamar, N. 2018. "OF-EC: A novel energy consumption aware objective function for RPL based on fuzzy logic". *Journal of Network and Computer Applications* 117: 42–58. https://doi.org/10.1016/j.jnca.2018.05.015

Li, C., Hsu, N. C., & Tsay, S.-C. 2011. "A study on the potential applications of satellite data in air quality monitoring and forecasting". *Atmospheric Environment* 45 (22): 3663–3675. https://doi.org/10.1016/j.atmosenv.2011.04.032

Li, X., Peng, L., Hu, Y., Shao, J., & Chi, T. 2016. "Deep learning architecture for air quality predictions". *Environmental Science and Pollution Research International* 23 (22): 22408–22417. https://doi.org/10.1007/s11356-016-7812-9

Li, X., Peng, L., Yao, X., Cui, S., Hu, Y., You, C., & Chi, T. 2017. "Long short-term memory neural network for air pollutant concentration predictions: method development and evaluation". *Environmental Pollution* 231: 997–1004. https://doi.org/10.1016/j.envpol.2017.08.114.

Li, Z., Halang, W., & Chen, G. 2006. *Integration of Fuzzy Logic and Chaos Theory*. Springer. https://doi.org/10.1007/3-540-32502-6

Luna, A. S., Paredes, M. L., Oliveira, G. C., & Corrêa, S. M. 2014. "Prediction of ozone concentration in tropospheric levels using artificial neural networks and support vector machine at Rio de Janeiro, Brazil". *Atmospheric Environment* 98: 98–104. https://doi.org/10.1016/j.atm.osenv.2014.08.060.

Maltare, N. N., & Vahora, S. 2023. "Air Quality Index prediction using machine learning for Ahmedabad city." *Digital Chemical Engineering* 7: 100093. https://doi.org/10.1016/j.dche.2023.100093

Masood, A., & Ahmad, K. 2021. "A review on emerging artificial intelligence (AI) techniques for air pollution forecasting: Fundamentals, application and performance". *Journal of Cleaner Production* 322: 129072. https://doi.org/10.1016/j.jclepro.2021.129072

Mishra, D., & Goyal, P. 2015. "Development of artifcial intelligence based NO2 forecasting models at Taj Mahal, Agra". *Atmospheric Pollution Research* 6 (1): 99–106. https://doi.org/10.5094/APR.2015.012.

Mlakar, P., Boˇznar, M., & Lesjak, M. 1994. "Neural networks predict pollution". In Sven-Erik Gryning, Millán M. Millán (Eds.), *Air Pollution Modeling and its Application X* (pp. 659–660). Springer, Boston, MA. https://doi.org/10.1007/978-1-4615-1817-4_93.

Moazami, S., Noori, R., Amiri, B. J., Yeganeh, B., Partani, S., & Safavi, S. 2016. "Reliable prediction of carbon monoxide using developed support vector machine". *Atmospheric Pollution Research* 7 (3): 412–418. https://doi.org/10.1016/j.apr.2015.10.022.

Morio, J., & Balesdent, M. 2015. *Estimation of Rare Event Probabilities in Complex Aerospace and Other Systems: A Practical Approach.* Woodhead Publishing.

Nieto, P. J., Combarro, E. F., Díaz, J. J., & Montañés, E. 2013. "A SVM-based regression model to study the air quality at local scale in Oviedo urban area (Northern Spain): A case study". *Applied Mathematics and Computation* 219: 8923–8937. https://doi.org/10.1016/j.amc.2013.03.018.

Nieto, P. J., Lasheras, F. S., García-Gonzalo, E., & Juez, F. J. 2018. "PM10 concentration forecasting in the metropolitan area of Oviedo (Northern Spain) using models based on SVM, MLP, VARMA and ARIMA: A case study". *The Science of the Total Environment* 621: 753–761. https://doi.org/10.1016/j.scitotenv.2017.11.291.

Noh, M. J., & Park, S. C. 2020. "A prediction model of asthma diseases in teenagers using artificial intelligence models." *Journal of Information Technology Applications and Management* 27 (6): 171–180. https://doi.org/10.21219/jitam.2020.27.6.171

Noori, R., Hoshyaripour, G., Ashraf, K., & Araabi, B. N. 2010. "Uncertainty analysis ofdeveloped ANN and ANFIS models in prediction of carbon monoxide dailyconcentration". *Atmospheric Environment* 44 (4): 476–482. https://doi.org/10.1016/j.atmosenv.2009.11.005.

Novak, V., Perfiljeva, I., & Mockor, J. 1999. *Mathematical Principles of Fuzzy Logic*. Kluwer Academic Publishers. https://doi.org/10.1007/978-1-4615-5217-8

Ramirez, M. C. V., de Campos Velho, H. F., & Ferreira, N. J. 2005. "Artificial neural network technique for rainfall forecasting applied to the Sao Paulo region". *Journal of Hydrology* 301 (1-4): 146–162. https://doi.org/10.1016/j.jhydrol.2004.06.028

Ravikiran, U., Sarah, P., Anand, G., & Zacharias, E. (2019). Influence of Na, Sm substitution on dielectric properties of SBT ceramics. *Ceramics International* 45 (15): 19242–19246.

Roy, K., Kar, S., & Das, R. N. 2015. *Understanding the Basics of QSAR for Applications in Pharmaceutical Sciences and Risk Assessment*. Academic Press.

Sahiner, B., Pezeshk, A., Hadjiiski, L. M., Wang, X., Drukker, K., Cha, K. H., Summers, R. M., & Giger, M. L. 2019. "Deep learning in medical imaging and radiation therapy". *Medical Physics* 46 (1): e1–e36. https://doi.org/10.1002/mp.13264

Sapra, R. L., Mehrotra, S., & Nundy, S. 2015. "Artificial Neural Networks: Prediction of mortality/survival in gastroenterology". *Current Medicine Research and Practice* 5 (3): 119–129. https://doi.org/10.1016/j.cmrp.2015.05.007

Shirali, E., Nikbakht Shahbazi, A., Fathian, H., Zohrabi, N., & Mobarak Hassan, E. 2020. "Evaluation of WRF and artificial intelligence models in short-term rainfall, temperature and flood forecast (case study)". *Journal of Earth System Science* 129 (1): 188. https://doi.org/10.1007/s12040-020-01450-9

Slini, T., Karatzas, K. D., & Moussiopoulos, N. 2003. "Correlation of air pollution and meteorological data using neural networks". *International Journal of Environment and Pollution* 20: 218–229. https://doi.org/10.1504/IJEP.2003.004279.

Soh, P. W., Chang, J. W., & Huang, J. W. 2018. "Adaptive deep learning-based air quality prediction model using the most relevant spatial-temporal relations". *IEEE Access* 6: 38186–38199. https://doi.org/10.1109/ACCESS.2018.2849820.

Sohn, S. H., Oh, S. C., & Yeo, Y. K. 1999. "Prediction of air pollutants by using an artificial neural network". Korean *Journal of Chemical Engineering* 16: 382–387. https://doi.org/10.1007/BF02707129.

Song, Y., Qin, S., Qu, J., & Liu, F. 2015. "The forecasting research of early warning systems for atmospheric pollutants: A case in Yangtze River Delta region". *Atmospheric Environment* 118: 58–69. https://doi.org/10.1016/j.atmosenv.2015.06.032

Subramanyam, V. 2007. Evolution of Artificial Neural Network Controller for a Boost Converter. M.E. Thesis, National University of Singapore. https://core.ac.uk/download/pdf/48630874.pdf

Tandon, A., Awasthi, A., & Pattnayak, K. C. 2023. "Comparison of different Machine Learning methods on Precipitation dataset for Uttarakhand". 2023 2nd International Conference on Ambient Intelligence in Health Care (ICAIHC), Bhubaneswar, India, 2023, pp. 1–6. doi: 10.1109/ICAIHC59020.2023.10431402

Thanki, R., & Borra, S. 2019. Chapter 11—Application of machine learning algorithms for classification and security of diagnostic images. In N. Dey, S. Borra, A. S. Ashour, & F. Shi (Eds.), *Machine Learning in Bio-Signal Analysis and Diagnostic Imaging* (pp. 273–292). Academic Press. https://doi.org/10.1016/B978-0-12-816086-2.00011-4

Verma, R., Vinoda, K. S., Papireddy, M., & Gowda, A. S. 2016. "Toxic pollutants from plastic waste- A review". *Procedia Environmental Sciences* 35: 701–708. https://doi.org/10.1016/j.proenv.2016.07.069

Wang, J., & Song, G. 2018. "A deep spatial-temporal ensemble model for air quality prediction". *Neurocomputing* 314: 198–206. https://doi.org/10.1016/j.neucom.2018.06.049

Wang, P., Liu, Y., Qin, Z., & Zhang, G. 2015. "A novel hybrid forecasting model for PM10 and SO2 daily concentrations". *Science of the Total Environment* 505: 1202–1212. https://doi.org/10.1016/j.scitotenv.2014.10.078

Wang, P., Zhang, H., Qin, Z., & Zhang, G. 2017. "A novel hybrid-Garch model based on ARIMA and SVM for PM2. 5 concentrations forecasting". *Atmospheric Pollution Research* 8 (5): 850–860. https://doi.org/10.1016/j.apr.2017.01.003

Wang, Q. 2018. "Urbanization and global health: The role of air pollution". *Iranian Journal of Public Health* 47: 1644–1652.

Wardah, W., Khan, M. G. M., Sharma, A., & Rashid, M. A. 2019. "Protein secondary structure prediction using neural networks and deep learning: A review". *Computational Biology and Chemistry* 81: 1–8. https://doi.org/10.1016/j.compbiolchem.2019.107093

Wong, I. C., Ng, Y. K., & Lui, V. W. 2014. "Cancers of the lung, head and neck on the rise: perspectives on the genotoxicity of air pollution". *Chinese Journal of Cancer* 33 (10): 476–480. https://doi.org/10.5732/cjc.014.10093

Xu, R., Wang, D., Li, J., Wan, H., Shen, S., & Guo, X. 2023. "A hybrid deep learning model for air quality prediction based on the time–frequency domain relationship". *Atmosphere* 14 (2): 405. https://doi.org/10.3390/atmos14020405

Xu, Y., Du, P., & Wang, J. 2017. "Research and application of a hybrid model based on dynamic fuzzy synthetic evaluation for establishing air quality forecasting and early warning system: A case study in China". *Environmental Pollution* 223: 435–448. https://doi.org/10.1016/j.envpol.2017.01.043

Yadav, N., Rajendra, K., Awasthi, A., Singh, C., & Bhushan, B. (2023). Systematic exploration of heat wave impact on mortality and urban heat island: A review from 2000 to 2022. *Urban Climate* 51: 101622.

Yasaka, K., Akai, H., Kunimatsu, A., Kiryu, S., & Abe, O. 2018. "Deep learning with convolutional neural network in radiology." *Japanese Journal of Radiology* 36 (4): 257–272. https://doi.org/10.1007/s11604-018-0726-3

Yeganeh, B., Motlagh, M. S. P., Rashidi, Y., & Kamalan, H. 2012. "Prediction of COconcentrations based on a hybrid partial least square and support vector machinemodel". *Atmospheric Environment* 55: 357–365. https://doi.org/10.1016/j.atmosenv.2012.02.092

Yildirim, Y., & Bayramoğlu, M. 2006. "Adaptive neuro-fuzzy based modelling for prediction of air pollution daily levels in city of Zonguldak". *Chemosphere* 63 (9): 1575–1582. https://doi.org/10.1016/j.chemosphere.2005.08.070

Zadeh, L. A. 1965. "Fuzzy sets." *Information and Control* 8 (3): 338–353. https://doi.org/10.1016/S0019-9958(65)90241-X

Zenkner, G., & Navarro-Martinez, S. 2023. "A flexible and lightweight deep learning weather forecasting model." *Applied Intelligence* 53 (21): 24991–25002. https://doi.org/10.1007/s10489-023-04824-w

Zhang, B., Zhang, H., Zhao, G., & Lian, J. 2020. "Constructing a PM2.5 concentration prediction model by combining auto-encoder with Bi-LSTM neural networks." *Environmental Modelling & Software* 124: 104600. https://doi.org/10.1016/j.envsoft.2019.104600

Zhou, Y., Chang, F. J., Chang, L. C., Kao, I. F., & Wang, Y. S. 2019. "Explore a deep learning multi-output neural network for regional multi-step-ahead air quality forecasts". *Journal of Cleaner Production* 209: 134–145. https://doi.org/10.1016/j.jclepro.2018.10.243.

6 Integrating AI into Air Quality Monitoring

Precision and Progress

Rakesh Kumar, Ayushi Sharma, and Siddharth Swami

6.1 INTRODUCTION

Global monitoring of air quality has grown significantly in the last few years. The government's newly established or extended monitoring networks, as well as the vital contributions of nongovernmental organizations and concerned citizens worldwide, are responsible for these major advancements in the sphere of air quality monitoring. Despite advancements, many nations and areas continue to lack air quality monitoring, which prevents huge populations from having access to the data they need to manage pollution and make wise health decisions. West Asia, Africa, and Latin America have the least extensive monitoring networks worldwide (Almalawi et al., 2019). Nonetheless, there is still a serious problem with air pollution brought on by human activity, such as burning fossil fuels, and agriculture residue (Agarwal et al., 2014; Singh et al., 2010).

Revolutionary improvements in air quality monitoring are promised by inexpensive air pollution sensors and monitors. Government agencies and research teams create initiatives to evaluate and apply new technologies, yet some of their results are underwhelming. Due to the topic's complexity and multifaceted nature, it is challenging to keep track of every project in detail. A growing number of people, not only specialists, will be able to buy and install the appropriate type of ubiquitous sensors/monitors for their intended use as technology advances, albeit there may be difficulties with data interpretation (Morawska et al., 2018). Regulations, social needs, scientific developments, and high-performance computing power have all contributed to the evolution of RT-AQF, a subject that combines meteorology, atmospheric chemistry, mathematics, physics, and computer sciences, from weather forecasting. This review offers a thorough synopsis (Zhang et al., 2012). The current status of air quality sensors in cities is covered in this chapter, with an emphasis on wireless node network applications. The use of functional materials for chemical sensing and the possible uses of wireless sensor networks in smart cities are covered in the text (Penza et al., 2020). The color of the sky or the margins of far-off buildings can be used to visually identify air pollution. This results from atmospheric scattering, which is the loss of information and color variation caused by light reflecting off of

DOI: 10.1201/9781032683805-6

objects and scattering. Particle size, morphology, and concentration are important parameters (Anand et al., 2017; Awasthi et al., 2013; Singh et al. 2020; Ravikiran et al., 2019). These are the factors that researchers use to predict air quality levels from camera photos. Research suggests effective techniques for measuring air quality with camera-captured visual pictures. These approaches can be divided into three categories: simplified models, which employ simplified models but still don't provide accurate estimations, and physical model-based, which uses physical theories to compute atmospheric reflectance. Conventional machine learning techniques use atmospheric scattering theory to estimate air quality by analyzing the link between visual image features and data. Research studies develop models between picture features and particulate matter indicators such as $PM_{2.5}$, PM_{10}, and air quality index (AQI) using techniques such as decision trees, random forests, support vector regression (SVR), and linear regression. While subjective selection of picture attributes affects the effectiveness of machine learning-based techniques. The estimation of air pollution has gained approval for deep learning, which has the ability to learn image attributes on its own. Deep learning is being used by researchers to predict pollution from ultrafine particles, quantify particulate matter concentrations, and accurately classify air quality levels (Wang et al., 2024). According to uncertainties in atmospheric chemistry, physics, and emission mechanisms, managing air quality (AQ) in cities is difficult. A multifaceted system, the urban environment consists of the architecture, geometry, traffic patterns, green spaces, city layout, thermal characteristics, and microclimate. Developing efficient modelling and forecasting techniques is difficult due to the dynamic and non-linear nature of photochemical pollutants. Automobiles and industrial pollution react to produce ozone, a secondary pollutant. Nitrogen oxides and volatile organic molecules are its precursors; in the presence of sunshine and appropriate meteorological circumstances, they undergo photochemical reactions that alter ambient concentrations (Athanasiadis et al., 2006).

In developing nations, air pollution is a serious problem that contributes to health problems, ozone depletion, and decreased agricultural output. The factors contributing to the increase in pollution levels in metropolitan areas are population expansion, urbanization, automobile ownership, energy demand, and industrialization. Significant mortality rates have resulted from the worrisome levels of pollution encountered by Indian cities. Pollution projections for the near future can aid in lowering acute exposures. For the purpose of managing air quality, issuing health advisories, enhancing emission control initiatives, and supporting operational planning, air quality predictions are essential. Chemical transport models, weather patterns, and artificial neural networks (ANN) are some of the methods used in short-term forecasting. Unfortunately, because of the tremendous changes and high levels of pollution in India, there isn't much research available. Though its scope is restricted by its high cost, India's SAFAR program attempts to address this problem. An efficient forecasting model is required because there are more than 474 urban centers and a large amount of pollution. In order to predict ambient pollutants in highly contaminated areas, this study proposes a forecasting model in an urban center that uses ANN and real-time correction (RTC). The effectiveness of the model is evaluated and further study is needed to see how well it can be applied to various land use categories. The usefulness of an ANN model for predicting air pollution in a megacity is investigated in this work. It includes 32 sites and a variety of contaminants such as NO_2, ozone, and particles. The approach, which was created in partnership with the Central Pollution Control Board, is very reliable and replicable in areas with high pollution levels and requires little resource usage. An AI-based air quality forecasting method for controlling air pollution in India and other cities is presented in the article. For regulatory bodies, the instrument can function as an early warning system that makes proactive control measures possible. It also gives people the ability to realign their outside exposures according to estimated values of air quality, allowing them to make daily decisions and affecting train and aircraft itineraries. Additionally, the concept encourages individual contributions to emissions mitigation by increasing public knowledge (Agarwal, et al., 2020).

6.2 LITERATURE REVIEW

A rising climate has caused wildfires and dust storms, which have led to extremely high levels of pollution in Australia, South America, Siberia, and California. Predictive models can be built using past data on environmental contamination thanks to advancements in AI technology. AI has the capability to collect a wide range of ecological environment monitoring data, including pollution traces, environmental incidents, and eco-friendly characteristics. AI is also competent in creating an intellectual brainpower for environmental monitoring that supports all aspects of the industry, including all perpendicular different level analysis viewpoints and all horizontal monitoring elements. Mobile phones, laptops, and big screens are a number of stations to which it can be applied (Li et al., 2022). The different AI applications that are now being used to track and forecast air quality will be consolidated in this chapter. Additionally, the chapter discusses the challenges and prospects of integrating AI into air quality monitoring systems, emphasizing the potential for enhanced decision-making, public awareness, and environmental sustainability. The symbiosis of AI and air quality monitoring presents a promising path towards a cleaner and healthier environment.

The term "air pollution" refers to the release of chemicals into the atmosphere or other natural resources that can harm human health or that of other living things (Cioffi et al., 2020). It is the outcome of industrialization, population expansion, and human activity. While there may be short-term health problems like coughing and sneezing, chronic exposure can lead to serious health problems. The economy and standard of living are both impacted by the alarming rise in air pollution. Industrialization and economic growth enhance the quality of human life, but energy use and emissions, especially in developing nations, are major causes of air pollution. Seven million people die each year from breathing in high-pollution air, according to the World Health Organization. The father of AI, John McCarthy, described it as the science and engineering of building intelligent devices, mostly computer programs, that mimic human abilities like learning and problem-solving (Subramaniam et al., 2022).

The five main pollutants that cause pollution are ozone, particulate matter, carbon monoxide, sulfur dioxide, and nitrogen dioxide. These pollutants are measured by the

AQI. Every year, poor ambient air causes 4.2 million premature deaths, necessitating early intervention and preventative measures. The multidisciplinary area of environmental toxicology examines the detrimental impacts that biological, chemical, and physical agents have on ecosystems, including human populations. It is important for dealing with issues related to public health and the impact of contaminants on microorganisms and ecosystem services. Lung cancer, cardiovascular illness, and respiratory disorders can all be brought on by high air pollution levels.

There are two types of previous models used to anticipate air pollutants: statistical forecast models and physical forecast models. Whereas physical forecast models use atmospheric physics, chemistry, and aerodynamics to evaluate pollution diffusion, statistical forecasting uses past data to anticipate future events. In order to anticipate air quality and save computations and expenses, the statistical forecast model makes use of statistics and historical time sequence data. But it could not be as precise as the actual prognosis (Asha et al., 2022).

Artificial intelligence (AI) systems are able to monitor and forecast pollution levels by identifying factors such as industrial productivity and automobile emissions. These techniques can enhance public health, sustainable urban environments, and the environment while providing insights on environmental intervention. In order to reduce expenses and cover more ground, the study makes fresh recommendations for how to use the Internet of Things (IoT) to monitor air pollution. Strong predictive potential is provided by AI learning approaches including the gradient boosted decision tree ensemble model, linear regression, and SVR. In addition to causing global warming, climate change has a big influence on people's lives, professions, and economic development. According to the World Health Organization, air pollution claims the lives of almost 7 million people each year, with South and East Asia having the highest death rates. The main causes of air pollution and climate change are human activities such as burning fossil fuels, deforestation, and emissions from transportation (VoPham, 2018).

Two techniques for extracting models characterizing significant data classes or forecasting future trends are classification and prediction in data analysis. Category labels are predicted by classification, whereas continuous-valued variables are predicted using prediction models. In both approaches, there are two stages to the process: the training stage applies a classification/prediction algorithm, and the testing stage applies the model to data to make decisions. Qualitative factors are predicted by classification, while quantitative variables are forecast by prediction. Distinguishing themselves from deterministic models that match the real environment through the application of physical and chemical rules, data-driven forecasting models in air quality utilize distinct prediction techniques. Models for numerical and observational data assimilation should be differentiated from one other. Air quality forecasting models have been reviewed and categorized. In order to forecast current air quality in Athens, Greece, this work compares statistical techniques and classification systems. Air quality data from January 1999 to June 2002 is analyzed using environmental data, including observations of the ozone concentration level. With a focus on an urban monitoring station in the Marousi suburbs of the northeast, the study emphasizes the value of qualitative data in functional air quality control systems

(Agarwal et al., 2020). There hasn't been much prior study on AI-based air quality forecasting; most studies have relied on inductive analysis and subjective screening. With a top-down approach to literature evaluation, this work aims close this gap by using a top-down method to evaluate the literature. The chapter maps cooperation networks, co-citation networks, and co-occurrence networks using the scientometric program CiteSpace. Its objective is to review the field of air quality forecasting using AI techniques accurately and objectively from the standpoints of research hotspot evolution and specific analysis. Productive nations, organizations, writers, and their networks of collaboration are all included in the analysis, along with journals with high publication counts and articles with high citation counts. The outcomes of the reference co-citation analysis are used to derive the research hotspots and frontier evolution.

6.3 VARIOUS AIR POLLUTANTS FORECASTING BASED ON AI TECHNIQUES

There are a number of techniques based on ANI, machine learning, Support Vector Machines (SVM), and Deep Neural Networks that researchers are using to develop AI-based air pollution prediction systems. The advantages of fuzzy logic make it a popular tool for forecasting atmospheric air pollution. To anticipate air pollutants including CO, SO_2, O_3, and particulate matter, SVM have been employed in conjunction with other machine learning techniques. A strong machine learning technique called SVM is employed for a number of applications, such as regression, outlier identification, and classification (Kaur et al., 2018). With higher geographical and temporal data resolution, inexpensive air pollution instruments offer ground-breaking improvements in air quality monitoring. Due to the topic's complexity and multifaceted nature, it is challenging to keep track of every project in detail. Whether it's ambient air or interior monitoring, low-cost air monitoring equipment needs to be carefully characterized to fulfill specific requirements. Because it can handle high-dimensional data and nonlinear interactions, it is flexible and effective. With the largest margin between the closest points of various classes, the ideal hyperplane that divides data points in separate classes is sought after by SVMs in an N-dimensional space.

6.3.1 Ozone Concentration Forecasting Using AI Techniques

Unhealthy gases like ozone can impair agricultural yields and cause respiratory problems, skin, eye, and nose discomfort. Using hourly pollution concentrations and four climate factors, Murillo et al. employed ANN and SVM to predict ozone levels. NO_2 and O_3 have a major impact on the total AQI, according to research done by Chattopadhyay et al. (2007) on air pollution in Kolkata, India.

To forecast the concentrations of NO_2, O_3, SO_2, and CO, the scientists employed ANFIS models using semi-experimental regression. The ANFIS is a modeling technique for inference systems that uses fuzzy logic and data-driven adaptive networks to solve function approximation problems. For time series data forecasting, ANFIS fared better than regression models. To forecast the levels of O_3 in metropolitan city zones, models such as neuro-fuzzy, ANN, and MLR were employed. For prediction accuracy, ANN models performed better than Multiple Linear Regression Models (MLR) models. In terms of R^2 and RMSE values, ANFIS models surpass MLR in the prediction of gaseous pollutants, demonstrating the advancement of fuzzy logic-based algorithms.

6.3.2 Carbon Monoxide Forecasting Using AI Techniques

When dangerous gas concentrations in the blood are high enough, carbon monoxide exposure can be fatal. This is something that early warning systems must stop. Partial least squares (PLS) and SVM-based prediction models have been demonstrated in studies to provide improved CO concentration estimation. When forecasting CO, models like ANFIS, ANN, and SVR have acceptable levels of uncertainty. Delhi's CO levels were reliably predicted using a neuro-fuzzy model with an agreement index ranging from 0.88 to 0.93. More accurate than ANFIS models, ANN models forecast daily CO concentrations. Based on the research, the most accurate and dependable models for estimating carbon monoxide concentration are those based on AI, notably ANN, SVR, and fuzzy logic.

6.3.3 Carbon Dioxide Forecasting Using AI Techniques

The main contributor to global warming, carbon dioxide (CO_2), can harm human health by reducing blood serum pH, producing acidosis, and triggering respiratory and cardiovascular disorders. Thus, it is essential to create an accurate model for forecasting CO_2 emissions. Major pollutants like carbon monoxide, methane, nitrous oxide, sulfur dioxide, ozone, and total hydrocarbons have all been predicted using ANN techniques. While the ANFIS model forecasts emissions from clinical waste incineration plants with accuracy, the GMDH ANN model has demonstrated accuracy in estimating CO_2 emissions.

Research demonstrates that nonlinear autoregressive neural network time series models outperform other models in CO_2 emission predictions. To assess carbon emission intensity, Wei and his team have used an improved extreme learning machine along with factor analysis. Even in the case of nonlinear links, neural networks prove to be dependable methods for prediction and forecasting, particularly when it comes to schedule forecasting.

6.3.4 NO_2 and SO_2 Forecasting Using AI Techniques

Emissions of NO_2, mostly from burning fossil fuels, cause acid rain and have a negative impact on ecosystems and human health. The two main air pollutants that cause acid rain are SO_2 and NO_2, which have an impact on buildings, flora, water, land, and human health. It is essential to create efficient models for foreseeing and alerting about these pollutants. The concentrations of CO, NO_2, and ozone have all been predicted using ANN approaches, some of which have demonstrated excellent accuracy. For daily maximum concentration forecasting, hybrid adaptive forecasting models that combine ANN and SVM approaches as well as recurrent neural networks have been utilized.

AI approaches have revolutionized research on air quality predictions. This study uses sci-entometric and text analysis to examine the evolution of AI forecasting from 2000 to 2019. It looks at current topical interests, frontier evolution, and research hotspots. The results point to research gaps and suggest future directions, offering insights into the future of AI-based air quality forecasting.

6.3.5 Particulate Matter Forecasting Using AI Techniques

AI approaches have revolutionized research on air quality predictions. This study uses sci-entometrics and text analysis to examine the evolution of AI forecasting from 2000 to 2019. It looks at current topical interests, frontier evolution, and research hotspots. The results point to research gaps and suggest future directions, offering insights into the future of AI-based air quality forecasting. XGBoost-RF-ARIMA and other ensemble machine learning techniques are used to forecast $PM_{2.5}$ concentrations. A networked, scalable gradient-boosted decision tree framework is called XGBoost.

6.4 RELATIONSHIP BETWEEN AI AND ENVIRONMENTAL HEALTH

Cities' technological and economic expansion has resulted in an increase in pollution, including air, water, and noise. Air pollution, which is mostly produced by solid and liquid particles as well as some hydrocarbons, has been linked to respiratory diseases, cardiovascular illness, and lung cancer. Agriculture, the use of fossil fuels, domestic heating, natural calamities, and industrial emissions are the primary contributors. For 30 years, the US Clean Air Act program has researched air quality. Predicting air pollution levels accurately is critical for population protection and air quality management (Asha et al., 2022). Spatial science, also known as geographic information science, is critical for comprehending, analyzing, and visualizing real-world events based on their location. It identifies patterns in space using technologies like as GIS and remote sensing. With the emergence of big data, 80% of all data is geographic, necessitating the organization and strategy to creating new knowledge from geographical big data (VoPham et al., 2018). Pesticide exposure endangers human health, resulting in roughly 200 million deaths and 3 million poisonings globally each year. Pesticide exposure affects tens of thousands of farmers, particularly in underdeveloped nations. Pesticide exposure can cause skin irritation, eye irritation, headache, nausea, dizziness, diarrhea, vomiting, cancer, asthma, diabetes, and other health issues. Sustainable practices, such as the use of biochemicals, biofertilizers, and efficient nutrient usage, may help decrease agricultural output losses and the negative externalities associated with chemical use (Elahi et al., 2019). Table 6.1 proposes a different model for AI application for air quality measurement.

TABLE 6.1
Different AI Application for Air Quality Measurement

Reference	Proposed study	Model	Region	Parameter
T. Chiwewe and J. Ditsela,(2016)	To determine the ozone level	This model is used to find out level of ozone in a particular region of a selected country.	South Africa	To determine ozone level
Y. Zheng, X. Yi, M. Li, R. Li, Z. Shan, E. Chang, and T. Li,(2015)	To determine air quality by using of big data technology.	This model used for prediction of air quality for a future time i.e., next 48 hours.	China	Air quality measurement
E. Kalapanidas et al., 1999	This technique is used for prediction of air quality measurement for a particular time and in a particular region.	Artificial neural network.	Athens, Greece	Air quality measurement
R. Yu, Y. Yang, L. Yang, G. Han, and, O. A. Move (2016)	A random forest approach for predicting air quality.	Random forest model.	Shenyang city, China	Prediction of air quality
J. Y. Zhu, C. Sun, and V. Li,(2015	This technique is based on big data and is used for city wide air quality measurement	This model is used for city wide air quality measurement, but there is time complexity in this model.	China	meteorology and traffic

6.5 POTENTIAL APPLICATION USED IN AI FOR AIR QUALITY MONITORING

6.5.1 Real-Time Data Analysis

In the arena of air condition examining, real-time data analysis is crucial because it provides prompt insights into the complex dynamics of atmospheric conditions. There are various key points in this context where real-time data analysis is applied. One of the main advantages is quick pollution identification, which makes it possible to identify high pollutant levels right away. Quick detection is essential for reacting quickly to pollution incidents, putting preventative measures in place, and alerting the public in a timely manner so they may take the appropriate safety measures (Safiullin et al., 2019).

Manufacturers are adopting Industry 4.0 innovations, which provide full control over supply and material flow through modular automation optimized by data-driven input. Competitiveness, performance, adaptability, resilience, profitability, and customer service are all improved by these technologies. Industry 4.0 is centered around Smart Factory technology, robotics, machine-to-machine networking, and over-the-counter manufacturing. These tools interface with Industry 4.0 software, democratize data, and offer insights. Industry 4.0 envisions data, networked devices, capacity creation, and the upskilling of individuals in analytics and future technologies. One detrimental environmental factor that contributes to human existence and development is air pollution. For pollution management measures to be effective, accurate forecasting is essential. Numerical, statistical, and AI techniques are used in predicting. For increased accuracy, hybrid models have also been developed. This chapter compares the benefits and drawbacks of several forecasting models, reviewing their theory and practical applications. It seeks to give researchers and upcoming studies an overview. The utilization of these advances is essential for inter-organizational supply chain networks and the full value chain. To develop solutions that are both affordable and accessible, a robust network of partners is required, comprising of start-ups and tech companies. Figure 6.1 indicates real-time data analysis and early detection of air quality. The figure explores a decision process through conceptual framework with help of AI. Dynamic reaction to variations in air quality brought on by elements such as industrial activity, traffic, or weather is made possible by real-time data analysis. This adaptability enables decision-makers to mitigate environmental hazards in an informed manner. Notifying the public of pollution surges, real-time continuous analysis facilitates the construction of early warning systems (Javaid et al., 2022).

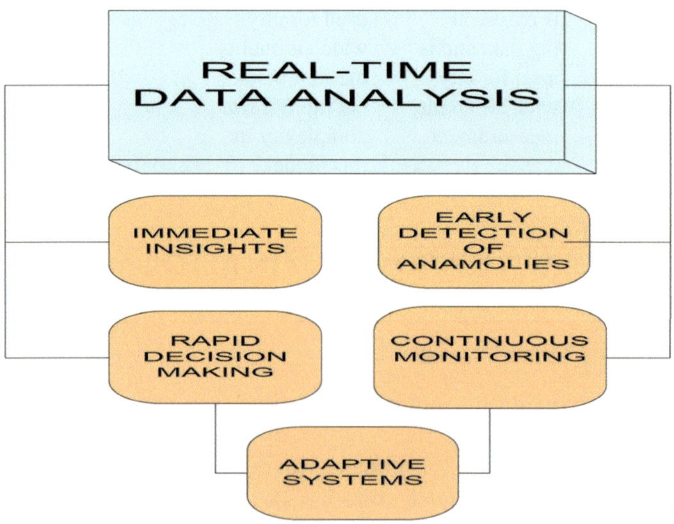

FIGURE 6.1 Real-time data analysis of air quality.

A proactive strategy for reducing the impact on the environment and public health entails real-time source attribution analysis. This makes it easier for regulatory bodies to enforce laws prohibiting polluting operations by identifying the causes of pollution. Additionally, it supports adaptive network optimization, facilitating focused interventions and policy initiatives (Mokrani et al., 2019,). The automation, self-cognition, and cutting edge technologies such as additive, robotics, and AI that make up Industry 4.0 are all combined. It can be accomplished with a variety of tools and technology and can yield creative answers to global problems. Because robotics stores and uses information wisely, it improves human-computer interactions. The physical world is better represented by modelling and simulation technologies, and vertical and horizontal systems integrate different result and practical aspects with non-functional layers. IoT strengthens the relationship between people and computers by bridging disparate fields. Sustainable development, which emphasizes people living, working, and building societies, can result from Industry 4.0. By carefully placing sensors according to pollution trends, adaptive optimization improves the efficacy and efficiency of air quality monitoring systems and offers comprehensive coverage in important locations (Sangwan et al., 2020).

6.5.2 Predictive Modelling

Because unmanned aerial vehicles (UAVs) are dependable and reasonably priced, they are being utilized more frequently to monitor air pollution. Even in situations with strong winds, these instruments are capable of measuring air quality in both vertical and horizontal dimensions. In addition, they can track air emissions and function as independent measuring stations, which offers several benefits for evaluating air quality. UAVs can be remotely operated and provide flexible wireless connectivity. They do, however, have certain drawbacks, such as the requirement for multi-pollutant approaches and the challenge of monitoring air pollutants due to their combination. Coordinating sensor data with GPS data is necessary for real-time monitoring, and in urban areas, safety precautions are essential. Furthermore, there are limitations on the use of UAVs for private, commercial, and research purposes. Deep learning (DL) algorithms have been created to enhance the forecasting of future air quality (Hemamalini et al., 2022). A vital tool for monitoring air quality is predictive modeling, which projects future meteorological conditions using sophisticated statistical techniques and mathematical algorithms. Figure 6.2 shows a predictive modelling of air quality measurement. The figure proposed air quality trend, early detection of air quality, resource allocation, palling for compliance and public health responses. It helps with long-term planning and regulatory decision-making by offering insights on patterns in air quality. When pollution events occur, it can send out early alerts so that authorities can take preventative measures. The efficacy and impact of air quality control are increased when authorities can deploy resources to pre-identified polluted areas or sectors thanks to predictive models, which help improve resource allocation (Guo et al., 2021).

There has been a notable surge in COVID-19 cases globally, leading to approximately 1.06 million fatalities. The death rate is between 3 and 4%, and people with

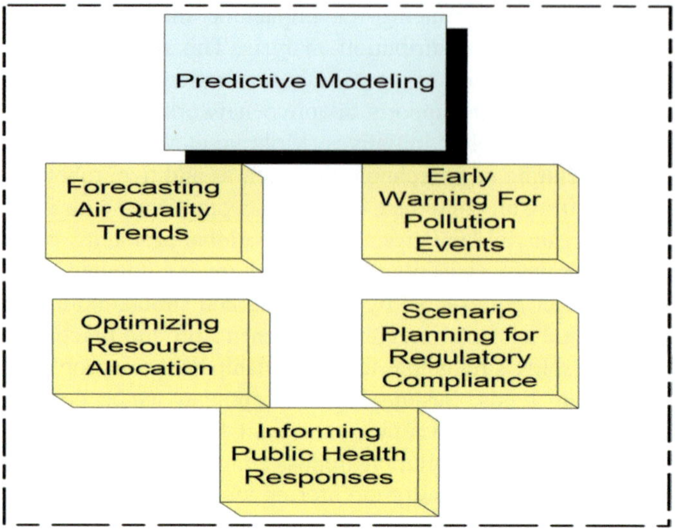

FIGURE 6.2 Important parameters to consider for predictive modelling of air quality.

underlying medical issues are at higher danger. Experts in public health attribute the high fatality rates from COVID-19 to prolonged exposure to air pollution, including $PM_{2.5}$ and NO_2. The development of real-time monitoring systems is crucial for promoting public health and wellness, as well as bettering individual health and well-being (Saini et al., 2021).

The planning and creation of economical device for an air quality monitoring system installed in cars, complete with cloud-based storage and air quality forecasting for the future. To lessen the number of people killed and injured in traffic accidents, the system is crucial for smart mobility applications and smart cities in the future. This device is important for identification of emission of harmful gases from vehicle. The chapter focuses on machine learning techniques and regression algorithms in the creation of sensor hardware and cloud-based predictive analysis for an in-vehicle air quality monitoring system. The prediction system combines the sensor and hardware. This system enables a real-time prediction system as well future prediction (Goh et al., 2021). By projecting the outcomes of potential changes to regulations or new emission control measures, predictive modelling is a proactive method to environmental management that helps ensure regulatory compliance. Additionally, it directs how the public health responds to variations in air quality by offering advice and health alerts to individuals who are more vulnerable. In addition to optimizing resource allocation, this proactive approach directs public health initiatives for more effective air quality management (Bock et al., 2019).

6.5.3 Pattern Recognition

Increased reliance on fossil fuels for transportation infrastructure results from urbanization and community growth, which raises emissions of pollutants linked to traffic.

Because atmospheric pollutants have a substantial impact on human health, including lung cancer, and other environmental issues like acid rain and the greenhouse gas effect, the problem of urban air pollution is widespread in both developed and developing countries. To reduce pollution and shield people from dangerous air pollutants, real-time data on air pollution, including $PM_{2.5}$, PM_{10}, and NO_2, is essential. Beijing has fewer than one-fourth of London's number of monitoring stations while having a greater area than London; as a result, many locations do not have enough precise air quality reports (Yang et al., 2017). A key technique in the monitoring of air quality is pattern recognition, which uses advanced algorithms to find intricate patterns in data. It supports regulatory bodies in enforcing penalties and assists in identifying emission patterns that indicate the origins of pollution. Additionally, pattern recognition speeds up the early detection of anomalies, allowing for prompt mitigation and intervention efforts (Mahariya et al., 2023). By finding relationships between variables influencing air quality, such as weather, traffic patterns, and industrial activity, pattern recognition aids in correlation analysis. This aids in the classification of air quality events, such as smog episodes and industrial pollutants, and the development of targeted pollution management strategies. Advanced pattern recognition algorithms also make adaptive monitoring tactics possible, which increase the effectiveness of air quality control by modifying the placements of monitoring devices and data collection based on trends that are identified (Anuşlu et al., 2019).

6.6 DISCUSSION

Sustainable nature is an important factor to achieve SDG-2030. AI played huge role in pollution management in term of prediction of real AQI and analysis futuristic measures thereto. This chapter find out various artificial quality prediction models in different regions which are effective and durable. These models are based on AI, big data, and machine learning. Real-time analysis, prediction models, and pattern recognition are some important applications which are based on technology. Different study models are used for particular regions like China, south Africa and Greece. These models predicted real-time data for selected region and analysis future result on the base of previous data.

This study was conducted on the basis of various datasets, applications, and models available in various regions. According to the analysis, smart manufacturing can address the scarcity of resources by enhancing resource management and industry efficiency. By lowering energy use and pollutant emissions, AI and machine learning are essential to sustainable development and green manufacturing. By incorporating stakeholders in production, predictive maintenance, inventory management, and other areas, AI and machine learning algorithms present prospects for sustainable development.

6.7 CONCLUSION

AI played a huge role in measures and prediction of AQI. It is essential to know about the real-time AQI supported in a sustainable environment. The chapter analyzes different applications used for measurement of AQI. The study also focused on applications used in particular geographical areas like China, South Africa, Greece, etc. The study examines how real-time AQI indexing can be used to maintain air quality at a satisfactory level in a specific area.

This chapter focused on the application of AI on the AQI and the future scope of AI in sustainable nature. The chapter finds different models of air quality measurement in different regions and their future scope. The study emphasized the potential application of air quality measurement like prediction pattern, recognition, etc. The chapter finds that AI applications help to find air quality and predict results. To achieve SDG-2030, AI can fill the gap of reality to use different applications for sustainable nature.

This chapter examines the application of AI in air quality monitoring, with particular emphasis on forecasting models, pollutant concentration estimation, and sensor data processing. It draws attention to the possibilities for better environmental sustainability, public awareness, and decision-making. Regression algorithms and artificial neural networks are two machine learning techniques used in this chapter to develop sensor hardware and cloud-based prediction analysis. It also emphasizes how AI affects people's life and climate change, and it suggests using IoT to monitor air pollution.

AI-based methods have demonstrated a great deal of promise for very accurate pollution forecasting. While statistical models are simpler to compute and apply, they still require a significant amount of historical data. Physical models, on the other hand, necessitate huge computations and a vast amount of data. Because they can simulate intricate nonlinear interactions and have less statistical restrictions, artificial neural networks, or ANNs, are a popular alternative for predicting air pollution. Hybrid artificial neural network models incorporate many techniques into a unified computational model, augmenting its capacity to manage intricate temporal variables and elevating precision. Because fuzzy logic can construct non-linear functions of arbitrary complexity and has a higher fault tolerance, it is commonly employed in air pollution forecasting.

6.8 LIMITATION AND FUTURE SCOPE

The study limited within AI used to measurement of AQI. The chapter analyzed different applications that exist to measure AQI in a particular geographical area. The topic keywords in this study are "air quality," "forecasting," "AI techniques," etc. These keywords are used to suggest efficient search tactics for air quality forecasting. In addition, it offers data and illustrations of notable nations, organizations, writers, publications, and industry players. The database used for this study is Scopus and web of science. There are number of areas which can be used for future aspects. The main limitation of the study was to find out the accuracy of different applications used to predict AQI. This can be improved through further research.

REFERENCES

Agarwal, R., Awasthi, A., Mital, S. K., Singh, N., & Gupta, P. K. (2014). Statistical model to study the effect of agriculture crop residue burning on healthy subjects. *Mapan*, 29, 57–65.

Agarwal, S., Sharma, S., Suresh, R., Rahman, M. H., Vranckx, S., Maiheu, B., … & Batra, S. (2020). Air quality forecasting using artificial neural networks with real time dynamic error correction in highly polluted regions. *Science of the Total Environment*, 735, 139454

Alfakeeh, A. S. (2022). An IoT based system for magnify air pollution monitoring and prognosis using hybrid artificial intelligence technique. *Environmental Research*, 206, 112576.

Almalawi, A., Alsolami, F., Khan, A. I., Alkhathlan, A., Fahad, A., Irshad, K., … & Anuşlu, M. D., & Fırat, S. Ü. (2019). Clustering analysis application on Industry 4.0-driven global indexes. *Procedia Computer Science*, 158, 145–152.

Anand, G., Nippani, S. K., Kuchhal, P., Vasishth, A., & Sarah, P. (2017). Dielectric, ferroelectric and piezoelectric properties of Ca0. 5Sr0. 5Bi4Ti4O15 prepared by solid state technique. *Ferroelectrics*, 516(1), 36–43.

Asha, P., Natrayan, L. B. T. J. R. R. G. S., Geetha, B. T., Beulah, J. R., Sumathy, R., Varalakshmi, G., & Neelakandan, S. (2022). IoT enabled environmental toxicology for air pollution monitoring using AI techniques. *Environmental Research*, 205, 112574.

Athanasiadis, I. N., Karatzas, K. D., & Mitkas, P. A. (2006, August). Classification techniques for air quality forecasting. In *Fifth ECAI Workshop on Binding Environmental Sciences and Artificial Intelligence, 17th European Conference on Artificial Intelligence*, Riva del Garda, Italy. 4:1–7.

Awasthi, A., Wu, B. S., Liu, C. N., Chen, C. W., Uang, S. N., & Tsai, C. J. (2013). The effect of nanoparticle morphology on the measurement accuracy of mobility particle sizers. *Mapan*, 28, 205–215.

Bakshi, B. R., & Paulson, J. A. (2022, June). Sustainability and industry 4.0: Obstacles and opportunities. In *2022 American Control Conference (ACC)* (pp. 2449–2460). IEEE.

Bock, M., Wiener, M., Gronau, R., & Martin, A. (2019). Industry 4.0 enabling smart air: Digital transformation at KAESER COMPRESSORS. In *Digitalization Cases: How Organizations Rethink Their Business for the Digital Age* (pp. 101–117).

Chattopadhyay, B. P., Mukherjee, A., Mukherjee, K., & Roychowdhury, A. (2007). Exposure to vehicular pollution and assessment of respiratory function in urban inhabitants. Lung, 185(5), 263–270.

Chiwewe, T., & Ditsela, J. (2016). Machine learning based estimation of Ozone using spatio-temporal data from air quality monitoring stations. *Presented at 2016 IEEE 14th International Conference on Industrial Informatics (INDIN)* (pp. 58–63). IEEE.

Cioffi, R., Travaglioni, M., Piscitelli, G., Petrillo, A., & De Felice, F. (2020). Artificial intelligence and machine learning applications in smart production: Progress, trends, and directions. *Sustainability*, 12(2), 492.

Elahi, E., Weijun, C., Zhang, H., & Nazeer, M. (2019). Agricultural intensification and damages to human health in relation to agrochemicals: Application of artificial intelligence. *Land Use Policy*, 83, 461–474.

Goh, C. C., Kamarudin, L. M., Zakaria, A., Nishizaki, H., Ramli, N., Mao, X., ... & Elham, M. F. (2021). Real-time in-vehicle air quality monitoring system using machine learning prediction algorithm. *Sensors*, 21(15), 4956.

Guo, D., Li, M., Lyu, Z., Kang, K., Wu, W., Zhong, R. Y., & Huang, G. Q. (2021). Synchroperation in industry 4.0 manufacturing. *International Journal of Production Economics*, 238, 108171.

Hemamalini, R. R., Vinodhini, R., Shanthini, B., Partheeban, P., Charumathy, M., & Cornelius, K. (2022). Air quality monitoring and forecasting using smart drones and recurrent neural network for sustainable development in Chennai city. *Sustainable Cities and Society*, 85, 104077.

Javaid, M., Haleem, A., Singh, R. P., Suman, R., & Gonzalez, E. S. (2022). Understanding the adoption of Industry 4.0 technologies in improving environmental sustainability. *Sustainable Operations and Computers*, 3, 203–217.

Kalapanidas, E. & Avouris, N. (1999, September). Applying machine learning techniques in air quality prediction. Proc. ACAI, 99, 58–64.

Karatas, M., Eriskin, L., Deveci, M., Pamucar, D., & Garg, H. (2022). Big data for healthcare Industry 4.0: Applications, challenges and future perspectives. *Expert Systems with Applications*, 200, 116912.

Kaur, P., Chaudhary, P., Bijalwan, A., & Awasthi, A. (2018). Network traffic classification using multiclass classifier. In *Advances in Computing and Data Sciences: Second International Conference, ICACDS 2018, Dehradun, India, April 20-21, 2018, Revised Selected Papers, Part I 2* (pp. 208–217). Springer Singapore.

Li, Y., Guo, J. E., Sun, S., Li, J., Wang, S., & Zhang, C. (2022). Air quality forecasting with artificial intelligence techniques: A scientometric and content analysis. *Environmental Modelling & Software*, 149, 105329.

Mahariya, S. K., Kumar, A., Singh, R., Gehlot, A., Akram, S. V., Twala, B., ... & Priyadarshi, N. (2023). Smart Campus 4.0: Digitalization of university campus with assimilation of industry 4.0 for innovation and sustainability. *Journal of Advanced Research in Applied Sciences and Engineering Technology*, 32(1), 120–138.

Mokrani, H., Lounas, R., Bennai, M. T., Salhi, D. E., & Djerbi, R. (2019, August). Air quality monitoring using iot: A survey. In *2019 IEEE International Conference on Smart Internet of Things (SmartIoT)* (pp. 127–134). IEEE.

Morawska, L., Thai, P. K., Liu, X., Asumadu-Sakyi, A., Ayoko, G., Bartonova, A., ... & Williams, R. (2018). Applications of low-cost sensing technologies for air quality monitoring and exposure assessment: How far have they gone?. *Environment International*, 116, 286–299.

Penza, M. (2020). Low-cost sensors for outdoor air quality monitoring. *Advanced Nanomaterials for Inexpensive Gas Microsensors*, 235–288.

Ravikiran, U., Sarah, P., Anand, G., & Zacharias, E. (2019). Influence of Na, Sm substitution on dielectric properties of SBT ceramics. *Ceramics International*, 45(15), 19242–19246.

Rollo, F., Bachechi, C., & Po, L. (2023). Anomaly *Detection and Repairing for Improving Air Quality Monitoring*. *Sensors*, 23(2), 640.

Safiullin, A., Krasnyuk, L., & Kapelyuk, Z. (2019, March). Integration of Industry 4.0 technologies for "smart cities" development. In *IOP Conference Series: Materials Science and Engineering* (Vol. 497, No. 1, p. 012089). IOP Publishing.

Saini, J., Dutta, M., & Marques, G. (2021). Indoor air quality monitoring systems and COVID-19. In *Emerging Technologies During the Era of COVID-19 Pandemic* (pp. 133–147). Springer International Publishing.

Sangwan, S. R., & Bhatia, M. P. S. (2020). Sustainable development in industry 4.0. In *A Roadmap to Industry 4.0: Smart Production, Sharp Business and Sustainable Development* (pp. 39–56). Springer International Publishing.

Sassanelli, C., Arriga, T., Zanin, S., D'Adamo, I., & Terzi, S. (2022). Industry 4.0 driven result-oriented PSS: an assessment in the energy management. *International Journal of Energy Economics and Policy*, 12(4), 186–203.

Singh, K., Thakur, A., Awasthi, A., & Kumar, A. (2020). Structural, morphological and temperature-dependent electrical properties of BN/NiO nanocomposites. *Journal of Materials Science: Materials in Electronics*, 31, 13158–13166.

Singh, N., Agarwal, R., Awasthi, A., Gupta, P. K., & Mittal, S. K. (2010). Characterization of atmospheric aerosols for organic tarry matter and combustible matter during crop residue burning and non-crop residue burning months in Northwestern region of India. *Atmospheric Environment*, 44(10), 1292–1300.

Subramaniam, S., Raju, N., Ganesan, A., Rajavel, N., Chenniappan, M., Prakash, C., ... & Dixit, S. (2022). Artificial intelligence technologies for forecasting air pollution and human health: A narrative review. *Sustainability*, 14(16), 9951.

VoPham, T., Hart, J. E., Laden, F., & Chiang, Y. Y. (2018). Emerging trends in geospatial artificial intelligence (geoAI): Potential applications for environmental epidemiology. *Environmental Health*, 17(1), 1–6.

Wang, X., Wang, M., Liu, X., Mao, Y., Chen, Y., & Dai, S. (2024). Surveillance-image-based outdoor air quality monitoring. *Environmental Science and Ecotechnology*, 18, 100319.

Yang, Y., Zheng, Z., Bian, K., Song, L., & Han, Z. (2017). Real-time profiling of fine-grained air quality index distribution using UAV sensing. *IEEE Internet of Things Journal*, 5(1), 186–198.

Yu, R., Yang, Y., Yang, L., Han, G., & Move, O. A. (2016). RAQ–A random forest approach for predicting air quality in urban sensing systems. *Sensors*, 16 (1), 86.

Zhang, Y., Bocquet, M., Mallet, V., Seigneur, C., & Baklanov, A. (2012). Real-time air quality forecasting, part I: History, techniques, and current status. *Atmospheric Environment*, 60, 632–655.

Zheng, Y., Yi, X., Li, M., Li, R., Shan, Z., Chang, E., & Li, T. (2015 August 10). Forecasting fine-grained air quality based on big data. In *Proceedings the 21st ACM SIGKDD International Conference on Knowledge Discovery and Data Mining* (pp. 2267–2276).

Zhu, J. Y., Sun, C., & Li, V. (2015). Granger-Causality-based air quality estimation with spatio-temporal (ST) heterogeneous big data. *Presented at 2015 IEEE Conference on Computer Communications Workshops (INFOCOM WKSHPS)*. IEEE.

7 Application of AI-Based Tools in Air Pollution Study

Sashi Yadav, Abhilasha Yadav, Asha Singh, Gunjan Goyal, Anita Sagwan, and Sunil Kumar Chhikara

7.1 INTRODUCTION

Air pollution significantly contributes to the global burden of environmental health costs, impacting healthcare systems and economies worldwide. Some 92% of people globally, as per estimates from the World Health Organization (WHO), are exposed to pollution at levels higher than those that are deemed harmful to their health (WHO, 2017). Numerous studies have indicated that extended and short contact with low-quality air raises the risk of heart attacks, lung cancer, and strokes and other serious diseases that pose a hazard to life (Agarwal et al., 2014; Wang et al., 2019; Yadav et al., 2023). Due to, this, the scientific community is in dire need of precise and dependable methods to counteract the negative impacts of air pollutants on living beings (Bublitz et al., 2019). Therefore, it's critical to provide trustworthy details on the origins of these contaminants to design effective approaches for mitigating environmental pollution. One of the most often used receptor models is the positive matrix factorization (PMF) which academics have used to assign source categories to better comprehend these sources. Additional receptor models comprise UNMIX, a multivariate receptor model, main component analysis, and chemical mass balance (CMB). PMF is a unique quantitative receptor model that takes into account the "uncertainty" of the data and incorporates this information into the final output, unlike traditional models. In recent years, AI has emerged as the most widely used technological instrument for reducing and limiting the harmful effects of different air contaminants. The atmospheric and medical sciences have shown a great deal of interest in this topic (Mo et al., 2019). For the diagnosis, management, and treatment of disorders linked to air pollution, AI approaches have been employed by several researchers as instruments for clinical decision support. For instance, Heuvelmans et al. (2021) developed a model based on deep learning to forecast the occurrence of lung cancer with inputs from 10,368 patient CT-scan datasets. The classification model performed admirably in classifying benign pulmonary tumors, according to the authors, who also stressed the importance of using standardized screening techniques to help increase the model's accuracy. A distinctive methodology for using fuzzy analysis and an ensemble model based on support vector machines to diagnose heart diseases

 DOI: 10.1201/9781032683805-7

(FSVM) was proposed by Nilashi et al. (2020). In another paper observations from the suggested FSVM model were fed data from the historical heart disease database. The ensemble model showed good accuracy for categorizing heart illness, according to the results, and it significantly sped up disease diagnostic computation time.

Additionally, AI is recognized as an essential instrument that regional preservation of the environment authorities should employ when deciding on mitigation strategies for reducing air pollution (Jerrett et al., 2005). AI has the ability to accurately predict pollution episodes by analyzing complex and non-linear relationships between various air quality indicators (Fernández and Vico, 2013). It has been demonstrated to be a precise and quick method to locate pollution hot spots and offers an exceptional capacity to monitor the present baseline for pollution. Furthermore, by using a variety of elements from the meteorological and data on air quality this approach generates analytical algorithms that help with forecasting and analyzing visibility, haze, smog, and other meteorological interventions in addition to improving air quality management. To anticipate ambient air pollution concentrations, interest in AI-based methods for identifying air pollution sources has recently increased. In view of the quick technical advancements in big data analytics including better computing platforms, fast parallel processing units, and scalable storage systems, AI has caused scholars to take notice of the development of sophisticated and precise sources for identifying air pollution (Masood and Ahmad, 2020; Rahimi, 2017;Bai et al., 2018). Prior research on air pollution identification contrasted AI-based systems with alternative approaches. McKendry (2002) contrasted the statistical technique known as multiple linear regression (MLR), which uses several explanatory factors to forecast a response variable's values, with artificial neural networks (ANN) for the goal of modeling particles in the air ($PM_{2.5}$ and PM_{10}) (Masoos and Ahmad, 2021). AI has the ability to maximize pollution prevention measures. Because intelligent systems are capable of assessing a wide range of air quality-influencing factors, they can be used to formulate context-specific and adaptive strategies. In urban planning, where focused strategies can be developed to reduce pollution from certain sources, such as traffic jams or industrial emissions and heat waves (Yadav et al., 2023), the use of AI-driven technologies is especially pertinent. The literature recognizes the enormous promise of AI in air pollution research, but it also points out certain difficulties, including issues with data quality, the interpretability of complicated models, and ethical issues. Establishing a balance between interpretability and model complexity is essential to building stakeholder confidence and comprehension.

7.1.1 Fundamentals of Air Pollution Source Identification

In the context of source identification, common air pollutants include sulfur dioxide, nitrogen oxides, and particulate matter, and volatile organic compounds are thoroughly researched. Because these pollutants come from several sources, including burning operations, industrial operations, and vehicle emissions, it is important to identify and mitigate each source using a tailored strategy (Yadav et al., 2022b). Problems still exist despite advances in methodology and technology. Because urban areas have complicated emission profiles, identifying individual sources requires advanced analysis. Accurate identification of temporal variability, including seasonal variations or event-driven emissions, necessitates ongoing monitoring (Kelp et al., 2020). Consolidating information from many sources—such as satellite images, emission inventories, and monitoring stations—makes it difficult to develop a cohesive picture of the causes of pollution (Mo et al., 2019). It is well recognized that most significant human activities, such as transportation, power plants, combustion engines, and industrial machines, discharge environmental contaminants into the atmosphere (Liu et al., 2022. These activities are the main sources of air pollution because of how widespread they are; at the moment, vehicles are thought to be responsible for more than 80% of it (Kumar et al., 2020). In addition, the ecosystem is somewhat impacted by natural phenomena including soil and volcanic eruptions, forest fires, and human activities such as field cultivation methods, gas stations, heaters for fuel tanks and cleaning practices (Kang et al., 2018). The sources of pollution are the main component taken into account when classifying air pollutants. It is crucial to discuss the four primary sources as a result, and they are as follows: primary, regional, cellular, and natural sources (Sahiner et al., 2019). Major sources of pollution include the plants engaged in metallurgy and other industries, refineries, power plants, petrochemical, chemical, and fertilizer sectors, related to municipal burning (Pal et al., 2015). Among the interior area sources are dry cleaning, gas stations, print businesses, and household cleaning duties. Mobile sources include vehicles such as cars, planes, trains, and other forms of transportation. The natural sources also include physical disasters such as dust storms, agricultural burning, forest fires, and volcanic erosion (Singh et al., 2010; Sun et al., 2020). Particle pollution, ground-level ozone, lead, and nitrogen oxides, and sulfur oxides pollution are the six principal air contaminants under observation by the World Health Organization (WHO). Air pollution can seriously damage a variety of environmental components, including soil, groundwater, and the atmosphere (Manisalidis et al., 2020). Moreover, it puts all living organisms in grave peril.

This is why these pollutants are the major focus of the present chapter as they are associated with further significant and all-encompassing problems with human health and the environment. A significant ecological influence on the pollution of the air is caused by factors such as acid rain, global warming, the greenhouse effect, and climate change (Manisalidis et al., 2020). Ever since the United States EPA released PMF, which Paatero invented, it has grown to be the most popular source identification technique. In January 2016, April 2016, July 2016, and October 2016 Liu et al.'s (2020) study used continuous online measurements to evaluate the concentrations of 99 volatile organic compounds (VOCs) at the urban site of Beijing, China. An examination of the composition and origins of VOCs was conducted, with a special emphasis on days with high levels of ozone (O_3) pollution. TVOCs, or total observed concentration of VOCs, were found to be 44.0 ± 28.9 ppbv. The amount of VOC contamination in Beijing has dropped, but it is still greater than other Chinese cities. Alkanes accounted for the greatest percentage of the seven main VOC categories that were sampled. There were clear temporal differences in the sources and concentrations of ambient VOCs. By PMF, six emission sources were found including burning of wood, combustion of coal, use of solvent biogenic and secondary emissions, and petrol and diesel vehicles (Anand and Kuchhal, 2018). The combustion source was found to be the main controlling element affecting Beijing's VOC reduction. Inner Mongolia, Shanxi, Hebei, Tianjin, Beijing, Henan, and Shandong were shown to be the primary potential source locations of surrounding VOCs using the potential source contribution function and the concentration-weighted trajectory (CWT) model (PSCF).

The 8:1 threshold for the VOC/NOx ratio demonstrated how sensitive Beijing's O_3 production is to VOCs. Subsequent examination of July's high- and low-O_3 days showed that higher VOC emissions and ideal weather circumstances for O_3 generation were the causes of the elevated O_3 levels. These results accentuate the noteworthy consequences of regional transportation and fuel burning with relation to the sources and amounts of volatile substances found in organic materials in Beijing's urban area. Furthermore, a variety dispersion models have been utilized in order to a considerable amount of time to evaluate the possible hazards related to the sources of air pollution release that are currently in place or that are proposed (Chakraborty et al., 2019). On the one hand, to accurately replicate variability, this procedure requires a sizable amount of data. However, because source lists contain obsolete data, the accuracy of numerical models associated with emission inventories, such WRF-Chem, may not be as good as it could be (Xi et al., 2015). Additionally, using these models to illustrate the effects of intricate geophysical features and dynamic climatic circumstances is difficult due to the complexities of air flows and topography features (Jiménez and Dudhia, 2012). Still, given the amount of contaminants, the wide variation in weather, and the variety of source and receptor parameter combinations provide significant obstacles to the creation of source inventories (Ma et al., 2012). According to data compiled the Environmental Protection Agency (EPA) of the United States, there are about 530 hazardous air pollutants that are linked to health hazards and can cause both acute and chronic health issues (USEPA, 2012).

The utilization of AI-based tools in air pollution research represents a significant leap forward in environmental science, fundamentally transforming traditional approaches with state-of-the-art technology. Harnessing sophisticated machine learning algorithms, these tools proficiently analyze extensive datasets gathered from a wide array of sensors, satellites, and monitoring stations, providing a detailed insight into the dynamics of air quality. Through the intricate examination of atmospheric data, AI models can accurately predict pollution levels, facilitating timely interventions and the formulation of policies.

7.2 IMPORTANCE OF SOURCE IDENTIFICATION FOR AIR QUALITY MANAGEMENT

Accurate identification of possible sources of air pollutant is the prerequisite for developing effective policies and control actions to enhance the quality of the air. One of the most urgent issues in contemporary public health is both the extended and brief exposure to ambient PM and its elements (Brook et al., 2017; Faridi et al., 2022). According to Querol et al. (2019), there are many different sources of ambient PM, including motor vehicle emissions, industrial facilities, power plants, burning crops, Re-suspended dust, wildfires, and dust storms (Yadav et al., 2022a). In order to identify the sources of ambient PM, notably $PM_{2.5}$ and related components, source apportionment (SA) studies have been done more frequently (Hopke et al., 2020; Munzel et al., 2021). One of the most important steps in supplying data that directs the creation of air quality management strategies is the recognition and numerical allocation of surrounding PM and its components (such as metal (loid)s, carbon species, sulphates, nitrates, and unidentified components) to their original sources (Hopke et al., 2020; Münzel et al., 2021). Researchers are studying the health effects of specific sources of PM concentrations to highlight their harmful impacts (Hopke et al., 2020; Münzel et al., 2021). Identification of the sources of air pollution is crucial for comprehending where the contaminants in the atmosphere come from. Traditional approaches have focused on methods like receptor modeling, which examine pollutant data statistically and chemically. On the other hand, AI-based methods that detect pollution sources make use of machine learning, data-driven algorithms, and the power of data analytics.

7.3 TRADITIONAL METHODS VS. AI-BASED APPROACHES

Traditional approaches have a proven track record in the industry, incorporating PMF receptor models and CMB. To determine the inputs from diverse sources, they rely on monitoring stations and chemical studies. These techniques are useful in a variety of settings because they offer useful insights into the local source and emission characteristics. Singh et al. (2017) have looked at how ambient $PM_{2.5}$ and PM_{10} originate from quantitatively in South Asia in studies conducted at the regional level (Almeida et al., 2020). Using the established multivariate receptor model, PMF, there have been studies on the composition and SA of PM mass concentrations in a number of nations worldwide (Taghvaee et al., 2018a,b; Soleimanian et al., 2021). Additional receptor simulations, such PCA and CMB, along with techniques for creating emission inventories have been employed to identify PM sources in great detail (Arhami et al., 2017; Esmaeilirad et al., 2020). These types of models and techniques are able to establish links between the bulk concentration of PM in the surrounding air and its constituents and a multitude of various resources, including secondary aerosols, emissions from transportation, industry, ships, soil, dust from the road, burning biomass, and sea salt, to name a few.

7.3.1 PMF Model

The traditional factor analysis (FA) served as the foundation for the PMF technique to particle source analysis. Afterwards, PMF has been extensively utilized to determine the origin of anthropogenic contaminants in a variety of media, including soil, air and water. Using PMF as a receptor model might be used with restrictions that are non-negative and define the matrix of concentrations to determine the origin locations as the result of multiplying a residue matrix by the matrices of contribution factors and source composition. Several environmental contaminants, including PAHs, POPs, volatile organic compounds (VOCs) (Yadav et al., 2022b), and heavy metals, have been source-apportioned using PMF. Numerous studies have been carried out in the last ten years on the origins of the particle size distribution (PNSD) and particle number concentration (PNC) using PMF, and source resolution of the particle chemical composition using filter sampling. China may have a high frequency of applications of volatile organic compounds (VOCs), PAHs, and heavy metal PMFs as a result of environmental pollution resulting from the nation's swift developing industrial and commercial sectors. In contrast, oOver the previous 20 years, the US has seen a considerable reduction in emissions, which has resulted in a drop in primary gaseous pollutants (CO, SO_2, and NOx) and $PM_{2.5}$. Nonetheless, it is commonly recognized that submicron particles (SMPs), which are smaller than 1 micron, have a closer link to negative effects on human health than other pollutants such as ultrafine particles (UFPs), which are smaller than 100 nanometers. PNC studies have been quite high in the USA during the past ten years because of the absence of information regarding the sources of ultrafine particles (Sun et al., 2020). AI-based methods, such as deep learning and machine learning models, present an alternative viewpoint. They are appropriate for source identification across huge geographic regions because they can handle enormous volumes of data, including real-time and geographical data. These technologies are able to spot correlations and patterns that conventional approaches can miss.

7.3.2 LCS (Low-Cost Sensor)

LCS are offering novel chances to fill in the gaps in observation data and gain insight into sub-scale emissions due to their lower cost and simpler installation. The viability and capacity of LCS as an additional monitoring resource to reference grades have been assessed by several academics (EIONET, 2019). These include comparing several sensor manufacturers, assessing the performance of a single sensor, and assessing a network of sensors. Studies have shown that its in-situ measuring calibration is necessary before deployment (Zheng et al., 2019; Simmhan et al., 2019). Studies have shown that LCS can be used to create heat maps of city pollution (Zheng et al., 2015; Yi et al., 2015) and characterize sources by understanding emissions from both nearby and distant sources (Hagan et al., 2019; Caubel et al., 2019). Additionally, they improve prediction accuracy through covariate data analysis, including weather and traffic data, and enhance dispersion models for better spatiotemporal health exposure evaluations (Mukherjee et al., 2019; Engel-Cox et al., 2013). These advancements raise awareness and benefit sensitive populations, such as students and individuals with asthma (Bulot et al., 2019; Roychowdhury et al., 2016).

Due to the extremely spatiotemporal nature of pollution levels and emission source contributions throughout the city, mobile sensors are becoming more and more common for health exposure assessment, source estimation, and awareness-raising. Many vehicle platforms, including those for buses, taxis, bikes, and pedestrians, have seen the pursuit of LCS (Ramos et al., 2018; Popoola et al., 2018). For instance, GSMA (2018) developed the "Smog mobile," a specialized car equipped with gas and particle sensors that was used to produce fine-grained pollution heat maps. In a different study, it was discovered that LCS equipped with image processing algorithms on public buses was appropriate for choosing routes that minimize pollution (Kaivonen and Ngai, 2020). These kinds of applications ought to be included in autonomous car navigation systems in the future. LCS is a useful instrument for evaluating health exposure because of its portability and cost-effectiveness. It has been applied, for instance, to the monitoring of activity-exposure correlation for trip distance and mobility alternatives (Steinle et al., 2015). In different research, Google Street View automobiles equipped with LCS discovered notable differences in traffic exposure and pollution levels among streets and cardiovascular disease incidences in an Oakland, USA, neighborhood (Alexeeff et al., 2018). Similarly, Zalakeviciute et al. (2018) used LCS and Google traffic data to construct cycle lanes in Quito by analyzing the $PM_{2.5}$ exposure of cyclists on the roadside. These cases imply that, in addition to research, properly calibrated LCS can offer instruments for urban planning and facilitate individual choices like going for a walk outside or picking a place to stay. For LCS to be used operationally, practical achievability and planning of operations are just as important as pollution-related evaluation. This covers aspects of data security, sensor network architecture, and multi-sensor interoperable data protocols (Al-Ali et al., 2010; Sanchez et al., 2014). However, significant obstacles to its widespread real-time applications include all-weather maintenance, data accuracy, and sensor life expectancy. Studies have indicated that low cost PurpleAir sensors have caused 90% improvement in Air Quality Index (AQI) by using Machine Learning (ML) techniques due to the addition of the temperature and humidity correlation correction. The "fit for purpose" strategy, which includes studying locating hotspots, identifying local sources of emissions, developing pollution maps, raising awareness, involving citizens, and providing health exposure assessments and advice, adds value to LCS even though its practical application is still in its infancy. Additionally, improved urban planning and citizen advisory can be created by learning how emissions relate to other factors including neighborhood socioeconomic features, traffic flow, attitudes, and weather conditions.

7.4 INTRODUCTION TO MACHINE LEARNING AND AI MODELS FOR SOURCE IDENTIFICATION

It is incredibly challenging to directly forecast and track the temporal and spatial changes in pollution due to the complexity and volatility of the environment. Recent developments in high-performance computers, different ML methodologies, and unprecedented data accumulation have opened up new possibilities for environmental contamination study. To calculate ground-level concentrations of air pollutants, allocate pollution sources, and predict the spatial distribution of water contaminants, satellite data processing has used ML methodology. Advanced algorithms like deep neural networks, in contrast to the active ML practices in chemical toxicity prediction, are still lacking in research of the environmental processes that lead to pollution (Liu et al., 2022). Deep learning methods for source detection and air quality forecasting are depicted in Table 7.1.

TABLE 7.1
Deep Learning Methods for Source Detection and Air Quality Forecasting

Scientific task	Conventional approaches	Deep learning approaches	Potential advantages
Ground-level pollution inversion estimates derived from satellite data	Regression on aerosol optical depth (AOD): linear or nonlinear	CNN (Hong et al., 2019), DBN (Li et al., 2017a)	Prevent intrinsic retrieval problems and appropriate for areas without stations
Spatiotemporal air quality forecasts	Statistical techniques (like ARIMA and MLR) and CTM simulations (like CMAQ and NAQPMS)	Temporal deep networks (e.g., RNN, LSTM, GRU (Bui et al., 2018; Freeman et al., 2018) STDL models (e.g., CNN-LSTM, CNN-GRU (Wang et al., 2018)	Identify nonlinear characteristics in high-dimensional data
Data filling and interpolation to create a dataset for pollution projections	The interpolation of linear or cubic spline, and inverse distance weighting (IDW)	Back propagation deep learning neural network (Li et al., 2017b), spatiotemporal semi-supervised deep learning neural network (Qi et al., 2018)	Appropriate for areas lacking surveillance stations
Improve CTM forecasts	Improve parameterization scheme	Deep learning model to replace and reproduce the chemical mechanism (Kelp et al., 2018; Keller et al., 2019), encoder-operator-decoder neural networks (Kelp et al., 2020)	Enhance, accelerate, and construct a more reliable numerical model
Estimating sources for air pollution dispersion projections	Computational fluid dynamics (CFD); Gaussian dispersion model	Mask-RCNN (region convolution neural network) (Kumar et al., 2020) and back propagation network coupled with Gaussian model (Ma et al., 2016).	Meet emergency needs with utmost precision and efficiency.

7.4.1 Deep Learning Methodology (For Instance, Neural Networks and Convolutional Neural Networks

Deep learning, a general-purpose technology that uses enormous amounts of data and gigantic neural networks to extract complicated knowledge, has recently surfaced. As a result, it has the ability to overcome the shortcomings of the air quality forecasting techniques used today. The area of deep learning, a subset of machine learning, has increased dramatically since 2010. The key to the accomplishment is the ability to effectively train enormous neural networks with several hidden layers to extract fundamental patterns as well as characteristics from enormous quantities of data. The ability of deep networks to learn knowledge representations in extremely high-dimensional areas, such as of establishing connections between network weights neurons is demonstrated. Hence, the way the networks are set up, or their mapping of the knowledge spaces across the several disciplines, is called architecture.

Deep learning has found widespread success, initially in computer science domains like speech recognition, computer vision, and natural language processing, and subsequently in basic sciences like chemistry and physics. Forecasts of air quality should likewise be successful because the evolution of contaminants and the prediction of image sequences are comparable (Donnelly et al., 2015; Esteva et al., 2017).

Over the past 20 years, there has been a significant rise in the amount of data gathered on air quality. For instance, satellite remote sensors are constantly producing fresh worldwide photos. However, more computing power allows for routine numerical model simulations. But the speed at which we can generate and gather data exceeds the speed at which we can understand it (Reichstein et al., 2019). These massive datasets allow us to develop deep learning-based air quality systems and anticipate air quality.

In spite of the noteworthy dynamic, spatially expansive, and behavioral variability in these systems (Kang et al., 2018), deep learning can automatically identify low-dimensional features from high-dimensional data, reduce the dimensionality of the data, and learn from it (Chin et al., 2000).

Deep networks' capacity to extract complicated nonlinear properties across scales from high-dimensional datasets is determined by their architectural designs. In contrast to shallow artificial neural networks, deep networks consist of many layers of neurons (Perez and Reyes, 2006). The connections between these neurons provide unique deep network designs suitable for many kinds of applications. profound networks for air quality forecasting that are more adept at extracting spatiotemporal information are more suitable. We quickly introduce a few deep network designs based on their spatiotemporal properties. The long short-term memory network, the recurrent neural network (RNN), and the gated recurrent unit (GRU) network are among the architectures that may adapt to temporal series predictions.

Among the designs that may adjust to spatial feature extractions are stacked autoencoders, deep belief networks (DBN), and convolutional neural networks (CNN). To build spatiotemporal deep learning (STDL) systems, temporal and spatial modules can be linked or integrated.

As STDL designs may reflect intricate spatiotemporal characteristics at various pollutant evolution sizes, they ought to be better equipped to respond to predictions

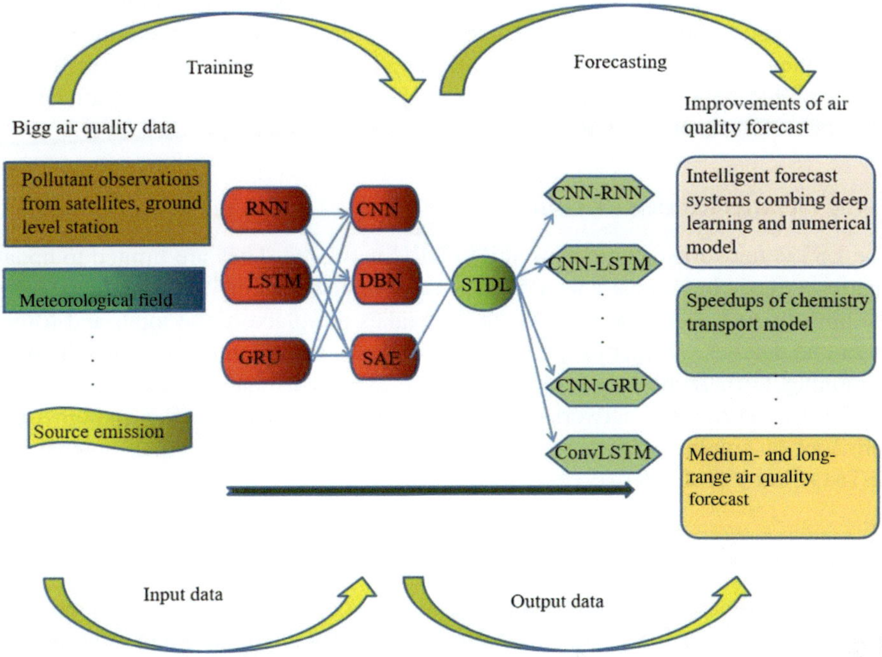

FIGURE 7.1 Various deep network designs.

about the state of the air. Figure 7.1 presents a deep learning framework for air quality along with different deep network architectures utilized in forecasting.

7.4.2 Recurrent Neural Networks

Variants of feed forward neural networks (FNNs) are called RNNs. Signals can only flow in a single direction because of FNNs, from input to output. FNNs are simple networks that connect inputs and outputs without the need for loops (Aishwarya and Aneena, 2018). RNNs, which are based on FNN, add a neuron self-connection and cyclic structure inside the network. As a result, input data can be learned, and through self-connected neurons, data sequences can affect network outputs. In many cases, RNNs perform better than FNNs by utilizing their memory features. Nevertheless, when the training period is too lengthy, RNNs may experience issues with failing to identify long-term relationships in the input data and experiencing disappearing and ballooning gradients (Kolen and Kremer, 2001).

7.4.3 Memory for Long–Short Term

Better RNNs are called LSTM networks (Hochreiter et al., 1997). Memory blocks are introduced to solve the vanishing gradient issue. Remember, output and input gates are the three different types of nonlinear multiplicative gates that make up the memory blocks. The multiplicative gates decide if the incoming data has to be remembered and regulate how the memory block operates. The flow of cell activation from input into a memory cell is managed by the input gate, and the flow of output from a memory cell into other nodes is managed by the output gate. In many situations, LSTM networks outperform typical RNNs due to their ability to learn lengthy sequences (Zaytar et al., 2016).

7.4.4 Gated Recurrent Unit

The LSTM networks are condensed into GRU networks. They are limited to update and reset gates, but they are nevertheless capable of balancing the data flows within the unit. In an LSTM, the gate for updates takes the place of the input and forgets gates and decides whether or not information has to be remembered. When it comes to training, GRU requires less computing power because it has fewer parameters than LSTM. However, GRU networks have demonstrated comparable or even higher performance on smaller datasets when it comes to voice and music signals compared to LSTM (Chung et al., 2014).

7.4.5 Artificial Neural Networks

An algorithm-based cognitive computing system is an ANN technique that mimics the surface ability using the nervous system in humans to simulate and examine non-linear data (Sapra et al., 2015; Olivieri and Allegrini, 2019). These are the classic instances of newly developed AI methods that are the dynamics and the structural components of a biological brain. They are made up of interconnected neurons, which are non-linear processing units (Ghritlahre and Prasad, 2018; Chatterjee and Pandya, 2016). Together, these neurons can address a wide range of issues, including control, function approximation, clustering, prediction, pattern classification, and content-based retrieval. Creation, instruction, and verification are the three stages that typically comprise an ANN modeling scheme. The optimal ANN design, the number of learning modes, the number of layers, the number of neurons in each layer, and the parameters for the input and output are all chosen during the design process (Subramanyam, 2008). During the training phase, synaptic weights are modified iteratively to lessen the discrepancy between the intended and anticipated values for every training set that the artificial neural network receives. The ANN's generalization performance is assessed for the unknown data at the end of the validation stage. After a satisfactory validation procedure, the ANN model could be further refined or taken into consideration for the prediction of unknowns (Anand et al., 2017; Dorohoi and Dimitriu, 2017; Hasanuzzaman and Rahim, 2019). According to Agirre et al. (2006), the most popular ANN configuration for predicting different types of air pollution occurrences is the multilayer perceptron (MLP). Three layers make up an MLP's design in general: input, concealed, and output. Additionally, contingent upon the intricacy and character these networks' architectures could have one or more hidden layers, depending on the assignment.

This allows the networks to discover complex relationships between the input and output data vectors. One important component of neural architecture that significantly affects the output of the network at the end involves determining how many neurons these buried layers. If there are either too many or too few nodes in the hidden layer, the network will have an over-fitting or under-fitting issue (Silva et al., 2017). In order to get around this problem, numerous strategies to increase the number of neurons have studied. These methods include the sequential orthogonal method, two-phase method, rule of thumb approach, trial and error method, and heuristics-based approach.

7.4.6 DNN (Deep Convolutional Networks)

Deep neural networks (DNNs), which combine deep architecture with cutting-edge training methods, are more advanced forms of artificial neural networks that achieve higher degrees of data abstraction and complexity (Abiodun et al., 2018). From a structural perspective, a neural network possesses more than three layers. That is, two or more strata in plain sight are considered "deep," while networks that have three layers or fewer are considered shallow (Sahiner et al., 2019; Wardah et al., 2019).

The hierarchy types of an input layer, an output layer, and several hidden (deep) layers comprise deep neural network are what give these networks their "deep" quality.

These networks have the ability to create due to their structural depth, which allows for more concise representation of input-output relationships compared to the neural network of our ancestors. The capacity of DNNs to carry out autonomous extracting features from unprocessed inputs, sometimes referred toas feature acquisition, sets them apart from conventional ANNs (Ganapathy et al., 2018). Furthermore, the scalability of the system, allowing DNN to improve its conclusions with additional data and a wider network, is another important distinction between DNN and ANN (Vieira et al., 2017). Advances in GPU processing power methods and data size are necessary for the very compute-intensive process of training DNNs (Sergeev et al., 2018). In the early DNNs, the standard back-propagation algorithm was the training method of choice by default. However, contemporary deeper networks trained with this method typically exhibit subpar performance and a disappearing gradient and overfitting problem. More complex algorithms, such as random, dropout, Support Vector Machines, and greedy layer-wise are available. The algorithms random forest, dropout, assistance vector systems, and greedy layer-wise have been developed and employed for numerous deep learning tasks to get around these problems (Kaur et al., 2018). Numerous deep learning architectures, including autoencoders, long short term memory networks (LSTM), deep belief nets, convolutional neural networks, recurrent neural networks, etc., have shown exceptional benchmark performances and a broad variety of uses in modern times.

The most effective and popular deep neural architectures for air pollution forecasting are the CNN, autoencoders, and LSTM. These models differ in terms of computational complexity and effectiveness (Kim et al., 2018; Zhang et al., 2020a, Zhang et al., 2020b).

7.4.7 Random Forest

A decision tree-based strategy for group learning is called random forest because of its resilience and capacity to manage complicated datasets, it is frequently used for source identification and classification. This is how it operates:

Using a random selection of the data and features for training, random forest generates a group of decision trees. This variety enhances generalization and lessens over fitting.

Voting Mechanism: Every decision tree in the ensemble contributes its classification for prediction-making. The class that is most frequently predicted wins out in a majority vote, which usually determines the final outcome.

For source discovery tasks including both structured and unstructured data, random forests work well. They are renowned for their resilience to noise and outliers and have the ability to handle highlight spaces with great dimensions.

7.4.8 Support Vector Machines (SVM)

Support Vector Machines (SVM) is a widely used machine learning technique that is frequently applied to source discovery (Kaur et al., 2018), especially in problems involving binary or multi-class classification. This is how SVM functions:

Finding a hyperplane that maximizes the margin between classes is what SVMs aim to do. The ideal hyperplane to divide the data points into distinct classes is this one.

Kernel Trick: SVM may translate the data entering a space with more dimensions, where it transforms linearly separable by employing kernel functions (such as polynomial and radial basis function). This allows SVM to tackle nonlinear source detection jobs.

One-Versus-All (OvA) Strategy: By employing an OvA strategy, which involves each class being trained with a different binary classifier, SVM can be expanded to handle multi-class situations. The class with the highest confidence score is the one that gets classified in the end.

7.5 ESTIMATING THE SOURCE FOR DISPERSION FORECASTS

A dispersion prediction is essential for preventing crises and shielding people from air pollution. The Gaussian dispersion model (Lodge, 1983; Seaman, 2000), the stochastic Lagrangian model (Flesch et al., 2004; Wilson et al., 1996), and the Computational Fluid Dynamics (CFD) model (Pontiggia et al., 2009; Mazzoldi et al., 2008) are the three primary dispersion models used today. Precise dispersion projections require precise localization of the pollutant sources. An algorithm for parameter estimation is therefore required. In an emergency, the source estimation and dispersion projections must be accurate and timely. In essence, dispersion forecasts and pollution source estimation need the establishment of fast-forward models. Such forward models could be based on deep networks or neural networks. For example,

So et al. (2010) estimated the combination of a neural network with data from optical sensors to calculate the dangerous rate of gas release. To estimate the concentration of released gases at certain off-site locations, Wang et al. (2015) suggested a method that combines gas dispersion, neural networks, and gas detectors. However, source estimates could not be carried out in these studies absent an air dispersion mechanism.

To make an estimate the source, combined machine intelligence algorithms for gas dispersion prediction with a traditional Gaussian model. The PSO, or particle swarm optimization, was then used to determine the parameters of the emission source. This method was used by Qiu et al. (2018), who also examined maximizing expectations (EM) technique for estimating source parameters. Using deep learning algorithms for source estimate is feasible when remote sensing picture datasets are accessible. To represent and detect the Mask-Region Convolution Neural Network (Mask-RCNN), which is hyperspectral for analyzing vast area hyperspectral imageries, and the discharges of methane plumes.

7.6 ARTIFICIAL NEURAL NETWORK UTILIZATION IN AIR POLLUTION SOURCE IDENTIFICATION

Conventional techniques for identifying sources, like chemical investigations and dispersion modelling, have yielded important information about the origins of pollution. However, when dealing with complicated metropolitan areas, these methodologies may have limits and often involve considerable computing hurdles requiring large amounts of data (Faridi et al., 2022). ANNs provide a creative way around a few of these restrictions. Neural nets seen in the human brain serve as the model for ANNs, which are computer models. Built from interconnected layers of artificial neurons, they are capable of deriving complicated relationships and patterns from data (Zheng et al., 2019).

An important benefit of utilizing artificial neural networks (ANNs) for source identification is its capacity to handle massive datasets quickly and adjust to non-linear correlations in the data (Abiodun et al., 2018). The data pertaining to air quality, encompassing input from many sensors, monitoring stations, and satellite imaging, is intrinsically intricate and frequently displays non-linear trends. Because ANNs are able to detect these subtle patterns, they are a good choice for identifying pollution sources in a variety of environments, such as cities, industrial areas, and areas where several sources of pollution are present (Chatterjee and Pandya, 2016). When identifying by employing ANNs, the cause of air pollution, several air quality measures (such as the concentration of pollutants like $PM_{2.5}$, NOx, SO_2, etc.) and meteorological data are usually collected (Yadav et al., 2022b).

7.6.1 Steps for Source Identification Using ANNs

Data Collection: It is crucial to collect thorough and representative data on air quality, including measurements of different contaminants and climatic factors. The ANN model is built on the basis of this data.

Data Pre-processing: To make sure the input data is appropriate for ANN training, it is essential to do feature selection, normalization, and data cleaning.

The advancement of the ANN model involves determining its design, which includes the quantity of layers, each layer's neurones, and activation functions. Using historical data, the model is trained to identify trends and connections.

To ensure the model can generalize and make correct predictions, it is trained on part of the data and validated on another.

The application potential of ANNs has been further increased by their integration with remote sensing and Geographic Information Systems (GIS). ANNs can offer insights into the geographic distribution of pollution sources by integrating spatial data. This has been shown to be crucial for developing policies and urban planning that try to lessen localized pollution (Da Silva et al., 2017).

7.7 DEEP NEURAL NETWORK (DNN) APPLICATION FOR SOURCE IDENTIFICATION

The study of air pollution source detection has observed an increase in the usage of Deep Neural Networks (DNN) in recent years. A subclass of artificial neural networks known as DNNs has proven to be exceptionally adept at processing complex data and resolving challenging issues (Poppla et al., 2018).

DNNs have been used to pinpoint pollution sources and forecast air quality. To produce real-time air quality forecasts, they can analyze complicated information such as meteorological, geographic, and emission data (Esteva et al., 2017). Additionally, by linking variations in pollutant concentrations with probable emission events, DNNs can link pollution spikes to particular sources, improving source identification capabilities (Kim et al., 2018).

7.7.1 Process Involved in Source Identification

Collecting Data and Preprocessing: Gathering thorough information on air quality is the first stage in utilizing DNN to identify the cause of air pollution. Typically, this data comes from meteorological, geographic, and air quality monitoring station sources. Preprocessing is preparing the data for DNN analysis by cleaning, combining, and formatting it.

The process of selecting relevant features or variables for analysis involves considering factors including geographic information, meteorological conditions, concentrations of pollutants (such as $PM_{2.5}$, NO_2, and SO_2), and wind speed and direction. These factors can all have an impact on air pollution. For the DNN to be properly trained to identify pollution sources, these properties are essential (Rahimi et al., 2017).

DNN Model Training: Next, utilizing the gathered and preprocessed data, a DNN is created and trained. Typically, a DNN's architecture consists of several layers of neurons, and to optimize model parameters, it uses techniques such stochastic gradient descent and back propagation. The input features are mapped by the model to the source of pollution.

Pollution Source Identification: The DNN can be used to detect pollution sources if it has received sufficient training. The program forecasts possible pollution sources when it receives fresh data on air quality. The DNN can offer details on the probable causes of elevated pollutant concentrations in a certain region, which may include of natural sources, traffic emissions, and industrial facilities (Esteva et al., 2017).

Validating the accuracy and dependability of the DNN model is crucial. This is known as model validation and accuracy assessment. Usually, to do this, the model's projections are compared with historical pollution source data or ground truth data. Common statistical metrics include root mean square error (RMSE) and mean absolute error (MAE). employed to assess the performance of models.

Constant Monitoring and Adjustment: The origins and circumstances of air pollution are ever-changing, making it a dynamic and ever-evolving issue. As a result, the DNN model needs to be updated and modified often to account for shifting patterns of pollutant sources. Frequent retraining of the model guarantees the accuracy and applicability of the pollutant source detection (Reichstein et al., 2019).

Policy and Management Recommendations: The DNN model's findings can help guide judgments on air quality management tactics and policy. Regulations and environmental organizations can provide focused interventions to lower pollution, like zoning laws, traffic management plans, or emission control measures, by identifying the origins of the pollution.

7.8 FUTURE TRENDS AND EMERGING TECHNOLOGIES

7.8.1 Advancements in AI for Air Pollution Source Identification

The fields of machine learning and artificial intelligence are always changing, and the identification of the sources of air pollution is probably not an exception. More advanced AI models and algorithms that can handle ever-more-complex data are possible future developments. Recurrent neural networks (RNNs) and convolutional neural networks (CNNs), two deep learning methods, may be tailored for image-based source identification and time series analysis, respectively. More precise and detailed source detection will be possible as a result of these developments.

7.8.2 Integration with Internet of Things (IoT) and Real-Time Monitoring

Real-time environmental monitoring is becoming more and more common with IoT technology. Air quality, weather, and pollutant levels can all be measured in real-time by sensors and devices, and AI systems will work in tandem with IoT infrastructure to process and analyze this continuous data stream. This will allow for quick identification of pollution sources, quick reaction to incidents, and the possibility of automated alerts and warnings to safeguard public health.

7.8.3 Potential for AI-Driven Policy and Decision Support Systems

The creation of sophisticated decision-support tools for lawmakers and environmental agencies can greatly benefit from artificial intelligence. AI can provide practical approaches for pollution prevention and control by evaluating both historical and current data. This involves making recommendations for zoning laws, traffic control strategies, and pollution reduction techniques. AI-powered systems can model the effects of various policies, assisting decision-makers in selecting the most practical and long-lasting solutions.

7.8.4 Better Data Fusion and Multimodal Analysis

Multimodal data fusion will be used more and more in future artificial intelligence systems for pollution source identification. To provide a more complete picture of pollution incidents, this entails integrating data from other sources, including social media, ground-based sensors, and satellite photography. AI algorithms will be created to efficiently combine and examine these various kinds of data, resulting in more precise source characterization and identification.

7.8.5 Explainable AI (XAI) and Regulatory Compliance

Transparency and interpretability will become increasingly important as AI systems become more integral to regulatory compliance and enforcement. Source identification models will use Explainable AI (XAI) methodologies, which will facilitate the understanding and validation of the reasoning behind AI-generated judgements. Ensuring trust and accountability in environmental policies is crucial.

7.8.6 International Collaboration and Data Sharing

Since air pollution is a global problem, international collaboration and data sharing can be advantageous for AI-driven source identification. This field will advance more quickly with the creation of shared datasets, open-source AI models, and cooperative research projects. Improved knowledge of the sources of transboundary pollution and more practical solutions can result from cross-border collaboration.

7.8.7 AI for Source Attribution in Emerging Economies

In emerging economies, where manufacturing is often accelerating faster than environmental legislation, AI-driven source identification can have a significant influence. AI systems that can detect sources of pollution and direct the expansion of sustainable industries can aid these areas by lessening the negative environmental effects of development.

7.9 CONCLUSION

AI—machine learning and deep learning in particular—has shown an impressive potential to increase the precision and promptness of air pollution forecasts and source identification. These systems are extremely useful for managing the environment and protecting public health. AI tools like support vector machines and random forests are crucial for locating the sources of pollution. Better source detection in both industrial and urban environments is made possible by their ability to manage complicated datasets and non-linear connections. AI-enabled IoT sensors that monitor air quality and pollutant levels in real time allow for quick reaction to pollution occurrences. Because of this convergence, pollution sources are rapidly recognized, allowing for prompt action to reduce hazards to the environment and public health.

AI is anticipated to empower communities and governments to take preventative action to protect the environment, public health, and air quality. AI facilitates more sustainable urban development, efficient industrial operations, and informed environmental policy by providing accurate and timely information.

REFERENCES

Abiodun, O. I., Jantan, A., Omolara, A. E., Dada, K. V., Mohamed, N. A., & Arshad, H. (2018). State-of-the-art in artificial neural network applications: A survey. *Heliyon*, *4*(11), 1–41.

Agarwal, R., Awasthi, A., Mital, S. K., Singh, N., & Gupta, P. K. (2014). Statistical model to study the effect of agriculture crop residue burning on healthy subjects. *Mapan*, 29, 57–65.

Agirre-Basurko, E., Ibarra-Berastegi, G., & Madariaga, I. (2006). Regression and multilayer perceptron-based models to forecast hourly O3 and NO2 levels in the Bilbao area. *Environmental Modelling & Software*, *21*(4), 430–446.

Aiswarya, B., & Aneena, A. A. (2018). A review on various techniques used in predicting pollutants. *IOP Conference Series: Materials Science and Engineering*, 396, 012016. https://doi.org/10.1088/1757-899X/396/1/012016.

Al-Ali, A. R., et al. (2010). A mobile GPRS-sensors array for air pollution monitoring. *IEEE Sensors Journal*, 10, 1666–1671. https://doi.org/10.1109/JSEN.2010.2045890.

Alexeeff, S. E., et al. (2018). High-resolution mapping of traffic related air pollution with Google street view cars and incidence of cardiovascular events within neighborhoods in Oakland. *CA Environmental Health*, 17, 38. https://doi.org/10.1186/s12940-018-0382-1

Almeida, S. M., Manousakas, M., Diapouli, E., Kertesz, Z., Samek, L., Hristova, E., & IAEA European Region Study GROUP. (2020). Ambient particulate matter source apportionment using receptor modelling in European and Central Asia urban areas. *Environmental Pollution*, *266*, 115199.

Anand, G., & Kuchhal, P. (2018). Experimental and theoretical investigation of viscosity, density and sound velocity of jatropha diesel and pure diesel blends. *Biofuels*, 9(1), 1–6.

Anand, G., Nippani, S. K., Kuchhal, P., Vasishth, A., & Sarah, P. (2017). Dielectric, ferroelectric and piezoelectric properties of Ca0. 5Sr0. 5Bi4Ti4O15 prepared by solid state technique. *Ferroelectrics*, 516(1), 36–43.

Arhami, M., Hosseini, V., Shahne, M. Z., Bigdeli, M., Lai, A., & Schauer, J. J. (2017). Seasonal trends, chemical speciation and source apportionment of fine PM in Tehran. *Atmospheric Environment*, *153*, 70–82.

Bai, L., Wang, J., Ma, X., & Lu, H. (2018). Air pollution forecasts: An overview. *International Journal of Environmental Research and Public Health*, *15*(4), 780.

Brook, R. D., Newby, D. E., & Rajagopalan, S. (2017). The global threat of outdoor ambient air pollution to cardiovascular health: Time for intervention. *JAMA Cardiology*, *2*(4), 353–354.

Bublitz, M., Oetomo, F., Sahu, A. S., Kuang, K., Fadrique, A., Velmovitsky, X. L., E., P., & Morita, P. P. (2019). Disruptive technologies for environment and health research: An overview of artificial intelligence, blockchain, and internet of things. *International Journal of Environmental Research and Public Health, 16*(20), 3847.

Bui, T. C., Le, V. D., & Cha, S. K. (2018). A deep learning approach for forecasting air pollution in South Korea using LSTM. *Machine Learning*, 1804.07891. https://arxiv.org/abs/1804.07891v3.

Bulot, F. M. J., et al. (2019). Long-term field comparison of multiple low-cost particulate matter sensors in an outdoor urban environment. *Science Report*, 9, 1–13. https://doi.org/10.1038/s41598-019-43716-3.

Caubel, J. J., et al. (2019). A distributed network of 100 black carbon sensors for 100 days of air quality monitoring in West Oakland, California. *Environmental Science Technology*, 53, 7564–7573. https://doi.org/10.1021/acs.est.9b00282.

Chakraborty, M., Bansal, S., Masiwal, R., & Awasthi, A. (2019). 5 air-pollution modelling aspects: An overview. In Pallavi Saxena & Vaishali Naik (Eds.), *Air Pollution: Sources, Impacts and Controls* (pp. 79–96). CABI.

Chatterjee, S. P., & Pandya, A. S. (2016). Artificial neural networks in drug transport modeling and simulation–II. In *Artificial Neural Network for Drug Design, Delivery and Disposition* (pp. 243–261). Academic Press.

Chin C, Brown DE. (2000). Learning in science: A comparison of deep and surface approaches. *Journal of Research in Science Teaching: The Official Journal of the National Association for Research in Science Teaching*, 37(2), 109–138. https://doi.org/10.1002/(SICI)1098-2736(200002)37:23.0.CO;2-7.

Chung, J., Gulcehre, C., Cho, K., & Bengio, Y. (2014). Empirical evaluation of gated recurrent neural networks on sequence modeling. *Computer Science*, http://arxiv.org/abs/1412.3555v1.

Da Silva, I. N., Hernane Spatti, D., Andrade Flauzino, R., Liboni, L. H. B., dos Reis Alves, S. F., da Silva, I. N., ... & dos Reis Alves, S. F. (2017). *Artificial Neural Network Architectures and Training Processes* (pp. 21–28). Springer International Publishing.

Donnelly, A., Misstear, B., & Broderick, B. (2015). Real time air quality forecasting using integrated parametric and non-parametric regression techniques. *Atmospheric Environmental*, 103(103), 53–65. https://doi. org/10.1016/j.atmosenv.2014.12.011.

Dorohoi, D. O., Barzic, A. I., & Aflori, M. (Eds.). (2017). *Electromagnetic Radiation in Analysis and Design of Organic Materials: Electronic and Biotechnology Applications*. CRC Press.

EIONET. (2019). ETC/ACM Report No. 2018/21 - Low cost Sensor Systems for Air Quality Assessment: Possibilities and Challenges [WWW Document]. Eionet Portal. www.eionet.europa.eu/etcs/etc-atni/products/etc-atni-reports/etc-acm-report-no-2018-21-low-cost-sensor-systems-for-air-quality-assessmentpossibilities-and-challenges

Engel-Cox, J., et al. (2013). Toward the next generation of air quality monitoring: particulate matter. *Atmospheric Environmental*, 80, 584–590. https://doi.org/10.1016/j. atmosenv.2013.08.016.

Esmaeilirad, S., Lai, A., Abbaszade, G., Schnelle-Kreis, J., Zimmermann, R., Uzu, G., & El Haddad, I. (2020). Source apportionment of fine particulate matter in a Middle Eastern Metropolis, Tehran-Iran, using PMF with organic and inorganic markers. *Science of the Total Environment*, *705*, 135330.

Esteva, A., Kuprel, B., Novoa, R. A., Ko, J., Swetter, S. M., Blau, H. M., et al. (2017). Dermatologist-level classification of skin cancer with deep neural networks. *Nature*, 542, 115–118. https://doi.org/10. 1038/nature21056.

Faridi, S., Brook, R. D., Yousefian, F., Hassanvand, M. S., Nodehi, R. N., Shamsipour, M., & Naddafi, K. (2022). Effects of respirators to reduce fine particulate matter exposures on blood pressure and heart rate variability: A systematic review and meta-analysis. *Environmental Pollution, 303*, 119109.

Fernández, J. D., & Vico, F. (2013). AI methods in algorithmic composition: A comprehensive survey. *Journal of Artificial Intelligence Research, 48*, 513–582.

Flesch, T. K., Wilson, J. D., Harper, L. A., Crenna, B. P., & Sharpe, R. R. (2004). Deducing ground-to-air emissions from observed trace gas concentrations: a field trial. *Journal of Applied Meteorology*, 43(3), 487–502. https://doi.org/10.1175/1520-0450(2004)0432.0.CO;2.

Freeman, B. S., Taylor, G., Gharabaghi, B., & Thé, J. (2018). Forecasting air quality time series using deep learning. *Journal of the Air & Waste Management Association*, 68(8), 866–886. https://doi.org/10.1080/10962247. 2018.1459956.

Ganapathy, N., Swaminathan, R., & Deserno, T. M. (2018). Deep learning on 1-D biosignals: A taxonomy-based survey. *Yearbook of Medical Informatics, 27*(01), 098–109.

Ghritlahre, H. K., & Prasad, R. K. (2018). Application of ANN technique to predict the performance of solar collector systems-A review. *Renewable and Sustainable Energy Reviews, 84*, 75–88.

GSMA. (2018). Report: Environment Monitoring: A Guide to Ensuring a Successful Mobile IoT Deployment [WWW Document]. www.gsma.com/iot/ resources/environmental-monitoring-guide-ensuring-successful-mobile-iot-deployment/ .

Hagan, D.H., et al. (2019). Inferring aerosol sources from low-cost air quality sensor measurements: A case study in Delhi, India. *Environ. Sci. Technol. Lett.* 6, 467–472. https://doi.org/10.1021/acs.estlett.9b00393.

Hasanuzzaman, M., & Abd Rahim, N. (Eds.) (2019). *Energy for Sustainable Development: Demand, Supply, Conversion and Management*. Academic Press.

Heuvelmans, M. A., van Ooijen, P. M., Ather, S., Silva, C. F., Han, D., Heussel, C. P., & Oudkerk, M. (2021). Lung cancer prediction by Deep Learning to identify benign lung nodules. *Lung Cancer, 154*, 1–4.

Hochreiter S, & Schmidhuber J. (1997). Long short-term memory. *Neural Computation*, 9(8), 1735–1780. https://doi.org/10.1162/neco.1997. 9.8.1735.

Hong, K.Y., & Pinheiro, P. O., & Weichenthal, S. (2019). Predicting global variations in outdoor PM2:5 concentrations using satellite images and deep convolutional neural networks: Image and Video Processing. https://arxiv.org/abs/1906.03975

Hopke, P. K., Dai, Q., Li, L., & Feng, Y. (2020). Global review of recent source apportionments for airborne particulate matter. *Science of the Total Environment, 740*, 140091.

Jerrett, M., Arain, A., Kanaroglou, P., Beckerman, B., Potoglou, D., Sahsuvaroglu, T., & Giovis, C. (2005). A review and evaluation of intra-urban air pollution exposure models. *Journal of Exposure Science & Environmental Epidemiology, 15*(2), 185–204.

Jiménez, P. A., & Dudhia, J. (2012). Improving the representation of resolved and unresolved topographic effects on surface wind in the WRF model. *Journal of Applied Meteorology and Climatology, 51*(2), 300–316.

Johnson, K., et al. (2020). PurpleAir PM2.5 Performance Across the U.S.#2. Meeting Between ORD, OAR/AirNow, and USFS, Research Triangle Park, NC, February 03, 2020. https://cfpub.epa.gov/si/si_public_record_report.cfm?Lab=CEMM&dirEntryId=348236.

Kaivonen, S., & Ngai, E. C.-H. (2020). Real-time air pollution monitoring with sensors on city bus. *Digital Communumication Network*, 6, 23–30. https://doi.org/10.1016/j.dcan.2019.03.003

Kang, G. K., Gao, J. Z., Chiao, S., Lu, S., Xie, G. (2018). Air quality prediction: Big data and machine learning approaches. *International Journal of Environmental Science and Development*, 9(1), 8–16. https://doi.org/10.18178/ijesd.2018.9.1.1066.

Kaur, P., Chaudhary, P., Bijalwan, A., & Awasthi, A. (2018). Network traffic classification using multiclass classifier. In *Advances in Computing and Data Sciences: Second International Conference, ICACDS 2018, Dehradun, India, April 20–21, 2018, Revised Selected Papers, Part I* 2 (pp. 208–217). Springer Singapore.

Keller, C. A., & Evans, M. J. (2019). Application of random forest regression to the calculation of gas-phase chemistry within the GEOS-Chem chemistry model v10. *Geoscientific Model Development*, 12(3), 1209–1225. https://doi.org/10.5194/gmd-12-1209-2019.

Kelp, M. M., Tessum, C.W., & Marshall, J. D. (2018). Orders-of-magnitude speedup in atmospheric chemistry modeling through neural network-based emulation. *Atmospheric Ocean Physics*. https:// arxiv.org/abs/1808.03874.

Kelp, M., Jacob, D. J., Kutz, J. N., Marshall, J. D., Tessum, C. (2020). Toward stable, general machine learned models of the atmospheric chemical system. *Journal of Geophysical Research: Atmospheres*, 125(23), e2020JD032759.

Kim, K., Kim, D. K., Noh, J., & Kim, M. (2018). Stable forecasting of environmental time series via long short term memory recurrent neural network. *IEEE Access*, *6*, 75216–75228.

Kolen, J. F., & Kremer, S. C. (2001). *Gradient Flow in Recurrent Nets: The Difficulty of Learning LongTerm Dependencies. A Field Guide to Dynamical Recurrent Networks* (pp. 237–243). IEEE. https:// doi.org/10.1109/9780470544037.ch14.

Kumar, S., Torres, C., Ulutan, O., Ayasse, A., Roberts, D., & Manjunath, B. S. (2020). Deep remote sensing methods for methane detection in overhead Hyperspectral imagery. In *The IEEE Winter Conference on Applications of Computer Vision*, 1776–1785. https:// doi.org/10.1109/WACV45572.2020.9093600.

Li, T., Shen, H., Yuan, Q., & Zhang, L. (2017b). A novel solution for remote sensing of air quality: from satellite reflectance to ground pm2.5. *Atmospheric Ocean Physics*.

Li, V., Lam, J., Chen, Y., & Gu, J. (2017a). Deep learning model to estimate air pollution using M-BP to fill in missing proxy urban data. In GLOBECOM 2017–2017 IEEE Global Communications Conference, 1–6. https://doi.org/10.1109/GLOCOM.2017.8255004.

Liu, X., Lu, D., Zhang, A., Liu, Q., & Jiang, G. (2022). Data-driven machine learning in environmental pollution: gains and problems. *Environmental Science & Technology*, *56*(4), 2124–2133.

Liu, Y., Song, M., Liu, X., Zhang, Y., Hui, L., Kong, L., Zhang, Y., Zhang, C., Qu, Y., An, J., Ma, D., Tan, Q., & Feng, M. (2020). Characterization and sources of volatile organic compounds (VOCs) and their related changes during ozone pollution days in 2016 in Beijing, China. *Environmental Pollution*, *257*, 113599. https://doi.org/10.1016/j.envpol.2019.113599

Lodge, J. (1993). Handbook on atmospheric diffusion. *Atmospheric Environmental*, 17(3), 673–675. https://doi.org/10.1016/0004-6981(83)90164-6.

Ma, D., & Zhang, Z. (2016). Contaminant dispersion prediction and source estimation with integrated Gaussian-machine learning network model for point source emission in atmosphere. *Journal of Hazardous Materials*, 311, 237–245. https://doi.org/10.1016/j.jhazmat.2016.03.022.

Ma, H. W., Shih, H. C., Hung, M. L., Chao, C. W., & Li, P. C. (2012). Integrating input output analysis with risk assessment to evaluate the population risk of arsenic. *Environmental Science & Technology*, *46*(2), 1104–1110.

Manisalidis, I., Stavropoulou, E., Stavropoulos, A., & Bezirtzoglou, E. (2020). Environmental and health impacts of air pollution: A review. *Frontiers in Public Health*, *8*, 505570. https://doi.org/10.3389/fpubh.2020.00014

Masood, A., & Ahmad, K. (2020). A model for particulate matter (PM2. 5) prediction for Delhi based on machine learning approaches. *Procedia Computer Science*, *167*, 2101–2110.

Masood, A., & Ahmad, K. (2021). A review on emerging artificial intelligence (AI) techniques for air pollution forecasting: Fundamentals, application and performance. *Journal of Cleaner Production*, *322*, 129072. https://doi.org/10.1016/j.jclepro.2021.129072

Mazzoldi, A., Hill, T., & Colls, J. J. (2008). CFD and Gaussian atmospheric dispersion models: A comparison for leak from carbon dioxide transportation and storage facilities. *Atmospheric Environmental*, 42(34), 8046–8054. https://doi.org/10.1016/j.atmos env.2008. 06.038.

McKendry, I. G. (2002). Evaluation of artificial neural networks for fine particulate pollution (PM10 and PM2. 5) forecasting. *Journal of the Air & Waste Management Association*, *52*(9), 1096–1101.

Mo, X., Zhang, L., Li, H., & Qu, Z. (2019). A novel air quality early-warning system based on artificial intelligence. *International Journal of Environmental Research and Public Health*, *16*(19), 3505.

Mukherjee, A., et al. (2019). Measuring spatial and temporal PM2.5 variations in Sacramento, California, communities using a network of low-cost sensors. *Sensors,* 19, 4701. https://doi.org/10.3390/s19214701.

Münzel, T., Sørensen, M., Lelieveld, J., Hahad, O., Al-Kindi, S., Nieuwenhuijsen, M., & Rajagopalan, S. (2021). Heart healthy cities: genetics loads the gun but the environment pulls the trigger. *European Heart Journal*, *42*(25), 2422–2438.

Nilashi, M., Ahmadi, H., Manaf, A. A., Rashid, T. A., Samad, S., Shahmoradi, L., & Akbari, E. (2020). Coronary heart disease diagnosis through self-organizing map and fuzzy support vector machine with incremental updates. *International Journal of Fuzzy Systems*, *22*, 1376–1388.

Olivieri, A. C., & Allegrini, F. (2019). *Chemometrics and Statistics: Neural Networks*. Elsevier.

Pal, S., Agrawal, A., Nippani, S. K., & Anand, G. (2015). Synthesis of NS doped graphene for hydrogen storage. *Procedia Materials Science*, 10, 103–110.

Perez, P., & Reyes, J. (2006). An integrated neural network model for PM10 forecasting. *Atmospheric Environmental*, 40(16), 2845–2851. https://doi. org/10.1016/j.atmosenv.2006.01.010.

Pontiggia, M., Derudi, M., Busini, V., & Rota, R. (2009). Hazardous gas dispersion: a CFD model accounting for atmospheric stability classes. *Journal of Hazardous Materials*, 171(1–3), 739–747. https://doi.org/10.1016/j. jhazmat.2009.06.064.

Popoola, O.A.M., et al. (2018). Use of networks of low cost air quality sensors to quantify air quality in urban settings. *Atmospheric Environmental*, 194, 58–70. https://doi.org/10.1016/j.atmosenv.2018.09.030

Qi, Z., Wang, T., Song, G., Hu, W., Li, X., Zhang, Z. (2018). Deep air learning: interpolation, prediction, and feature analysis of fine-grained air quality. *IEEE Transactions on Knowledge and Data Engineering*, 30(12), 2285–2297. https://doi.org/10.1109/TKDE.2018. 2823740.

Qiu, S., Chen, B., Wang, R., Zhu, Z., Wang, Y., & Qiu, X. (2018). Atmospheric dispersion prediction and source estimation of hazardous gas using artificial neural network, particle swarm optimization and expectation maximization. *Atmospheric Environmental*, 178, 158–163.

Querol, X., Tobías, A., Pérez, N., Karanasiou, A., Amato, F., Stafoggia, M., ... & Alastuey, A. (2019). Monitoring the impact of desert dust outbreaks for air quality for health studies. *Environment International*, 130, 104867.

Rahimi, A. (2017). Short-term prediction of NO_2 and NO_x concentrations using multilayer perceptron neural network: A case study of Tabriz, Iran. *Ecological Processes*, 6, 1–9.

Ramos, F., et al. (2018). Promoting pollution-free routes in smart cities using air quality sensor networks. *Sensors* 18, 2507. https://doi.org/10.3390/s18082507

Reichstein, M., Camps-Valls, G., Stevens, B., Jung, M., Denzler, J., Nuno Carvalhais, N., et al. (2019). Deep learning and process understanding for data-driven earth system science. *Nature*, 566(7743): 195–204. https://doi.org/10.1038/s41586-019-0912-1.

Roychowdhury, A., Chattopadhyaya, V., & Shukla, S., 2016. Reinventing Air Quality Monitoring -Potential of Low Cost Alternative Monitoring Methods [Internet]. http://cdn.cseindia.org/attachments/0.85392300_1505190810_reinventing-air-quality-monitoring-potential-of-low-cost-dec27.pdf.

Sahiner, B., Pezeshk, A., Hadjiiski, L. M., Wang, X., Drukker, K., Cha, K. H., & Giger, M. L. (2019). Deep learning in medical imaging and radiation therapy. *Medical physics*, 46(1), e1–e36.

Sanchez, L., et al. (2014). SmartSantander: IoT experimentation over a smart city Testbed. *Computer Networks. Special Issue on Future Internet Testbeds – Part I* , 61, 217–238. https://doi.org/10.1016/j.bjp.2013.12.020.

Sapra, R. L., Mehrotra, S., & Nundy, S. (2015). Artificial neural networks: prediction of mortality/survival in gastroenterology. *Current Medicine Research and Practice*, *5*(3), 119–129.

Seaman, N. L. (2000). Meteorological modeling for air-quality assessments. *Atmospheric Environmental*, 34(12–14), 2231–2259. https://doi.org/10.1016/S1352-2310(99)00466-5.

Sergeev, A., & Del Balso, M. (2018). Horovod: Fast and easy distributed deep learning in TensorFlow. *arXiv preprint: arXiv:1802.05799*.

Simmhan, Y., et al. (2019). SATVAM: Toward an IoT cyber-Infrastructure for low-cost urban air quality monitoring. In *2019 15th International Conference on EScience (EScience). Presented at the 2019 15th International Conference on eScience (eScience)* (pp. 57–66). https://doi.org/10.1109/eScience.2019.00014.

Singh, N., Agarwal, R., Awasthi, A., Gupta, P. K., & Mittal, S. K. (2010). Characterization of atmospheric aerosols for organic tarry matter and combustible matter during crop residue burning and non-crop residue burning months in Northwestern region of India. *Atmospheric Environment*, 44(10), 1292–1300.

Singh, N., Murari, V., Kumar, M., Barman, S. C., & Banerjee, T. (2017). Fine particulates over South Asia: Review and meta-analysis of PM2. 5 source apportionment through receptor model. *Environmental Pollution*, *223*, 121–136.

So, W., Koo, J., Shin, D., & Yoon, E. S. (2010). The estimation of hazardous gas release rate using optical sensor and neural network. *Computer Aided Chemical Engineering* (Elsevier), 28(10), 199–204. https://doi.org/10.1016/S1570-7946(10)28034-3.

Soleimani, M., Ebrahimi, Z., Mirghaffari, N., Moradi, H., Amini, N., Poulsen, K. G., & Christensen, J. H. (2021). Seasonal trend and source identification of polycyclic aromatic hydrocarbons associated with fine particulate matters (PM 2.5) in Isfahan City, Iran, using diagnostic ratio and PMF model. *Environmental Science and Pollution Research*, 9, 1–16.

Steinle, S., et al. (2015). Personal exposure monitoring of PM2.5 in indoor and outdoor microenvironments. *Science Total Environmental*, 508, 383–394. https://doi.org/10.1016/j.scitotenv.2014.12.003 .

Subramanyam, V. (2008). Evolution of artificial neural network controller for a boost converter.

Sun, X., Wang, H., Guo, Z., Lu, P., Song, F., Liu, L., & Wang, F. (2020). Positive matrix factorization on source apportionment for typical pollutants in different environmental media: A review. *Environmental Science: Processes & Impacts*, *22*(2), 239–255.

Taghvaee, S., Sowlat, M. H., Hassanvand, M. S., Yunesian, M., Naddafi, K., & Sioutas, C. (2018a). Source-specific lung cancer risk assessment of ambient PM2. 5-bound polycyclic aromatic hydrocarbons (PAHs) in central Tehran. *Environment International*, *120*, 321–332.

Taghvaee, S., Sowlat, M. H., Mousavi, A., Hassanvand, M. S., Yunesian, M., Naddafi, K., & Sioutas, C. (2018b). Source apportionment of ambient PM2. 5 in two locations in central Tehran using the Positive Matrix Factorization (PMF) model. *Science of the Total Environment*, *628*, 672–686.

USEPA (2012). *Regulatory Impact Analysis for the Final Revisions to the National Ambient Air Quality Standards for Particulate Matter.* www3.epa.gov/ttnecas1/regdata/RIAs/finalria.pdf

Vieira, S., Pinaya, W. H., & Mechelli, A. (2017). Using deep learning to investigate the neuroimaging correlates of psychiatric and neurological disorders: Methods and applications. *Neuroscience & Biobehavioral Reviews*, *74*, 58–75.

Wang, B., Chen, B., & Zhao, J. (2015). The real-time estimation of hazardous gas dispersion by the integration of gas detectors, neural network and gas dispersion models. *Journal of Hazardous Materials*, 300, 433–442. https://doi.org/10.1016/j.jhazmat.2015.07.028.

Wang, H., Zhuang, B., Chen, Y., Li, N., & Wei, D. (2018). Deep inferential spatial-temporal network for forecasting air pollution concentrations. *Machine Learning*. https://arxiv.org/abs/1809.03964v1.

Wang, J., Bai, L., Wang, S., & Wang, C. (2019). Research and application of the hybrid forecasting model based on secondary denoising and multi-objective optimization for air pollution early warning system. *Journal of Cleaner Production*, *234*, 54–70.

Wardah, W., Khan, M. G., Sharma, A., & Rashid, M. A. (2019). Protein secondary structure prediction using neural networks and deep learning: A review. *Computational Biology and Chemistry*, *81*, 1–8.

Wilson, J. D., & Sawford, B. L. (1996). Review of lagrangian stochastic models for trajectories in the turbulent atmosphere. *Bound-Layer Meteorology*, 78(1–2), 191–210. https://doi.org/10.1007/BF00122492

World Health Organization. (2017). *Inheriting a Sustainable World? Atlas on Children's Health and the Environment*. World Health Organization.

Xi, X., Wei, Z., Xiaoguang, R., Yijie, W., Xinxin, B., Wenjun, Y., & Jin, D. (2015, November). A comprehensive evaluation of air pollution prediction improvement by a machine learning method. In *2015 IEEE International Conference on Service Operations and Logistics, and Informatics (SOLI)* (pp. 176–181). IEEE.

Yadav, N., Rajendra, K., Awasthi, A., Singh, C., & Bhushan, B. (2023). Systematic exploration of heat wave impact on mortality and urban heat Island: A review from 2000 to 2022. *Urban Climate*, 51, 101622.

Yadav, S., Dhankhar, R., & Chhikara, S. K. (2022a). Significant changes in urban air quality during Covid-19 pandemic lockdown in Rohtak City, India. *Asian Journal of Chemistry,* 34(12), 3189–3196. https://doi.org/10.14233/ajchem.2022.23950

Yadav, S., Dhankhar, R., & Chhikara, S. K. (2022b). Volatile organic compounds in ambient air: Potential sources, distribution and impact. *Ambient Science*, 9(3), 22–28. https://doi.org/10.21276/ambi.2022.09.3.ta01

Yi, W.Y., et al. (2015). A survey of wireless sensor network based air pollution monitoring systems. *Sensors,* 15, 31392–31427. https://doi.org/10.3390/s151229859.

Zalakeviciute, R., et al. (2018). *Urban air pollution mapping and traffic intensity: active transport application*. intechOpen.

Zaytar, M. A., & Amrani, C. E. (2016). Sequence to sequence weather forecasting with long short-term memory recurrent neural networks. *International Journal Computer Application*, 143(11), 7–11. https://doi.org/ 10.5120/ijca2016910497.

Zhang, B., Zhang, H., Zhao, G., & Lian, J. (2020a). Constructing a PM2. 5 concentration prediction model by combining auto-encoder with Bi-LSTM neural networks. *Environmental Modelling & Software*, *124*, 104600.

Zhang, X., Zhang, T., Zou, Y., Du, G., & Guo, N. (2020b). Predictive eco-driving application considering real-world traffic flow. *IEEE Access*, *8*, 82187–82200.

Zheng, T. et al. (2019). Gaussian process regression model for dynamically calibrating and surveilling a wireless low-cost particulate matter sensor network in Delhi. *Atmospheric Measuremnt Techniques*, 12, 5161–5181. https://doi.org/10.5194/amt-12-5161-2019

Zheng, Y. et al. (2015). Forecasting fine-grained air quality based on big data. In *Proceedings of the 21st ACM SIGKDD International Conference on Knowledge Discovery and Data Mining, KDD '15* (pp. 2267–2276). Association for Computing Machinery, Sydney, NSW, Australia. https://doi.org/10.1145/ 2783258.2788573.

8 Study of Extreme Weather Events in the Central Himalayan Region through Machine Learning and Artificial Intelligence

A Case Study

Alok Sagar Gautam, Aman Deep Vishwkarma, Yasti Panchbhaiya, Karan Singh, and Sanjeev Kumar

8.1 INTRODUCTION

The Central Himalayas region in Uttarakhand, India, is a diverse and ecologically significant area known for its towering mountain ranges, lush forests, and pristine rivers. This region is not only a hotspot for biodiversity but also home to numerous communities that rely on its resources for their livelihoods (Gautam et al., 2023). However, the Central Himalayas are increasingly experiencing extreme weather events, such as intense rainfall, snowfall, and flash floods, which have far-reaching impacts on both the environment and local communities (Mishra, 2019). Studying these extreme weather events is of paramount significance as it helps us understand the complex dynamics of climate change, assess vulnerability, and develop strategies to mitigate the risks and protect the delicate balance of this unique region. This knowledge is vital for conservation efforts, disaster preparedness, and the well-being of the people living in this vulnerable area. Artificial intelligence (AI) aids in climate change mitigation, employing Machine Learning (ML) for predict carbon removal and climate change. In the monsoon season (June–July–August–September), cloud bursts are very common. Every year, Uttarakhand experiences flash floods, landslides, damaged roads, fatalities, and losses of crops, property,

DOI: 10.1201/9781032683805-8

and livestock. Unforgettable major extreme weather events have been reported in Uttarakhand, such as the Kedarnath flash flood in 2013, and the Chamoli incident in 2020 (Khanaduri, 2020; Mishra, 2019; Naithani et al., 2011). Due to its short duration, scientists and researchers find it complex and challenging to understand and predict. Even the predictions of such events are very complex and costly when using Doppler radars. Only one unit of Doppler radar is installed in the state (Mukteshwar), and it is not possible to cover the entire region of Uttarakhand. In this condition, ML/deep learning approaches are essential to provide a reliable option to predict cloudbursts based on historical datasets and train models with similar datasets. In this chapter, our focus has been directed towards cultivating insights into deep learning models, encompassing the Random Forest, Artificial Neural Network (ANN), General Regression Neural Networks (GRNN), Support Vector Machine (SVM), Gated Recurrent Unit (GRU), and Long Short-Term Memory (LSTM) network models. This focus aims to anticipate the potential incidence of cloudburst events. Predictive Power Score (PPS) tools are used to understand dataset features, time series model inputs, and model optimization. In the past decades from 2010 to the present, various studies have been completed in the field of atmospheric physics and processes using machine learning and artificial intelligence. Gentine et al. (2018) used an ANN technique and found improvement in convection parameterizations using machine learning emulators. Krasnopolsky et al. (2005) also used the ANN technique to improve the speed of global climate model parameterizations. Wu et al. (2019) Implemented diverse methodologies, including the random forest model, SVM, ANN, and GRNN, to investigate the impacts of climate change on above-ground biomass. Tiwari et al. (2015) proposed an Arduino-based cost-effective cloudburst predetermination system based on real-time calculation of rainfall intensity. Samya et al. (2011) proposed forecasting cloudburst techniques using data mining as well as ANN, and they observed that Data Mining Techniques, ANN, fuzzy and logic result in better accuracy. Pabreja (2012) developed a clustering method using expected data from the Numerical Weather Prediction model and Leh cloudburst events, predicting the formation of cloudbursts up to five to six days in advance.

Identified gaps in the study: It is not possible to install an Automatic Weather Station (AWS), Optical Particle Counter, dust samplers, and other instruments overall monitoring stations due to extensive cost and manpower. Therefore, to resolve this problem, an ANN-based method can provide a better understanding of the dataset and extreme weather events. Indian researchers have focused on understanding extreme weather events and predictions across India (Tiwari et al., 2016). Pabreja (2012) tried to predict the cloudburst over Leh, but very few ML models were found to understand extreme events over Uttarakhand. We have not found any suitable literature that covers the modeling of meteorological parameters to understand aerosol particle size distribution, their chemistry, and cloudbursts. In this chapter, we have focused on developing an understanding of the random forest model based on meteorological datasets to comprehend extreme weather events (rainfall) over the Central Himalayan region.

8.2 METHODS AND METHODOLOGY

8.2.1 Study Region

Uttarakhand is characterized as a mountainous state in India, situated within the latitudinal range of 28° 43' to 31° 28'N and longitudinal span of 77° 34' to 81° 03' E. It encompasses a landscape predominantly composed of hills, with approximately 93% of its terrain classified as mountainous, and boasts substantial forest coverage of 64% (Tak et al., 2011; Prasad & Kumar, 2022; Farooquee & Maikhuri, 2007). Two major rivers, the Ganga and Yamuna, originate from Uttarakhand, along with significant Hindu and Sikh holy temples like Badrinath, Kedarnath, and Hemkund Sahib. Geographically located on the southern slope of the Himalayas range, the region exhibits varying climates and vegetation at different elevations, including alpine shrubs and meadows (3000 to 5000 m), subalpine conifer forests (3000 to 2600 m), broadleaf forests (2600 to 1500 m), and subtropical pine forests (below 1500 m) (Negi, 1991). Several extreme events have been reported in the Himalayan region, such as the Kedarnath flash flood in 2013, and the Rishi Ganga flash flood in 2021 (Rao et al., 2014; Singh et al., 2022; Nainwal et al., 2022). Based on elevation, geography, and climatic conditions influencing the occurrence of unprecedented weather phenomena, we will categorize distinct zones to signify varying severity levels. Further details on elevation and geography can be found in Figure 8.1.

8.2.2 Data Pre-processing

We extracted meteorological data from our installed Automatic Weather Station (AWS) in the Srinagar Garhwal Valley. We utilized a five-year minute-resolution

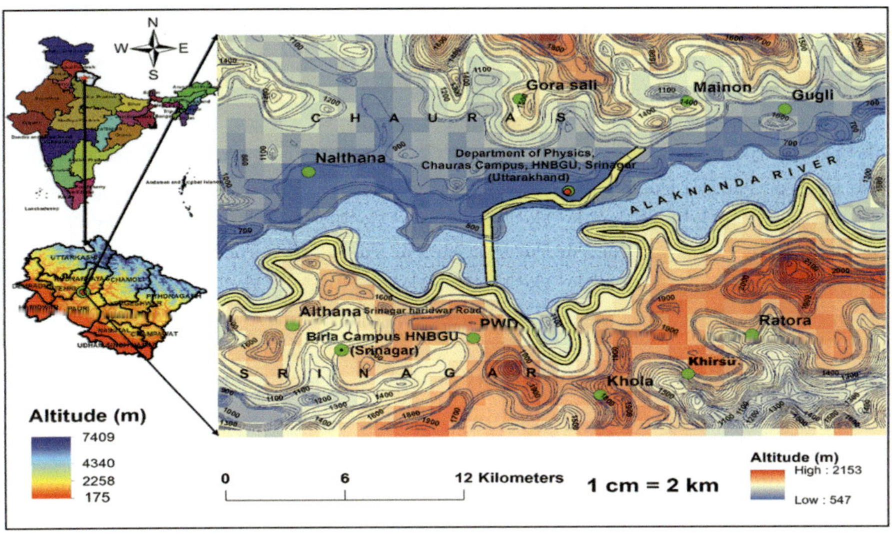

FIGURE 8.1 Location map of the sampling site in Srinagar, Central Himalaya region, Uttarakhand.

meteorological dataset from Ground-based AWS in Srinagar Garhwal. Our study focused on testing and training random forest models using exclusively one-hour resolution data spanning from August 15, 2018, to August 14, 2019. This approach enhances model precision and reliability within the chosen timeframe for our scientific investigation. To predict rainfall, we have used random forest in the R-Studio software and utilized the random forest package, caret, and ggplot2 in the study.

8.2.3 Models and Tools

8.2.3.1 Random Forest Model

The random forest approach is an ensemble learning method that combines numerous decision trees. It addresses overfitting by instructing each tree on a stochastic subset of both data and features, promoting diversity in the ensemble, a process known as bootstrap aggregating. Feature randomization at each split enhances diversity among trees, preventing dominance by specific features. The model's strength lies in aggregating predictions from individual trees, utilizing majority consensus for classification and averaging for regression. Recognized for its adaptability and resilience, random forest is widely employed across diverse domains, demonstrating efficacy in handling sizable datasets, and high-dimensional feature spaces, and providing robust solutions for classification and regression tasks.

8.2.3.2 Feedforward Neural Network (FFNN)

After data collection and pre-processing, the next step is model optimization, with ANN models often used for predicting air quality due to their ability to approximate real-valued target functions. Neural networks consist of nodes with activation functions determining output values. Sigmoid and Rectified Linear Units (ReLUs) may be used as activation functions. For aerosol size information, a FFNN is preferred over a Time Delay Neural Network (TDNN) due to data pre-processing removing rows and lacking time continuity. FFNN comprises layers, with the final layer producing the network's output.

8.2.3.3 Extreme Gradient Boosting (XG Boost) Algorithm (XG Boost Model)

This model addresses occasional boundary overflow issues in outputs, adopting a wavelet-based neural network for flood forecasting in MATLAB 2019a. To simplify, a one or two-layer model configuration is considered to manage computing requirements. Selected configurations are evaluated using testing data and performance metrics to determine the most suitable approach.

8.2.3.4 Long Short-Term Memory (LSTM)

LSTM, a breakthrough in deep learning ANN, features internal memory for long-term information retention. Sigmoid activation gates enable data updating or forgetting, vital for effective learning (Van Houdt et al. 2020; Graves 2012; Kalchbrenner et al. 2015; Malhotra et al. 2015; Cheng et al. 2019; Sivagami et al. 2021).

8.2.3.5 Gated Recurrent Unit (GRU)

GRU is a variant of recurrent neural networks (RNNs) with update and reset gates. These gates control the flow of information by filtering out irrelevant data, thereby enhancing training efficiency. The reset gate filters undesired information, while the update gate determines the amount of prior information passed through the network. The gated structure of GRU enables effective handling of large sequences, ensuring that relevant dependencies are captured without neglecting the sequential context (Dey & Salem, 2017; Chung et al., 2014; Sivagami et al., 2021; Niu et al., 2020; Agarap et al., 2018; Chen et al., 2019).

8.2.3.6 Predictive Power Score (PSS)

PPS is a superior tool for assessing non-linear relationships, handling categorical and nominal data. It overcomes the limitations of correlation matrices, offering asymmetry awareness. PPS values, ranging from 0 to 1, are computed using regression decision trees for numerical goals and weighted accuracy through Classification Decision Trees for categorical purposes.

8.3 RESULT AND DISCUSSION

8.3.1 Rainfall Prediction Based on Random Forest Model

In our study, we employed Ground-Based Automatic Weather Station (AWS) data spanning from August 15, 2018, to August 14, 2019, to train and test our rainfall prediction model for Srinagar. The dataset was divided into 80% for training and 20% for testing purposes. In evaluating the model's effectiveness, we analyzed pivotal metrics, incorporating the Root Mean Square Error (RMSE), Mean Absolute Error (MAE), and R-squared to gauge its performance and predictive accuracy as given in Table 8.1. The exploration of our random forest model unveiled intriguing discoveries when investigating the influence of the number of trees on its performance (as suggested by Figure 8.2). With 100 trees, the model demonstrated commendable results, achieving an RMSE of 2.74, an MAE of 1.39, and an R^2 of 0.57. This initial assessment suggested a promising level of predictive accuracy. However, as the number of trees increased to 1500, a notable shift in performance occurred. The RMSE rose to 3.02, indicating an increase in predictive errors, while the MAE expanded to 1.54, signaling larger discrepancies between predicted and actual values. Most notably, the R^2 decreased to 0.46, signifying a diminishing ability to explain

TABLE 8.1
Performance of the Random Forest Model

Number of Trees	RMSE	MAE	R^2
100	2.74	1.39	0.57
1500	3.02	1.54	

variance. The results indicated that both RMSE and MAE were minimized at 100 and 1500 trees. However, the R^2 values of 0.57 and 0.46 at these respective points highlight a trade-off, emphasizing the delicate balance between model complexity and predictive accuracy (Figure 8.3).

In addition to quantitative metrics, we employed scatter plots to visually assess the model's performance with 100 and 1500 trees. The graphs depict the correlation between observed and forecasted precipitation, providing valuable insights into the

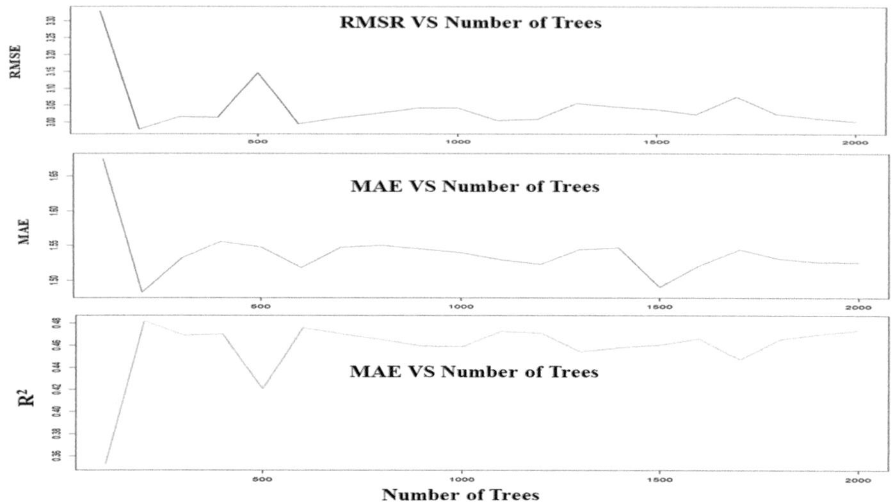

FIGURE 8.2 Variation of errors and number of trees.

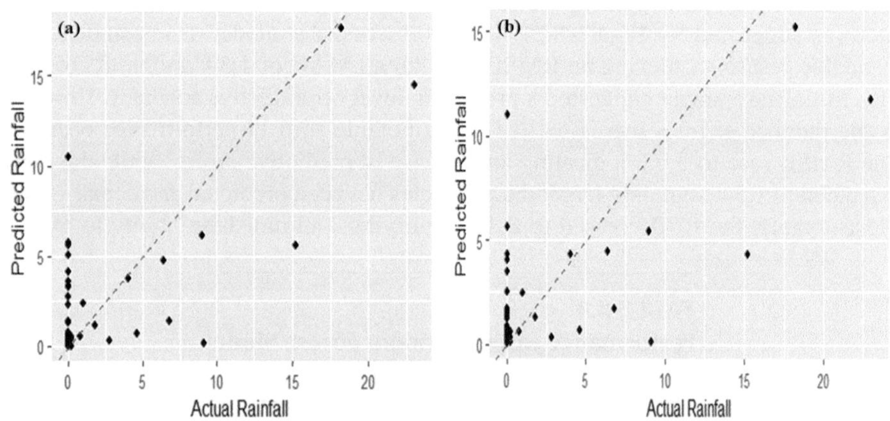

FIGURE 8.3 A scatter plot between actual and predicted rainfall at 100 (a) and 1500 (b) trees.

model's capacity to accurately represent fluctuations in the dataset. (Figure 8.3). In Wei et al.'s (2022) study, a random forest model was trained on historical atmospheric and oceanic indices spanning from 1951 to 2019. The model highlighted three key predictors for extreme events, forecasting an increased likelihood of extreme precipitation during the summer of 2020. Decision tree analysis unveiled three principal pathways conducive to such occurrences. Whereas in India, Shah et al. (2023) introduced a nowcasting model that integrated random forest with Principal Component Analysis, utilizing data from the Automatic Weather Station in Thiruvananthapuram. Their predictions, particularly during the peak July monsoon, demonstrated impressive performance with 90% accuracy and a 68% probability of detection. Notably, atmospheric pressure and wind speed emerged as significant predictors in this context.

In a related study by Shah et al. (2022), a data-driven algorithm for precipitation nowcasting was proposed, leveraging data from the Automatic Weather Station at Vikram Sarabhai Space Centre, Thiruvananthapuram. The study compared three ensemble AI models—random forest, Adaboost, and Xgboost—revealing that random forest and Xgboost achieved an 80% accuracy rate. Noteworthy predictors in this analysis included pressure and temperature, underscoring their importance in rainfall prediction. These findings collectively underscore the utility of machine learning approaches in enhancing weather forecasting capabilities, particularly in regions prone to extreme precipitation events.

Our study emphasizes the importance of carefully considering the number of trees in random forest models. While an increased number initially enhances predictive accuracy, our findings underscore the existence of a critical point where further complexity compromises the model's overall effectiveness in explaining variance in the target variable.

8.4 CONCLUSIONS

Based on the case study and application of the random forest model, we have concluded the following key facts. The case study provides a deeper understanding of advanced machine learning and artificial intelligence methodologies to predict and comprehend extreme weather events in Uttarakhand, India. Meteorological data is available from satellite and ground-based systems. We can clearly comprehend SVM, ANN, and GRNN. Going beyond traditional analysis, we proactively forecast events related to cloud bursts and extreme weather events. To achieve this, we have developed a random forest model specifically focused on predicting rainfall, to be modified for predicting extreme weather events and enhancing disaster preparedness in the ecologically delicate Central Himalayan region. This model integrates crucial meteorological factors including temperature, humidity, wind direction, and wind speed. The performance of our Srinagar rainfall prediction model was assessed through various metrics, including RMSE, MAE, and R^2. Notably, the model trained with 100 trees exhibited optimal performance, showcasing an RMSE of 2.74, MAE of 1.39, and an R^2 of 0.57. However, increasing the number of trees to 1500 resulted in higher errors (RMSE 3.02, MAE 1.54) and a reduction in explanatory power (R^2 0.46). This underscores the delicate balance between model complexity and performance, highlighting that 100 trees offer the best trade-off for accuracy and efficiency in this context. Looking ahead, our research roadmap involves incorporating additional meteorological parameters, such as solar radiation, pressure, dew points, and higher data resolution (1 minute). Furthermore, we plan to explore the applicability of other advanced techniques like SVM, ANN, and GRNN to gain a more comprehensive understanding of extreme weather events in Uttarakhand. This holistic approach contributes significantly to advancing the scientific understanding of these events, supporting sustainable environmental management, and bolstering disaster resilience efforts in the region.

ACKNOWLEDGMENT

The authors, ASG and YP, would like to thank to Science and Engineering Research Board (SERB), Department of Science and Technology New Delhi-110016 for their financial support for research work.

REFERENCES

Agarap, A. F. M. (2018, February). A neural network architecture combining gated recurrent unit (GRU) and support vector machine (SVM) for intrusion detection in network traffic data. In *Proceedings of the 2018 10th international conference on machine learning and computing* (pp. 26–30).

Cheng, J., Su, J., Cui, T., Li, X., Dong, X., Sun, F., Yang, Y., Tong, D., Zheng, Y., Li, Y., Li, J., Zhang, Q., & He, K. (2019). Dominant role of emission reduction in PM_{2.5} air quality improvement in Beijing during 2013–2017: A model-based decomposition analysis. *Atmospheric Chemistry and Physics*, *19*(9), 6125–6146. https://doi.org/10.5194/acp-19-6125-2019

Chung, J., Gülçehre, Ç., Cho, K., & Bengio, Y. (2014). *Empirical evaluation of gated recurrent neural networks on sequence modeling. arXiv (Cornell University).* https://arxiv.org/pdf/1412.3555

Dey, R., & Salem, F. M. (2017, August). Gate-variants of gated recurrent unit (GRU) neural networks. In *2017 IEEE 60th international midwest symposium on circuits and systems (MWSCAS)* (pp. 1597–1600). IEEE.

Graves, A. (2012). Supervised sequence labelling with recurrent neural networks. In *Studies in computational intelligence*. https://doi.org/10.1007/978-3-642-24797-2

Gautam, A. S., Kumar, S., Gautam, S., Singh, K. P., Ram, K., Siingh, D., Ambade, B., & Sharma, M. (2023). Regional air quality: Biomass burning impacts of SO2 emissions on air quality in the Himalayan region of Uttarakhand, India. *Air Quality, Atmosphere & Health*. https://doi.org/10.1007/s11869-023-01426-w

Gentine, P., Pritchard, M., Rasp, S., Reinaudi, G., & Yacalis, G. (2018). Could machine learning break the convection parameterization deadlock?. Geophysical Research Letters, 45(11), 5742–5751.

Haywood, J. (2021). Atmospheric aerosols and their role in climate change. In *Climate change* (pp. 645–659). Elsevier.

Huntingford, C., Jeffers, E. S., Bonsall, M. B., Christensen, H., Lees, T., & Yang, H. (2019). Machine learning and artificial intelligence to aid climate change research and preparedness. *Environmental Research Letters*, 14(12), 124007. https://doi.org/10.1088/1748-9326/ab4e55

Kalchbrenner, N., Danihelka, I., & Graves, A. (2015). Grid Long Short-Term Memory. *arXiv (Cornell University)*. https://arxiv.org/pdf/1507.01526.pdf

Krasnopolsky, V. M., Fox-Rabinovitz, M. S., & Chalikov, D. (2005). New approach to calculation of Atmospheric model physics: Accurate and fast neural network emulation of longwave radiation in a climate model. *Monthly Weather Review, 133*(5), 1370–1383. https://doi.org/10.1175/mwr2923.1

Malhotra, P., Vig, L., Shroff, G., & Agarwal, P. (2015). Long short-term memory networks for anomaly detection in time series. The European Symposium on Artificial Neural Networks. www.elen.ucl.ac.be/Proceedings/esann/esannpdf/es2015-56.pdf

Mishra, M. N. (2019). Active tectonic deformation of the Shillong plateau, India: Inferences from river profiles and stream-gradients. Journal of Asian Earth Sciences, 181, 103904.

Mukherjee, S. (2014). Extra terrestrial remote sensing and geophysical applications to understand Kedarnath cloudburst in Uttarakhand, India. *Journal of Geophysics and Remote Sensing*, 3(3).

Negi, G. C. S. (2017). *Impacts of hydropower projects in the Himalaya: A pilot study in Bhagirathi River Basin in Uttarakhand, India*. The Bookline Publishers.

Pabreja, K. (2012). Clustering technique to interpret Numerical Weather Prediction output products for forecast of Cloudburst. International Journal of Computer Science and Information Technologies (IJCSIT), 3(1), 2996–2999.

Rosenfeld, D., Fischman, B., Zheng, Y., Goren, T., & Giguzin, D., 2014. Combined 857 satellite and radar retrievals of drop concentration and CCN at convective cloud Journal Pre-proof 35 858 base. *Geophysical Research Letters*, 41, 3259–3265. 859 https://doi.org/10.1002/2014GL059453

Sainath, T. N., Vinyals, O., Senior, A., & Sak, H. (2015, April). Convolutional, long short-term memory, fully connected deep neural networks. In *2015 IEEE international conference on acoustics, speech and signal processing (ICASSP)* (pp. 4580–4584). IEEE.

Samya, R., & Rathipriya, R. (2016). Predictive analysis for weather prediction using data mining with ANN: a study. International Journal of Computational Intelligence and Informatics, 6(2), 150–154.

Shah, N. H., Priamvada, A., & Shukla, B. P. (2023). Random forest-based nowcast model for rainfall. Earth Science Informatics, 16(3), 2391–2403.

Sivagami, M., Radha, P. B., & Balasundaram, A. (2021). Sequence model-based cloudburst prediction for the indian state of uttarakhand. *Disaster Advances*, 1–9. https://doi.org/10.25303/f2512105

Thorn, P. D. (2015). Nick Bostrom: Superintelligence: Paths, dangers, strategies. *Minds and Machines*, 25(3), 285–289. https://doi.org/10.1007/s11023-015-9377-7

Tiwari, A., & Verma, S. K. (2015). Cloudburst predetermination system. *ISOR J. Comput. Eng*, 17, 44–56.

Tiwari, P. R., Kar, S. C., Mohanty, U. C., Dey, S., Kumari, S., & Sinha, P. (2016). Seasonal prediction skill of winter temperature over North India. Theoretical and *Applied Clim*atology, 124, 15–29.

Van Houdt, G., Mosquera, C., & Nápoles, G. (2020). A review on the long short-term memory model. *Artificial Intelligence Review*, 53(8), 5929–5955. https://doi.org/10.1007/s10462-020-09838-1

Wang, L., Liu, J., Gao, Z., Li, Y., Huang, M., Fan, S., ... & Yang, T. (2019). Observations of the atmospheric boundary layer structure over Beijing urban area during air pollution episodes. Atmospheric Chemistry and Physics Discussion. https://doi. org/10.5194/acp-2018-1184, in review.

Wang, Y., & Chung, S. H. (2019). Soot formation in laminar counterflow flames. *Progress in Energy and Combustion Science*, *74*, 152–238. https://doi.org/10.1016/j.pecs.2019.05.003

Zhang, Z., Sato, H., Matsubar, Y., Suetake, A., Wakasugi, N., Chen, C., Sakurai, Y., & Suganuma, K. (2022). (Digital Presentation) lifetime prediction of SiC power module by using time-series analysis of acoustic emission during power cycling test. *ECS Meeting Abstracts*, *MA2022-02*(37), 1356. https://doi.org/10.1149/ma2022-02371356mtgabs

Zhao, S., Yao, X., Yang, J., Jia, G., Ding, G., Chua, T., Schuller, B., & Keutzer, K. (2022). Affective image content analysis: Two decades review and new perspectives. *IEEE Transactions on Pattern Analysis and Machine Intelligence*, *44*(10), 6729–6751. https://doi.org/10.1109/tpami.2021.3094362

9 Machine Learning Applications in Air Quality Management and Policies

Abhishek Upadhyay, Puneet Sharma, and Sourangsu Chowdhury

9.1 INTRODUCTION

Air pollution represents a significant global environmental challenge, particularly pronounced in the developing world, with profound impacts on human health, climate, and ecosystems. Mitigation strategies aimed to address this issue are on the prime agenda for policymakers. Comprehensive air pollution data stands out as a vital cornerstone for formulating effective air quality policies. This includes diverse types of air quality data like air pollution levels, source and chemical species, air pollution predictions, and air pollutant emission pathways. Accurate prediction of air quality is indispensable for the development of effective air quality policies and management strategies. These predictions offer crucial insights into the future state of air quality, aiding policymakers in devising targeted interventions and regulatory measures to mitigate the adverse health impacts and environmental consequences associated with air pollution (Shrestha et al., 2021; Sofia et al., 2020). Anticipating potential episodes of high pollution enables timely implementation of control measures and issuance of public advisories, safeguarding public health and vulnerable populations. For instance, resolved data on pollutant concentrations, emission sources, and their spatial distribution assists policymakers in identifying pollution hotspots and understanding the magnitude of the problem. This information facilitates the formulation of targeted policies and interventions aimed at mitigating pollution in specific areas or sectors (Lelieveld et al., 2015; Upadhyay et al., 2018b). Furthermore, reliable air quality predictions facilitate the evaluation of the effectiveness of existing policies and support the assessment of various emission reduction scenarios, guiding evidence-based air quality management strategies aimed at achieving better public health outcomes and environmental sustainability. Moreover, accurate data serve as a foundation for modeling and forecasting future scenarios, enabling policymakers to anticipate potential environmental challenges and design proactive policies to

DOI: 10.1201/9781032683805-9

address them. By providing a reliable basis for decision-making, accurate data not only enhances the precision and effectiveness of policies but also fosters public trust and support for regulatory measures. Recent studies emphasize the significance of integrating advanced modeling techniques with comprehensive observational data to enhance the accuracy of air quality predictions, enabling policymakers to make informed decisions for effective pollution control and management.

Air quality prediction relies on various sophisticated tools and models that integrate atmospheric science, computer modeling, and data analytics. Chemical transport models (CTMs) serve as fundamental tools, employing numerical algorithms to simulate the transport, dispersion, and chemical transformation of pollutants in the atmosphere (Grell et al., 2005; Tuccella et al., 2012; Visconti, 2016). These models utilize complex equations derived from fluid dynamics, atmospheric chemistry, and thermodynamics to represent the intricate interplay of emission sources, chemical reactions, atmospheric transport processes, and deposition mechanisms. Additionally, statistical models, such as Machine Learning (ML) techniques, complement CTMs by leveraging advanced algorithms to analyze vast datasets, including meteorological observations, satellite imagery, and ground-level monitoring data, to enhance predictive accuracy (M. Wang et al., 2016; Xu et al., 2021). Satellite remote sensing technologies also play a crucial role, providing extensive spatial coverage and valuable information on atmospheric composition, aerosol properties, and pollutant concentrations, which are integrated into predictive models for a comprehensive understanding of air quality dynamics (Khattatov et al., 2000; Kumar et al., 2020). These diverse tools and approaches, coupled with advancements in computational capabilities and data assimilation techniques, continually improve air quality prediction accuracy, supporting informed decision-making in air quality management and policy formulation.

A range of statistical methods, spanning from basic to advanced, were employed to conduct various types of predictions, including air quality prediction. One primary limitation of basic methods lies in their bias toward the mean value and their limited capability to comprehend non-linear relationships (Mishra et al., 2015). These limitations can affect the accuracy and effectiveness of basic methods when dealing with complex or non-linear datasets. Conversely, advanced ML techniques exhibit greater proficiency in modeling non-linear relationships, thereby presenting an advantage over basic statistical methods (Liu et al., 2022). ML algorithms have a wide range of features such as the ability to understand the linear and non-linear relationships, to handle multi-dimensional and high-dimensional air quality datasets, their adeptness in capturing temporal and spatial interactions, reading complex patterns in datasets, etc. that enables them to be used in various air quality prediction applications.

This chapter delineates the fundamental principles of frequently employed ML methods within the realm of air quality, focusing on their theoretical aspects in the methodology section. It expounds on their typical applications, outlining the advantages and disadvantages inherent to each technique. Subsequently, the latter part of the chapter elucidates the practical applications of ML in air quality prediction. This section encompasses diverse air quality prediction challenges and delineates the utilization of ML for specific application scenarios. The chapter culminates in a conclusive summary, summarizing key findings and offering insights into potential future directions for research and application in this domain.

9.2 MACHINE LEARNING METHODOLOGIES IN THE FIELD OF AIR QUALITY

Air quality prediction involves using data analysis to understand connections between different factors in a dataset and make predictions based on these relationships and conditions. This has a wide range of applications like air quality forecasting, high temporally and spatially resolved air pollution datasets, understanding processes of the atmosphere, and many more. Depending on the specific application, scientists opt for data mining algorithms designed to make categorical predictions (classified as classifiers) or numeric predictions akin to regression. Commonly employed classification algorithms in this context encompass decision trees, Bayesian classifiers, support vector machines, and multilayer perceptrons. For numeric predictions in air quality forecasting, prevalent data mining methods include artificial neural networks, support vector regression, and regression trees.

In the realm of air quality prediction, regression methods have long been foundational in modeling pollutant concentrations. Linear regression, a cornerstone in traditional air quality predictions, offers a fundamental framework for understanding the linear relationships between predictor variables and pollutant concentrations. It provides insights into the magnitude and direction of associations, elucidating the impact of meteorological parameters such as temperature, humidity, wind speed, and direction on pollutant dispersion (Liu et al., 2022). Advancements within regression modeling have aimed to model non-linear relations of atmospheric processes. The integration of non-linear regression techniques, such as polynomial regression, to capture more intricate relationships between predictor variables and air quality indicators (Liu et al., 2009) has paved the way to better capture the response of perturbed predictors to regional air quality. Bayesian regression approaches have enriched the air quality forecasting landscape by accounting for uncertainties inherent in the data. Bayesian regression models noted for their ability to incorporate prior knowledge and update beliefs iteratively based on observed data, offer a probabilistic framework for capturing uncertainties in pollutant concentration predictions (Beloconi & Vounatsou, 2020). In essence, regression methods, encompassing linear, non-linear, and Bayesian approaches, form an integral component of air quality prediction endeavors. While linear regression models provide fundamental insights into the linear relationships between predictors and pollutants, non-linear regression techniques and Bayesian methodologies augment the predictive capacity by accommodating complexities and uncertainties within air quality datasets.

The ensemble learning approach of RF models, amalgamating multiple decision trees, has showcased proficiency in predicting particulate matter concentrations, leveraging the collective wisdom of constituent trees to deliver robust forecasts. Gradient Boosting Machines (GBM) is an ensemble method that has garnered attention for its versatility in handling multi-dimensional and high-dimensional air quality datasets.

GBM's ability to harness the strengths of diverse models enhances predictive accuracy and resilience, underpinning its relevance in real-world applications (Ma et al., 2020).

Deep learning architectures, notably Convolutional Neural Networks (CNNs) and Long Short-Term Memory networks (LSTMs), have emerged as formidable contenders owing to their adeptness in capturing temporal and spatial dependencies inherent in air quality data (J. Zhang & Li, 2022). CNNs excel in identifying spatial patterns, while LSTMs adeptly model temporal sequences, collectively enriching the predictive capacities in air quality forecasting. The utilization of neural networks, particularly deep learning architectures, in air quality prediction has surged owing to their capacity to capture intricate spatiotemporal patterns and nonlinear relationships within complex environmental datasets. CNNs, renowned for their proficiency in image recognition tasks, have found resonance in spatial analysis within air quality datasets. Their ability to extract hierarchical features from spatially structured data makes them adept at discerning intricate spatial patterns in pollutant distributions. Studies by Mao and Lee (2019) demonstrated the applicability of CNNs in air quality prediction by leveraging the spatial characteristics of air pollutant data, thereby improving forecasting precision. In parallel, LSTMs, a variant of recurrent neural networks designed to capture temporal dependencies, excel in modeling sequential data such as time series. In air quality forecasting, LSTMs are more favorably used to capture temporal dynamics and dependencies within pollutant concentration time series data. Their architecture facilitates the retention of long-term dependencies, enabling the assimilation of historical information crucial for accurate predictions (Drewil & Al-Bahadili, 2022).

Support Vector Machines (SVMs), noted for their ability to delineate intricate decision boundaries, have demonstrated efficacy in discerning patterns within air pollutants datasets. SVMs are established ML algorithms, that have demonstrated prowess in air quality prediction, offering robustness and generalization capabilities amidst complex and high-dimensional datasets. SVMs, originally devised for binary classification tasks, have been adeptly extended to handle multi-class and regression problems within air quality forecasting (Leong et al., 2020). One of the key advantages of SVMs lies in their ability to handle high-dimensional datasets, making them well-suited for air quality prediction tasks encompassing a multitude of input variables. Moreover, SVMs are equipped with regularization parameters that aid in preventing overfitting, thereby enhancing their generalization performance even in scenarios with limited training data (Müller et al., 1997). Cluster analysis, a cornerstone in unsupervised learning, offers a distinctive perspective in exploring patterns and discerning inherent structures within air quality datasets. This technique, aimed at uncovering latent groupings or clusters within data points, has been instrumental in uncovering spatial and temporal relationships among pollutant concentrations and environmental parameters (G. Huang, 1992). The application of cluster analysis in air quality research revolves around identifying homogeneous groups of monitoring stations or geographical regions exhibiting similar pollutant patterns or characteristics. Stolz et al. (2020) highlighted the significance of cluster analysis in categorizing monitoring stations based on similarities in air quality patterns, aiding in the identification of regions experiencing analogous pollution levels or exhibiting comparable meteorological influences (Stolz et al., 2020).

9.3 APPLICATION OF MACHINE LEARNING IN AIR QUALITY STUDIES AND PREDICTION

Comprehensive air quality data is fundamental for policy making. It encompasses detailed information on air quality across different locations and temporal scales, specific details about pollution sources and types, and a deeper insight into the underlying processes. The utmost priority lies in ensuring the accuracy of this information, as it is essential for developing precise and effective policies, fostering trust, and garnering support from the public and stakeholders. The features and methods of ML described in the previous section enable us to generate required air quality datasets and information with high accuracy and save on computational resources too. It provides a way for efficient and feasible policy making. The application of ML in generating policy-relevant air quality data is described below.

9.3.1 Spatial and Temporal Data Filling

Ground-based monitoring is considered as the most accurate source of air quality problems. Because of operational issues like power failure, instrument defects, sensor failure, and network outages, it can produce erroneous data or missing data. ML methods are utilized to understand the data pattern, detect outliers, and fill missing data in ground based observation. An artificial neural network (ANN) method denoising autoencoder is used in one study to fix missing data in air pollution measurements. This approach predicts missing information by studying adjacent locations, eliminating the necessity for any other additional data (Wardana et al., 2022). Another study has proposed a hybrid of linear interpolation and graph ML for the imputation of air pollutant time series (Betancourt et al., 2023) for determining missing values in hourly air pollution data. Where linear interpolation performs well for smaller gaps but for larger gaps a graph ML method which is a combination of random forest (RF) with correct and smooth method generally performs well (Betancourt et al., 2023).

Spatial air pollutant maps greatly aid in air quality decision-making. They're generated through a network of air quality monitors, using ML to predict missing data spatially and temporally. These methods employ ground-based observations, low-cost sensors, or other data like land use. For instance, NO_2 maps are created using station observations and distances, employing methods like Inverse Distance Weighting (IDW) and Least Absolute Shrinkage and Selection Operator (LASSO), often enhanced by an ANN ensemble for better accuracy (Van Roode et al., 2019). These techniques are effective in densely monitored smaller regions, while larger areas with fewer sites rely on additional data sources like land use, activity, population, or satellite maps. Approaches combining satellite data, chemical transport models, and ground observations through geographical weighted regression produce $PM_{2.5}$ distributions and fine particulate matter composition (Donkelaar et al., 2016; van Donkelaar et al., 2019). The precision of empirical models using statistical and

ML techniques heavily relies on input data quality rather than the methodology employed (Bechle et al., 2023). Hence, better and more comprehensive input data leads to improved predictions.

The global distribution of ground-based monitoring networks for air quality is uneven, especially in low and middle-income countries (LMICs), where there are high levels of air pollution (Martin et al., 2019). In LMICs, many monitoring stations operate manually and only sample data twice or thrice every week, resulting in a lack of continuous temporal data. To improve air quality management, significant investment from these countries' GDP is necessary to expand monitoring networks. However, the substantial health impacts and financial obstacles associated with this expansion can be mitigated by supplementing ground-based data with satellite information (Dey et al., 2012). Nevertheless, satellite retrievals conducted daily encounter spatial gaps caused by cloud cover and other challenges during data retrieval, potentially leading to biased estimates (Chowdhury et al., 2019; Dey et al., 2012).

Numerous efforts have been undertaken to address the gaps present in satellite data sampling. Recent studies have employed various statistical and interpolation methodologies to address these gaps, detailed elsewhere in the literature. Katoch and their colleagues highlighted how spatial gaps in satellite Aerosol Optical Depth (AOD) could lead to a substantial bias in annual $PM_{2.5}$ exposure estimation (Katoch et al., 2023). This bias in $PM_{2.5}$ exposure estimation can consequently result in a significant overestimation of both the health and economic impacts of air pollution. ML algorithms, particularly RF techniques, have only recently been employed to address these gaps; they have shown promise in providing a more comprehensive estimation with minimal errors compared to previous methodologies. Recent studies have delved into employing artificial intelligence techniques to tackle the issue of sampling gaps in satellite AOD and subsequently estimate $PM_{2.5}$ levels (Kianian et al., 2021; Tiwari et al., 2023; Y. Wang et al., 2022). Research has also leveraged ML approaches to address gaps in satellite-derived NO_2 data. Zhang and team employed a spatiotemporal neural network (STNN) model to predict surface NO_2 levels in China (Wei et al., 2022). Their approach incorporated various data sources such as satellite observations, ground measurements, meteorological and geographical data. Similarly, Qin et al. utilized geographically and temporally weighted regression (GTWR) and extremely randomized trees (ERT) models to estimate daily surface NO_2 concentrations in the central–eastern regions of China using OMI tropospheric NO_2 products alongside meteorological information (Qin et al., 2020).

9.3.2 Improving CTM Prediction

Chemical transport models (CTMs), commonly referred to as air quality models, simulate the dispersion, transport, and chemical transformation of various chemical species in the atmosphere, predicting their atmospheric concentrations (Grell et al., 2005; Kukkonen et al., 2012). The CTMs are built to solve mathematical equations for describing physical and chemical processes occurring in the atmosphere. These equations integrate principles from fluid dynamics, chemical kinetics, thermodynamics, and atmospheric chemistry. CTMs are used in wide applications like air quality forecasting (Grell et al., 2005; Kukkonen et al., 2012), understanding physical and chemical processes of air pollutants (Déméautis et al., 2022; Fast et al., 2006; Grell & Freitas, 2014; Jiang et al., 2019), source apportionment of air pollutants (Daellenbach et al., 2020; Lelieveld et al., 2015; Upadhyay et al., 2018a), future projections of air pollution and air quality mitigation policies (Kumar et al., 2018; Upadhyay et al., 2020; West et al., 2013).

However, CTMs inherit some uncertainties from numerical approximations, incomplete representation of intricate atmospheric processes to reduce computational requirements particularly regarding chemical reactions and their kinetics (Gunwani & Mohan, 2017; Z. Huang et al., 2019; Mallet & Sportisse, 2006). CTMs require inputs such as emissions, meteorological data, topography, land-use information, and chemical reaction rates to simulate air quality (Figure 9.1). Emission inventories entail gridded emission data of major air pollutants from diverse emission sources. These estimates are typically derived from activity data and emission rates, regridded to the domain of interest (Janssens-Maenhout et al., 2015; M. Li et al., 2017). The accuracy of emission estimation relies on the precision of activity data and the uncertainty associated with emission rates, leading to uncertainties in CTM simulations (Matthias et al., 2018; Upadhyay et al., 2019). Weather or climate models provide meteorological parameters for CTMs. Uncertainty is simulating meteorological parameters and inaccuracies in other input data like topography, and land use contribute to prediction uncertainties (Bei et al., 2017; Chatani & Sharma, 2018). These sources of uncertainty collectively result in model biases, typically assessed through model validation against observations or ground truth data (Z. Huang et al., 2019). Addressing these uncertainties necessitates continuous advancements in scientific comprehension, enhanced data collection methods, improved process parameterizations, and the

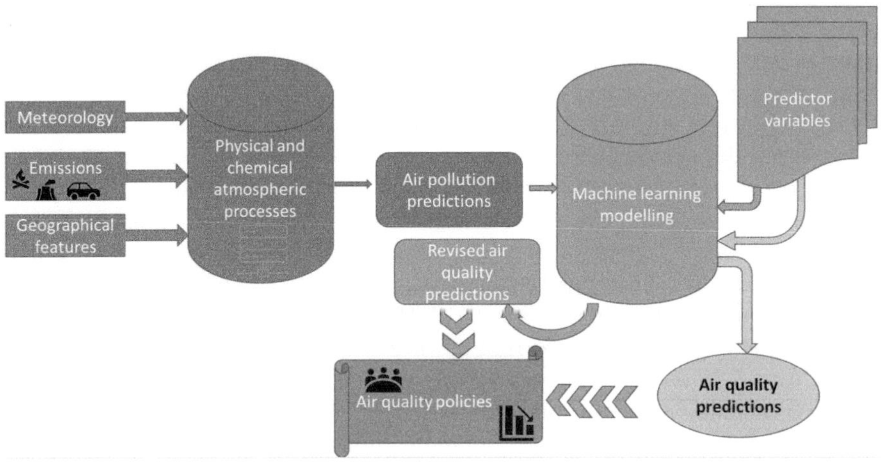

FIGURE 9.1 Schematic representation for application of machine learning models in (i) improving air quality predictions from chemical transport model (CTM) as well as describing framework for CTM; (ii) air quality predictions for air quality policies.

integration of observations and model simulations. Improvement have been made in model parametrization for chemical and physical processes (Dudhia, 1989; Grell & Freitas, 2014; Hong et al., 2006) and emission inventories (Crippa et al., 2020). Additionally, advancements in data assimilation techniques have been achieved (Khattatov et al., 2000; Kumar et al., 2020; Z. Li et al., 2013). However, implementing these improvements is time-consuming, computationally intensive, and demands substantial human resources.

Simple statistical models to advanced ML models are integrated with CTMs to improve model performance (Figure 9.1) and these are relatively less computationally intensive with easy incorporation. There are several instances where a wide variety of ML models have been used in improving CTM predictions, like multiplicative ratio correction and the Kalman filter method are used to improve air quality forecasting for Portugal (Borrego et al., 2011). Both these methods used to improve forecasts are site-specific approaches where ground-based measurements and model simulations at each monitoring site are used. They have shown significant improvement in hourly air quality forecasts for the studied period. Similarly, likelihood and spatiotemporal spline methods are used for bias-correction of chemical transport model predictions and this is further used for epidemiological study (Crooks & Özkaynak, 2014). ML methods are also used in conjugation with CTM, incorporating other dependent variables to provide more accurate predictions (Di et al., 2016; Ivatt & Evans, 2020; M. Wang et al., 2016).

ML methods can replace the need of fine-scale computationally intensive CTM simulation. This method uses coarser model output, meteorological data, and land use data as input and avoids the need for fine-resolution emission inventory and large computational resources (Dinkelacker et al., 2023). Accuracy and robustness of ML methods depends on input dataset. Ground based observation is one of the prime prerequisite for these ML applications where longer time series and spatially dense data produce a model with high accuracy and robustness. Other predictor variables which are correlated with target paramenter (air pollutant concentration) are also used in ML methods. Some commonly used variables are population density, urbanization, green area, roads, elevation and land use, etc.

ML methods are used to resolve complex atmospheric chemistry parameterizations within the model too. An application of ML methods to predict ozone has shown the potential to replace the need for fully interactive atmospheric chemistry schemes (Nowack et al., 2018). These chemical schemes are computationally expensive and therefore modeling communities often use climatological ozone fields to reduce computational cost. The representation of ozone doesn't align with the actual state of climate simulated and it doesn't correspond to any specific climate change scenario. The methodology proposed by Nowack et al. (2018), for parameterization, relies solely on the temperature from the preceding timestep. While demonstrating promising validation statistics, it necessitates rigorous testing to ascertain its robustness. Furthermore, its integration into a fully coupled model requires thorough exploration and evaluation. Keller et al. (2019) have shown potential of a ML model to replace computationally expensive complex gas phase chemistry (Keller & Evans, 2019). They proposed an alternative RF regression framework to represent the gas-phase chemistry within the GEOS-Chem model. Their RF predictor models did well in predicting changes in long lived pollutants such as O_3, CO, NOx, SO_2, Sulfate and many VOC species and absolute concentration of short-lived species such as NO_3 and N2O5. However, this implementation was predicted better for total NOx rather than its individual components, NO and NO_2. A pivotal outcome of their work involved segregating species based on their atmospheric lifetime, thereby significantly enhancing model performance through meticulous conservation of atoms, notably within the NOx family (Keller & Evans, 2019; Liao et al., 2020).

The utilization of neural networks to substitute an atmospheric chemical solver has demonstrated a noteworthy reduction in computational workload within global chemical transport models. While previous substitutions have been known to prompt rapid error amplification during simulations, recent advancements in ML techniques have exhibited a capacity not only to diminish computational demands but also to manage and control error growth effectively. This methodology has undergone testing within the framework of the global chemical transport model, GEOS-Chem (Kelp et al., 2020, 2022). Furthermore, its complete development holds the promise of conferring a significant advantage to air pollution and climate simulations, potentially offering a decisive edge in these domains. Gil et al. (2023) leverage advanced computational techniques rooted in deep neural networks to precisely calculate the mixing ratios of HONO, also known as nitrous acid (Gil et al., 2023). Developed and implemented in Python, this pioneering simulation technique, named the Reactive Nitrogen Species using a Deep Neural Network (RND), marks a significant advancement in accurately estimating HONO levels and affirm its potential as an invaluable supplementary model for accurately predicting HONO mixing ratios within densely polluted urban environments.

CTM provides a solid grounding in physical principles, the ML supplements this foundation by capturing nuanced dynamics and subtle interactions that may elude traditional modeling paradigms. This holistic approach facilitates a better understanding of air pollutants transport and distributions, support stakeholders with actionable insights for informed decision-making in pollution management, public health protection measures, and urban planning initiatives. Such interdisciplinary collaborations underscore the significance of bridging conventional and modern modeling methodologies to address the multifaceted challenges posed by urban air pollution.

9.3.3 Forecasting

Air quality forecasts are required to release timely public advisories to deal with high pollution events, to predict health impact in advance help in better preparedness, and help in predicting some economic activities. Accurate air quality forecasting is essential the development of efficient and reliable policies and subsequent actions. Forecasting systems entail analyzing different parameters associated with a particular location or timeframe within a comprehensive dataset to forecast the expected air quality levels. This process involves assessing various factors such as pollutant concentrations, meteorological conditions, and emission sources to estimate the likely air quality outcome for that specific instance or area. Accurate air quality forecasts are always challenging because the dynamics of air pollution is highly non-linear and with high dimensionality and it needed a wide range of source information. Traditional methodologies for forecasting air quality have historically relied on mathematical and statistical models. These methods assimilate historical data, emission inventories, and meteorological parameters to model pollutant dispersion and concentrations. However, their efficacy often encounters challenges in capturing intricate non-linear relationships inherent in air quality data, particularly in dynamic and heterogeneous environmental settings. In the recent past, the advent of ML algorithms has sparked a paradigm shift in air quality prediction. Pioneering studies employing SVM (Lu & Wang, 2008), RF models (Kim et al., 2023), and the integration of CNNs and LSTMs (J. Zhang & Li, 2022) for forecasting various air quality parameters have underscored the transformative potential of these methodologies. Post-processing air quality forecasts obtained from physical models utilizing ML techniques is an effective way to generate better forecasts. The following methods are some widely used methods to improve air quality forecast by postprocessing air quality forecasting model output (a) Running-mean bias correction; (b) Kalman filter; and (c) analogues (Bertrand et al., 2023). The running-mean bias correction method involves correcting systematic errors in forecast outputs by computing the average bias over a sliding time window and adjusting future predictions accordingly. This technique aims to mitigate consistent over or underestimations in forecasted values. The Kalman filter is a dynamic estimation method that updates forecast predictions based on the assimilation of real-time observations. It uses a recursive algorithm to continuously adjust forecasts using available measurements, thereby improving the accuracy of predictions by dynamically incorporating new information. It involves identifying past meteorological conditions that closely resemble the current atmospheric state and using historical air quality data associated with these analogous conditions to refine present forecasts. This method leverages similarities in past conditions to improve the accuracy of current air quality predictions.

Machine learning models for air quality prediction operate on a distinct paradigm, leveraging data-driven approaches to discern patterns and relationships from historical datasets. While they also require input datasets, these may extend beyond traditional meteorological and emissions data to include satellite imagery, remote sensing data, and other sources, enriching the breadth of information available for analysis. Unlike conventional models, machine learning algorithms do not explicitly require predefined boundary conditions; instead, they implicitly capture boundary effects through the input data. Historical data from neighboring regions or external sources indirectly inform the model about boundary influences on local air quality. Thus, while conventional models excel in capturing intricate physical processes and require explicit boundary specifications, machine learning models offer a more flexible framework for learning complex relationships directly from data, without relying on predefined equations or boundary conditions. Both approaches contribute significantly to advancing our understanding and prediction capabilities in air quality research, catering to the diverse needs and complexities of real-world atmospheric systems. While machine learning models excel in filling data gaps and enhancing the accuracy of simulations, they confront the challenge of data scarcity. The efficacy of these models is intricately tied to the quality and quantity of input data. Enhanced model performance and more accurate predictions are achieved through a substantial and high-quality input dataset. Thus, the optimization and refinement of machine learning models are contingent upon the availability of comprehensive and reliable input data.

Accurate air quality predictions hold the key to orchestrating targeted interventions, facilitating effective environmental management strategies, and informing public health policies. The convergence of interdisciplinary approaches, entwining domain expertise with computational prowess, holds the promise of creating a future where air quality forecasts become not just accurate but indispensable tools in preserving environmental equilibrium and human well-being.

9.4 CONCLUSIONS AND POLICY DIRECTIONS

Over the past five decades, numerous regions worldwide have established monitoring networks to track air pollution using a combination of fixed stations and mobile platforms. These networks supplement their data with information gathered from satellites. However, despite this seemingly comprehensive description of the global air pollution monitoring system, the reality is far from complete. Satellite instruments offer global coverage, yet their measurements lack the necessary frequency and precision, especially when assessing air quality near the Earth's surface. Many parts of the world, mostly LMICs, lack an adequate number of air quality monitoring stations. Even in Europe or the USA where the network of monitoring stations are seemingly dense, there can be considerable distance of hundreds of kilometers between neighboring monitoring stations. This spatial gap poses challenges in obtaining detailed and localized air quality information. ML methods offer a cost-effective and less computationally intensive way to fill in these gaps. This helps policymakers by giving them useful evidence about where and when air pollutants peak, even in areas that are beyond the reach of satellite data and monitoring stations.

Understanding and predicting air pollution presently relies on intricate numerical models. These models simulate both weather patterns and air pollution chemistry, demanding high computational power and significant costs. Operating on supercomputers, these models often take several days or even months to generate systems for decision-making. These systems determine the contribution of emissions from various sources to the exposure of air pollutants in a specific area. ML techniques offer a potential solution to accelerate these models. By statistically parameterizing multiple physical and chemical processes within these models, ML can create simplified models with reduced complexity. Such streamlined models can then be employed for more efficient decision-making processes regarding air quality and emission sources. Furthermore, when coupled with physiological sensors and sophisticated medical information systems, AI-driven pollution monitoring holds promise for conducting in-depth health impact assessments and comprehending the intricate effects of air pollution exposure on human health. These AI-based systems can serve as rapid tools for early warnings, enabling vulnerable individuals to better plan their outdoor activities and steer clear of potentially hazardous environments (*cleanairfund*, 2023; Y. Zhang et al., 2021) .

There are several encouraging examples showing the diverse use of ML in air quality management and policies. Agencies are extensively using machine learning techniques to develop operational products to tackle air quality and they are working in the collaboration with the policy makers to implement these for air quality management. United Nation Environment Protection (UNEP) and IQAir have worked in partnership to develop a large global platform for air quality monitoring and this utilize machine learning methods to observe impact of air quality on human health in real time and support in health impact control measures (UNEP, 2022). Similarly National Aeronautics and Space Administration (NASA) and World Resources Institute (WRI) have worked together for CityAQ project, intended to generate customized and localized air quality monitoring specifically for urban areas. It operates by assimilating even limited ground and air monitoring data, employing a machine-learning technique to enhance the precision of GEOS-CF model predictions. They have applied these in eight cities (Addis Ababa, Jakarta, Kigali, Leon-Salamanca-Celaya Metro, Monterrey Metro, Guadalajara Metro, Sao Paulo, and Bogotá, Colombia) and in the process aim to expand it to more cities (NASA, 2023). NASA SERVIR South east Asia has developed a tool for air quality (SERVIR SEA's air quality tool) with the aim to identify sources of heavy smoke plumes and high pollution events, transport of air pollutants with their trajectory, and direction for public health advisories concerning unhealthy level of air pollution based on machine learning methods. This is being used by the Royal Thai Government's Ministry of Natural Resources and Environment and the Lao People's Democratic Republic (Servirglobal, 2023).

Yet, the current reliability and effectiveness of such systems often raise doubts, with limited available information regarding their practical performance. Similar to other domains, the primary risk associated with AI solutions lies in their uncritical reliance. Consequently, it becomes crucial to gain a comprehensive understanding of the capabilities and constraints inherent in AI-based air quality monitoring systems. Equally important is maintaining control over our actions concerning their deployment and interpretation.

REFERENCES

Bechle, M. J., Bell, M. L., Goldberg, D. L., Hankey, S., Lu, T., Presto, A. A., Robinson, A. L., Schwartz, J., Shi, L., Zhang, Y., & Marshall, J. D. (2023). Intercomparison of Six national empirical models for PM2.5 air pollution in the contiguous US. *Findings*. https://doi.org/10.32866/001c.89423

Bei, N., Wu, J., Elser, M., Feng, T., Cao, J., El-haddad, I., Li, X., & Huang, R. (2017). Impacts of meteorological uncertainties on the haze formation in Beijing – Tianjin – Hebei (BTH) during wintertime: A case study. December 2015, 14579–14591.

Beloconi, A., & Vounatsou, P. (2020). Bayesian geostatistical modelling of high-resolution {NO} 2 exposure in {Europe} combining data from monitors, satellites and chemical transport models. *Environment International*, *138*, 105578. https://doi.org/10.1016/j.envint.2020.105578

Bertrand, J.-M., Meleux, F., Ung, A., Descombes, G., & Colette, A. (2023). Technical note: Improving the European air quality forecast of the Copernicus Atmosphere Monitoring Service using machine learning techniques. *Atmospheric Chemistry and Physics*, *23*(9), 5317–5333. https://doi.org/10.5194/acp-23-5317-2023

Betancourt, C., Li, C. W. Y., Kleinert, F., & Schultz, M. G. (2023). Graph machine learning for improved imputation of missing tropospheric ozone data. *Environmental Science & Technology*, *57*(46), 18246–18258. https://doi.org/10.1021/acs.est.3c05104

Borrego, C., Monteiro, A., Pay, M. T., Ribeiro, I., Miranda, A. I., Basart, S., & Baldasano, J. M. (2011). How bias-correction can improve air quality forecasts over Portugal. *Atmospheric Environment*, *45*(37), 6629–6641. https://doi.org/https://doi.org/10.1016/j.atmosenv.2011.09.006

Chatani, S., & Sharma, S. (2018). Uncertainties caused by major meteorological analysis data sets in simulating air quality over India. *x*, 6230–6247. https://doi.org/10.1029/2017JD027502

Chowdhury, S., Dey, S., Di, L., Smith, K. R., Pillarisetti, A., & Lyapustin, A. (2019). Tracking ambient PM 2 . 5 build-up in Delhi national capital region during the dry season over 15 years using a high-resolution (1 km) satellite aerosol dataset. *Atmospheric Environment*, *204*(December 2018), 142–150. https://doi.org/10.1016/j.atmosenv.2019.02.029

cleanairfund. (2023). www.cleanairfund.org/news-item/ai-wearables-innovative-tech/

Crippa, M., Solazzo, E., Huang, G., Guizzardi, D., Koffi, E., Muntean, M., Schieberle, C., Friedrich, R., & Janssens-Maenhout, G. (2020). High resolution temporal profiles in the emissions database for global atmospheric research. *Scientific Data*, *7*(1), 121. https://doi.org/10.1038/s41597-020-0462-2

Crooks, J. L., & Özkaynak, H. (2014). Simultaneous statistical bias correction of multiple PM2.5 species from a regional photochemical grid model. *Atmospheric Environment*, *95*, 126–141. https://doi.org/https://doi.org/10.1016/j.atmosenv.2014.06.024

Daellenbach, K. R., Uzu, G., Jiang, J., Cassagnes, L.-E., Leni, Z., Vlachou, A., Stefenelli, G., Canonaco, F., Weber, S., Segers, A., Kuenen, J. J. P., Schaap, M., Favez, O., Albinet, A., Aksoyoglu, S., Dommen, J., Baltensperger, U., Geiser, M., El Haddad, I., … Prévôt, A. S. H. (2020). Sources of particulate-matter air pollution and its oxidative potential in Europe. *Nature*, *587*(7834), 414–419. https://doi.org/10.1038/s41586-020-2902-8

Déméautis, T., Delles, M., Tomaz, S., Monneret, G., Glehen, O., Devouassoux, G., George, C., & Bentaher, A. (2022). Pathogenic mechanisms of secondary organic aerosols. *Chemical Research in Toxicology*, *35*(7), 1146–1161. https://doi.org/10.1021/acs.chemrestox.1c00353

Dey, S., Di, L., Donkelaar, A. Van, Tripathi, S. N., Gupta, T., & Mohan, M. (2012). Remote sensing of environment variability of outdoor fi ne particulate (PM 2 . 5) concentration in the Indian Subcontinent: A remote sensing approach. *Remote Sensing of Environment*, *127*, 153–161. https://doi.org/10.1016/j.rse.2012.08.021

Di, Q., Koutrakis, P., & Schwartz, J. (2016). A hybrid prediction model for PM2.5 mass and components using a chemical transport model and land use regression. *Atmospheric Environment*, *131*, 390 399. https://doi.org/https://doi.org/10.1016/j.atmosenv.2016.02.002

Dinkelacker, B. T., Garcia Rivera, P., Marshall, J. D., Adams, P. J., & Pandis, S. N. (2023). High-resolution downscaling of source resolved PM2.5 predictions using machine learning models. *Atmospheric Environment*, *310*, 119967. https://doi.org/https://doi.org/10.1016/j.atmosenv.2023.119967

Donkelaar, A. Van, Martin, R. V, Brauer, M., Hsu, N. C., Kahn, R. A., Levy, R. C., Lyapustin, A., Sayer, A. M., & Winker, D. M. (2016). Global Estimates of Fine Particulate Matter using a Combined Geophysical-Statistical Method with Information from Satellites, Models, and Monitors. https://doi.org/10.1021/acs.est.5b05833

Drewil, G. I., & Al-Bahadili, R. J. (2022). Air pollution prediction using {LSTM} deep learning and metaheuristics algorithms. *Measurement: Sensors*, *24*, 100546. https://doi.org/10.1016/j.measen.2022.100546

Dudhia, J. (1989). Numerical study of convection observed during the winter monsoon experiment using a mesoscale two-dimensional model. In *Journal of the Atmospheric Sciences* (Vol. 46, Issue 20, pp. 3077–3107). https://doi.org/10.1175/1520-0469(1989)046<3077:NSOCOD>2.0.CO;2

Fast, J. D., Jr, W. I. G., Easter, R. C., Zaveri, R. A., Barnard, J. C., Chapman, E. G., Grell, G. A., & Peckham, S. E. (2006). Evolution of ozone, particulates, and aerosol direct radiative forcing in the vicinity of Houston using a fully coupled meteorology-chemistry- aerosol model. *111*, 1–29. https://doi.org/10.1029/2005JD006721

Gil, J., Lee, M., Kim, J., Lee, G., Ahn, J., & Kim, C.-H. (2023). Simulation model of reactive nitrogen species in an urban atmosphere using a deep neural network: RNDv1.0. *Geoscientific Model Development*, *16*(17), 5251–5263. https://doi.org/10.5194/gmd-16-5251-2023

Grell, G. A., & Freitas, S. R. (2014). A scale and aerosol aware stochastic convective parameterization for weather and air quality modeling. *Atmospheric Chemistry and Physics*, *14*(10), 5233–5250. https://doi.org/10.5194/acp-14-5233-2014

Grell, G. A., Peckham, S. E., Schmitz, R., Mckeen, S. A., Frost, G., Skamarock, W. C., & Eder, B. (2005). Fully coupled '" online "' chemistry within the WRF model. *39*, 6957–6975. https://doi.org/10.1016/j.atmosenv.2005.04.027

Gunwani, P., & Mohan, M. (2017). Sensitivity of WRF model estimates to various PBL parameterizations in di ff erent climatic zones over India. *Atmospheric Research*, *194*(April), 43–65. https://doi.org/10.1016/j.atmosres.2017.04.026

Hong, S.-Y., Noh, Y., & Dudhia, J. (2006). A new vertical diffusion package with an explicit treatment of entrainment processes. *Monthly Weather Review*, *134*(9), 2318–2341. https://doi.org/10.1175/MWR3199.1

Huang, G. (1992). A stepwise cluster analysis method for predicting air quality in an urban environment. *Atmospheric Environment. Part B. Urban Atmosphere*, *26*(3), 349–357. https://doi.org/10.1016/0957-1272(92)90010-P

Huang, Z., Zheng, J., Ou, J., Zhong, Z., Wu, Y., & Shao, M. (2019). A feasible methodological framework for uncertainty analysis and diagnosis of atmospheric chemical transport models. *Environmental Science & Technology*, *53*(6), 3110–3118. https://doi.org/10.1021/acs.est.8b06326

Ivatt, P. D., & Evans, M. J. (2020). Improving the prediction of an chemistry transport model using gradient-boosted regression treesatmospheric. *Atmospheric Chemistry and Physics*, *20*(13), 8063–8082. https://doi.org/10.5194/acp-20-8063-2020

Janssens-Maenhout, G., Crippa, M., Guizzardi, D., Dentener, F., Muntean, M., Pouliot, G., Keating, T., Zhang, Q., Kurokawa, J., Wankmüller, R., Denier Van Der Gon, H., Kuenen, J. J. P., Klimont, Z., Frost, G., Darras, S., Koffi, B., & Li, M. (2015). HTAP-v2.2: A mosaic of regional and global emission grid maps for 2008 and 2010 to study hemispheric transport of air pollution. *Atmospheric Chemistry and Physics*, *15*(19). https://doi.org/10.5194/acp-15-11411-2015

Jiang, J., Aksoyoglu, S., Ciarelli, G., Oikonomakis, E., El-Haddad, I., Canonaco, F., O'Dowd, C., Ovadnevaite, J., Minguillón, M. C., Baltensperger, U., & Prévôt, A. S. H. (2019). Effects of two different biogenic emission models on modelled ozone and aerosol concentrations in Europe. *Atmospheric Chemistry and Physics*, *19*(6), 3747–3768. https://doi.org/10.5194/acp-19-3747-2019

Katoch, V., Kumar, A., Imam, F., Sarkar, D., Knibbs, L. D., Liu, Y., Ganguly, D., & Dey, S. (2023). Addressing biases in ambient PM2.5 exposure and associated health burden estimates by filling satellite AOD retrieval gaps over India. *Environmental Science & Technology*, *57*(48), 19190–19201. https://doi.org/10.1021/acs.est.3c03355

Keller, C. A., & Evans, M. J. (2019). Application of random forest regression to the calculation of gas-phase chemistry within the GEOS-Chem chemistry model v10. *Geoscientific Model Development*, *12*(3), 1209–1225. https://doi.org/10.5194/gmd-12-1209-2019

Kelp, M. M., Jacob, D. J., Kutz, J. N., Marshall, J. D., & Tessum, C. W. (2020). Toward stable, general machine-learned models of the atmospheric chemical system. *Journal of Geophysical Research: Atmospheres*, *125*(23), e2020JD032759. https://doi.org/https://doi.org/10.1029/2020JD032759

Kelp, M. M., Jacob, D. J., Lin, H., & Sulprizio, M. P. (2022). An online-learned neural network chemical solver for stable long-term global simulations of atmospheric chemistry. *Journal of Advances in Modeling Earth Systems*, *14*(6), e2021MS002926. https://doi.org/https://doi.org/10.1029/2021MS002926

Khattatov, B. V, Lamarque, J.-F., Lyjak, L. V, Menard, R., Levelt, P., Tie, X., Brasseur, G. P., & Gille, J. C. (2000). Assimilation of satellite observations of long-lived chemical species in global chemistry transport models. *Journal of Geophysical Research: Atmospheres*, *105*(D23), 29135–29144. https://doi.org/https://doi.org/10.1029/2000JD900466

Kianian, B., Liu, Y., & Chang, H. H. (2021). Imputing satellite-derived aerosol optical depth using a multi-resolution spatial model and random forest for PM2.5 prediction. *Remote Sensing*, *13*(1). https://doi.org/10.3390/rs13010126

Kim, B., Kim, E., Jung, S., Kim, M., Kim, J., & Kim, S. (2023). {PM2}.5 {Concentration} {Forecasting} {Using} {Weighted} {Bi}-{LSTM} and {Random} {Forest} {Feature} {Importance}-{Based} {Feature} {Selection}. *Atmosphere*, *14*(6), 968. https://doi.org/10.3390/atmos14060968

Kukkonen, J., Olsson, T., Schultz, D. M., Baklanov, A., Klein, T., Miranda, A. I., Monteiro, A., Hirtl, M., Tarvainen, V., Boy, M., Peuch, V.-H., Poupkou, A., Kioutsioukis, I., Finardi, S., Sofiev, M., Sokhi, R., Lehtinen, K. E. J., Karatzas, K., San José, R., … Eben, K. (2012). A review of operational, regional-scale, chemical weather forecasting models in Europe. *Atmospheric Chemistry and Physics*, *12*(1), 1–87. https://doi.org/10.5194/acp-12-1-2012

Kumar, R., Barth, M. C., Pfister, G. G., Delle Monache, L., Lamarque, J. F., Archer-Nicholls, S., Tilmes, S., Ghude, S. D., Wiedinmyer, C., Naja, M., & Walters, S. (2018). How will air quality change in South Asia by 2050? *Journal of Geophysical Research: Atmospheres*, *123*(3), 1840–1864. https://doi.org/10.1002/2017JD027357

Kumar, R., Ghude, S. D., Biswas, M., Jena, C., Alessandrini, S., Debnath, S., Kulkarni, S., Sperati, S., Soni, V. K., Nanjundiah, R. S., & Rajeevan, M. (2020). Enhancing accuracy of air quality and temperature forecasts during paddy crop residue burning season in Delhi via chemical data assimilation. *Journal of Geophysical Research: Atmospheres*, *125*(17), e2020JD033019. https://doi.org/https://doi.org/10.1029/2020JD033019

Lelieveld, J., Evans, J. S., Fnais, M., Giannadaki, D., & Pozzer, A. (2015). The contribution of outdoor air pollution sources to premature mortality on a global scale. *Nature*, *525*(7569), 367–371. https://doi.org/10.1038/nature15371

Leong, W. C., Kelani, R. O., & Ahmad, Z. (2020). Prediction of air pollution index ({API}) using support vector machine ({SVM}). *Journal of Environmental Chemical Engineering*, *8*(3), 103208. https://doi.org/10.1016/j.jece.2019.103208

Li, M., Zhang, Q., Kurokawa, J. I., Woo, J. H., He, K., Lu, Z., Ohara, T., Song, Y., Streets, D. G., Carmichael, G. R., Cheng, Y., Hong, C., Huo, H., Jiang, X., Kang, S., Liu, F., Su, H., & Zheng, B. (2017). MIX: A mosaic Asian anthropogenic emission inventory under the international collaboration framework of the MICS-Asia and HTAP. *Atmospheric Chemistry and Physics*, *17*(2), 935–963. https://doi.org/10.5194/acp-17-935-2017

Li, Z., Zang, Z., Li, Q. B., Chao, Y., Chen, D., Ye, Z., Liu, Y., & Liou, K. N. (2013). A three-dimensional variational data assimilation system for multiple aerosol species with WRF/Chem and an application to PM). A three-dimensional variational data assimilation system for multiple aerosol species with WRF/Chem and an application to PM$_{2.5}$ prediction. {2.5}$ prediction. *Atmospheric Chemistry and Physics*, *13*(8), 4265–4278. https://doi.org/10.5194/acp-13-4265-2013

Liao, Q., Zhu, M., Wu, L., Pan, X., Tang, X., & Wang, Z. (2020). Deep learning for air quality forecasts: A Review. *Current Pollution Reports*, *6*(4), 399–409. https://doi.org/10.1007/s40726-020-00159-z

Liu, Y., Paciorek, C. J., & Koutrakis, P. (2009). Estimating {Regional} {Spatial} and {Temporal} {Variability} of {PM}). Estimating {Regional} {Spatial} and {Temporal} {Variability} of {PM} $_{\textrm{2.5}}$ {Concentrations} {Using} {Satellite} {Data}, {Meteorology}, and {Land} {Use} {Information}. {\textrm{2.5}}$ {Concentrations} {Using} {Satellite} {Data}, {Meteorology}, and {Land} {Use} {Information}. *Environmental Health Perspectives*, *117*(6), 886–892. https://doi.org/10.1289/ehp.0800123

Liu, Y., Wang, P., Li, Y., Wen, L., & Deng, X. (2022). Air quality prediction models based on meteorological factors and real-time data of industrial waste gas. *Scientific Reports*, *12*(1), 9253. https://doi.org/10.1038/s41598-022-13579-2

Lu, W.-Z., & Wang, D. (2008). Ground-level ozone prediction by support vector machine approach with a cost-sensitive classification scheme. *Science of The Total Environment*, *395*(2–3), 109–116. https://doi.org/10.1016/j.scitotenv.2008.01.035

Ma, J., Yu, Z., Qu, Y., Xu, J., & Cao, Y. (2020). Application of the {XGBoost} {Machine} {Learning} {Method} in {PM2}.5 {Prediction}: {A} {Case} {Study} of {Shanghai}. *Aerosol and Air Quality Research*, *20*(1), 128–138. https://doi.org/10.4209/aaqr.2019.08.0408

Mallet, V., & Sportisse, B. (2006). Uncertainty in a chemistry-transport model due to physical parameterizations and numerical approximations: An ensemble approach applied to ozone modeling. *Journal of Geophysical Research: Atmospheres*, *111*(D1). https://doi.org/https://doi.org/10.1029/2005JD006149

Mao, Y., & Lee, S. (2019). Deep {Convolutional} {Neural} {Network} for {Air} {Quality} {Prediction}. *Journal of Physics: Conference Series*, *1302*(3), 32046. https://doi.org/10.1088/1742-6596/1302/3/032046

Martin, R. V, Brauer, M., Donkelaar, A. Van, Shaddick, G., Narain, U., & Dey, S. (2019). Atmospheric environment: X No one knows which city has the highest concentration of fi ne particulate matter. *Atmospheric Environment: X*, *3*(June), 100040. https://doi.org/10.1016/j.aeaoa.2019.100040

Matthias, V., Arndt, J. A., Aulinger, A., Bieser, J., Denier van der Gon, H., Kranenburg, R., Kuenen, J., Neumann, D., Pouliot, G., & Quante, M. (2018). Modeling emissions for three-dimensional atmospheric chemistry transport models. *Journal of the Air & Waste Management Association*, *68*(8), 763–800. https://doi.org/10.1080/10962247.2018.1424057

Mishra, D., Goyal, P., & Upadhyay, A. (2015). Artificial intelligence based approach to forecast PM<inf>2.5</inf> during haze episodes: A case study of Delhi, India. *Atmospheric Environment*, *102*. https://doi.org/10.1016/j.atmosenv.2014.11.050

Müller, K.-R., Smola, A. J., Rätsch, G., Schölkopf, B., Kohlmorgen, J., & Vapnik, V. (1997). Predicting time series with support vector machines. In G. Goos, J. Hartmanis, J. Van Leeuwen, W. Gerstner, A. Germond, M. Hasler, & J.-D. Nicoud (Eds.), *Artificial {Neural} {Networks} — {ICANN}'97* (Vol. 1327, pp. 999–1004). Springer Berlin Heidelberg. https://doi.org/10.1007/BFb0020283

NASA (2023). https://earthobservatory.nasa.gov/images/152131/filling-an-air-pollution-data-gap

Nowack, P., Braesicke, P., Haigh, J., Abraham, N. L., Pyle, J., & Voulgarakis, A. (2018). Using machine learning to build temperature-based ozone parameterizations for climate sensitivity simulations. *Environmental Research Letters*, *13*(10), 104016. https://doi.org/10.1088/1748-9326/aae2be

Qin, K., Han, X., Li, D., Xu, J., Loyola, D., Xue, Y., Zhou, X., Li, D., Zhang, K., & Yuan, L. (2020). Satellite-based estimation of surface NO2 concentrations over east-central China: A comparison of POMINO and OMNO2d data. *Atmospheric Environment*, *224*, 117322. https://doi.org/https://doi.org/10.1016/j.atmosenv.2020.117322

Servirglobal (2023). https://servirglobal.net/news/using-artificial-intelligence-boost-air-quality-southeast-asia

Shrestha, M., Upadhyay, A., Bajracharya, S., Sharma, M., & Maharjan, A. (2021). Meeting summary State of air pollution and potential mitigation mechanisms for the greater Punjab region. 1–19. https://doi.org/10.1175/BAMS-D-22-0132.1

Sofia, D., Gioiella, F., Lotrecchiano, N., & Giuliano, A. (2020). Mitigation strategies for reducing air pollution. *Environmental Science and Pollution Research*, *27*(16), 19226–19235. https://doi.org/10.1007/s11356-020-08647-x

Stolz, T., Huertas, M. E., & Mendoza, A. (2020). Assessment of air quality monitoring networks using an ensemble clustering method in the three major metropolitan areas of {Mexico}. *Atmospheric Pollution Research*, *11*(8), 1271–1280. https://doi.org/10.1016/j.apr.2020.05.005

Tiwari, S., Kumar, A., Mantri, S., & Dey, S. (2023). Modelling ambient PM2.5 exposure at an ultra-high resolution and associated health burden in megacity Delhi: exposure reduction target for 2030. *Environmental Research Letters*, *18*(4), 44010. https://doi.org/10.1088/1748-9326/acc261

Tuccella, P., Curci, G., Visconti, G., Bessagnet, B., Menut, L., & Park, R. J. (2012). Modeling of gas and aerosol with WRF/Chem over Europe: Evaluation and sensitivity study. *Journal of Geophysical Research: Atmospheres*, *117*(D3). https://doi.org/https://doi.org/10.1029/2011JD016302

UNEP (2022). www.unep.org/news-and-stories/story/how-artificial-intelligence-helping-tackle-environmental-challenges

Upadhyay, A., Dey, S., & Goyal, P. (2019). A comparative assessment of regional representativeness of EDGAR and ECLIPSE emission inventories for air quality studies in India. *Atmospheric Environment*, 117182. https://doi.org/https://doi.org/10.1016/j.atmosenv.2019.117182

Upadhyay, A., Dey, S., Chowdhury, S., & Goyal, P. (2018a). Expected health bene fi ts from mitigation of emissions from major anthropogenic PM 2 . 5 sources in India: Statistics at state level *. *Environmental Pollution*, *242*, 1817–1826. https://doi.org/10.1016/j.envpol.2018.07.085

Upadhyay, A., Dey, S., Chowdhury, S., & Goyal, P. (2018b). Expected health benefits from mitigation of emissions from major anthropogenic PM<inf>2.5</inf> sources in India: Statistics at state level. *Environmental Pollution*, *242*. https://doi.org/10.1016/j.envpol.2018.07.085

Upadhyay, A., Dey, S., Chowdhury, S., & Kumar, R. (2020). Tradeoffs between air pollution mitigation and meteorological response in India. *Scientific Reports*, 1–10. https://doi.org/10.1038/s41598-020-71607-5

van Donkelaar, A., Martin, R. V, Li, C., & Burnett, R. T. (2019). Regional estimates of chemical composition of fine particulate matter using a combined geoscience-statistical method with information from satellites, models, and monitors. *Environmental Science & Technology*, *53*(5), 2595–2611. https://doi.org/10.1021/acs.est.8b06392

Van Roode, S., Ruiz-Aguilar, J. J., González-Enrique, J., & Turias, I. J. (2019). An artificial neural network ensemble approach to generate air pollution maps. *Environmental Monitoring and Assessment*, *191*(12), 727. https://doi.org/10.1007/s10661-019-7901-6

Visconti, G. (2016). Simple Climate Models. In *Fundamentals of Physics and Chemistry of the Atmosphere* (pp. 461–502). Springer International Publishing. https://doi.org/10.1007/978-3-319-29449-0_14

Wang, M., Sampson, P. D., Hu, J., Kleeman, M., Keller, J. P., Olives, C., Szpiro, A. A., Vedal, S., & Kaufman, J. D. (2016). Combining land-use regression and chemical transport modeling in a spatiotemporal geostatistical model for ozone and PM2.5. *Environmental Science & Technology*, *50*(10), 5111–5118. https://doi.org/10.1021/acs.est.5b06001

Wang, Y., Du, Y., Fang, J., Dong, X., Wang, Q., Ban, J., Sun, Q., Ma, R., Zhang, W., He, M. Z., Liu, C., Niu, Y., Chen, R., Kan, H., & Li, T. (2022). A random forest model for daily PM2.5 personal exposure assessment for a Chinese Cohort. *Environmental Science & Technology Letters*, *9*(5), 466–472. https://doi.org/10.1021/acs.estlett.1c00970

Wardana, I. N. K., Gardner, J. W., & Fahmy, S. A. (2022). Estimation of missing air pollutant data using a spatiotemporal convolutional autoencoder. *Neural Computing and Applications*, *34*(18), 16129–16154. https://doi.org/10.1007/s00521-022-07224-2

Wei, J., Liu, S., Li, Z., Liu, C., Qin, K., Liu, X., Pinker, R. T., Dickerson, R. R., Lin, J., Boersma, K. F., Sun, L., Li, R., Xue, W., Cui, Y., Zhang, C., & Wang, J. (2022). Ground-Level NO2 surveillance from space across China for high resolution using interpretable spatiotemporally weighted artificial intelligence. *Environmental Science & Technology*, *56*(14), 9988–9998. https://doi.org/10.1021/acs.est.2c03834

West, J. J., Smith, S. J., Silva, R. A., Naik, V., Zhang, Y., Adelman, Z., Fry, M. M., Anenberg, S., & Horowitz, L. W. (2013). Co-benefits of mitigating global greenhouse gas emissions for future air quality and human health. *Nature Climate Change*, *3*(10), 885–889. https://doi.org/10.1038/nclimate2009

Xu, M., Jin, J., Wang, G., Segers, A., Deng, T., & Lin, H. X. (2021). Machine learning based bias correction for numerical chemical transport models. *Atmospheric Environment*, *248*, 118022. https://doi.org/https://doi.org/10.1016/j.atmosenv.2020.118022

Zhang, J., & Li, S. (2022). Air quality index forecast in {Beijing} based on {CNN}-{LSTM} multi-model. *Chemosphere*, *308*, 136180. https://doi.org/10.1016/j.chemosphere.2022.136180

Zhang, Y., Geng, P., Sivaparthipan, C. B., & Muthu, B. A. (2021). Big data and artificial intelligence based early risk warning system of fire hazard for smart cities. *Sustainable Energy Technologies and Assessments*, *45*, 100986. https://doi.org/https://doi.org/10.1016/j.seta.2020.100986

10 A Glimpse into Tomorrow's Air

Leveraging PM 2.5 with FP Prophet as a Forecasting Model

Kush Singla, Love Singla, and Mamta Bansal

10.1 INTRODUCTION

The 21st century is known to be the most developed century for India till now. The growth of social, economic, and urban development of several cities has led to the rise of several problems like pollution (air pollution, water pollution, soil pollution, electronic waste), global warming, hazardous waste deposition in the environment, natural resource depletion, and many more, in the last few years. Compared to other types of pollution, air pollution is one of the most critical and significant environmental issues, dramatically impacting every biotic component present in this world, including plants, animals, humans, and several others. Reports from the World Health Organization (WHO) state that air pollution is a significant cause of premature deaths all around the globe. It was estimated that around 7 million premature deaths were recorded annually by several diseases such as heart attack, stroke, pulmonary diseases, lung cancer, etc., whose root cause is air pollution, which mainly affects low and middle-income countries (Anand et al., 2017; Northey et al., 2018; "The Global Health Cost of $PM_{2.5}$ Air Pollution: A Case for Action Beyond 2021," 2022; Ravikiran et al., 2019; Vito et al., 2009; Vitousek, 1994; Yilmaz et al., 2017). Some significant pollutants include carbon monoxide, ground-level ozone, particulate matter, nitrogen dioxide, and sulfur dioxide. These lead to many diseases and several health hazards that cause major cause of early deaths in India (Glencross et al., 2020; Pozzer et al., 2020; Q. Zhang et al., 2017). According to the global burden of disease study (2019), Uttar Pradesh, Bihar, Rajasthan, and Madhya Pradesh are the major areas of the Indian subcontinent which has a maximum level (64% of total death) of children death under 5 years of age due to air pollution (Pandey et al., 2021). Primary reasons behind the growth of air pollution include a hike in population density, uncontrolled growth in the industrial sector, an increase in motor vehicle usage, etc. In addition, urbanization is also a significant reason behind the rise in pollution (Yadav et al.,

DOI: 10.1201/9781032683805-10

2023). This impacts human health to a greater extent than has led scientists to think about the future trends in air quality (Sarkar et al., 2022; Zou et al., 2019)

The Air Quality Index (AQI) is an essential and crucial indicator comprising statistical calculations used to determine the air quality in the surroundings. This involves consolidating various contaminants' concentrations into a single value (Zhu et al., 2017). According to the National Air Quality Monitoring Programme (NAQMP) standards initiated by the Central Pollution Control Board (CPCB), 11 parameters are considered regularly at different locations across different cities. These parameters include sulfur dioxide (SO_2), nitrogen dioxide (NO_2), particulate matter with less than 10 and 2.5 microns (PM_{10} and $PM_{2.5}$), carbon monoxide (CO), ammonia (NH_3), lead (Pb), ozone (O_3), benzene (C_6H_6), arsenic (As), nickel (Ni), Benzo(a)pyrene (Central Pollution Control Board, 2023). These societies have standardized parametric values for the AQI, as shown in Table 10.1. Lower values imply an excellent state of health and vice versa.

Particulate matter (PM) is the term used for a mixture of solid and liquid particles as these are minute in size, which can be seen with an electron microscope. These include dust particles, dirt, soot, or smoke. It is differentiated based on diameter as we primarily study PM_{10} and $PM_{2.5}$ (inhalable particles with diameters of 10 microns and 2.5 microns, respectively, even smaller than the diameter of a human hair (having 50–70 microns)) concerning the AQI. These molecules can originate from different sources, such as construction sites, smoke stacks, fires, open residue burning, etc. (Agarwal et al., 2014; Singh et al., 2010). These particulate matter are known to be a significant risk to human health as these are the leading cause of visibility reduction in several regions of the United States and their neighboring areas. The particulate matter having 2.5 microns ($PM_{2.5}$) is supposed to be the most dangerous molecule based on theoretical and practical reports. Theoretically, it was reported that smaller-sized particulate matter (PMs) having a more significant surface area volume ratio are more capable of penetrating alveolar capillary barriers and entering bloodstreams, causing damage to several body parts such as ovarian functions (X. Zhang et al., 2023). The latest information in November 2023 shows that particulate matter having 2.5 microns ($PM_{2.5}$) is a crucial and primary top threat across the Indian

TABLE 10.1
Air Quality Index Standard Values

Air Quality Index (AQI)	State of health
0–50	Good
50–100	Satisfactory
100–200	Moderate
200–300	Poor
300–400	Very poor
400–500	Severe

Source: Central Pollution Control Board (2023)

sub-continent, leading to a decrease in the life expectancy of children by five years (George et al., 2024).

Governance and solving the air pollution problem will take a long time and effort. To curb this problem, we have some techniques to predict the air quality of particular regions. These can help government departments and the public take proactive measures to protect people from the coming obstacles. This forecasting requires a precise and prompt manner. It will be beneficial to stop the originators of bad air quality as the Chinese government has stopped functioning coal-based power plants based on this kind of future prediction (H. Zhang et al., 2016). Table 10.2 shows some differences between conventional (traditional, statistical, or rule-based) and machine learning-based approaches (Liang et al., 2020; Pant et al., 2023).

Machine Learning (ML) is a part of artificial intelligence (AI), defined as using mathematical models/algorithms to train computer systems to make precise predictions/decisions without the help of direct instructions (Bishop & Nasrabadi, 2006). ML lets the computer grasp the data/past experience-based information, label the pattern, and make future trends. Due to this exceptional capability of calculating big data in such a précised and short period, machine learning is gaining immersive recognition in present-day research. Big data comprises massive data which can be structured, semi-structured, or unstructured. It isn't easy to collect and store according to the old traditional methods. Still, the help of ML makes it easy to store and calculate with great accuracy and quickly aids in gaining comprehensive knowledge of its hidden data (Chen et al., 2014). In contrast to great accuracy, some problem-based errors will be faced during future predictions using machine learning models, such as training of complex models requires significant resources and takes a longer time

TABLE 10.2
Differences between Conventional and ML-based Approaches

	Conventional approaches	Machine learning based
Data sources	Fixed-station monitoring, atmospheric models	Sensor networks, satellite imagery, weather data, historical records
Prediction model	ARIMA, linear regression, and dispersion models	Random Forest, XGBoost, LSTM networks, deep learning models
Complexity	Simpler	It can be higher complex
Accuracy	Moderate	Can be higher accuracy
Adaptability	Less adaptable	Continuously improves with new data
Real-time prediction	Limited capabilities	Optimized for real-time with an efficient algorithm
Spatial coverage	Limited to monitoring stations	Wider areas with interpolation
Explanatory power	Limited insights	More insights through feature analysis
Data requirements	Large historical data	Smaller datasets with techniques
Computational cost	Lower cost	It can be expensive for complex models
Development time	Faster development	Longer development for optimization

to be processed; some noise or irrelevant patterns in the training dataset also led to the poor generalization on unseen data, or inconsistency in data collection also makes a challenging situation while training the ML model (Liang et al., 2020; Neo et al., 2022).

Time series prediction/forecasting models are critical in analyzing time series problems and help different business and organization holders think about future predictions based on past trends and patterns. There are several models in the time-series forecasting model. FB Prophet is open source and one of the best algorithms based on its accuracy, user-friendliness, and flexibility. Facebook's data science team made this algorithm in 2017 (Taylor & Letham, 2017). Significant features of this algorithm include its scalability, accuracy, and ease of use in various applications. This algorithm is based on the time series data that functions based on the combination of the trend (overall movement of the prediction, i.e., increasing or decreasing; using piecewise linear regression model), seasonality (periodic pattern like week or month basis; using Fourier series) and several other noise components (some seasonal fluctuations). Its functionality is based on the Bayesian approach, which works as a determination of the posterior distribution of the model parameters in comparison to the only points taken into consideration (*Understanding FB Prophet: A Time Series Forecasting Algorithm | by Pratyush Khare | ILLUMINATION | Medium*, n.d.).

FB Prophet has several advantages over other models, which include (*When to Use Facebook Prophet – Crunching the Data*, n.d.):

1. It can count for the mean shifts/disrupted data.
2. It has the ability to withstand multiple seasonality.
3. It is easy to use and less technical in working data.
4. It is robust to outliners.
5. It can efficiently deal with the missing data
6. It can incorporate multivariate for different parameters
7. It doesn't require stationary data, which can be non-linearity in their trends
8. Results of this method don't require particular information or techniques to be read.
9. It produces the results relatively quicker in comparison to other models.

In this chapter, we have implemented the FB Prophet forecasting model to predict the future trends of the AQI using particulate matter having 2.5 microns ($PM_{2.5}$) as the variable against time series.

10.2 DATA COLLECTION

The Central Pollution Control Board (CPCB) corroborated the air pollution data with all the state authorities (*CPCB | Central Pollution Control Board*, n.d.). The list of all the authorized agencies that provide information about the particular centers is provided in supplementary data. This chapter focuses on future air quality prediction by taking ten significant cities around the Indian subcontinent as samples for evaluation (*CCR*, n.d.). The selected cities are Ahmedabad, Bangalore, Chandigarh, Chennai, Delhi, Gurugram, Hyderabad, Lucknow, Mumbai, and Patna. The input data is available on this platform (https://cpcb.nic.in/), which provides historical data on air quality by showing the concentration of different pollutants present in the air of a particular area. This data contains information from January 1, 2013, to December 26, 2023.

10.3 CODE RUN

The machine learning code for future prediction of the AQI (using $PM_{2.5}$ as the primary pollutant) is based on FB Prophet. This model is a versatile and powerful tool for time series forecasting tasks, offering flexibility, accuracy, and user-friendliness that caters to diverse forecasting needs, including air quality prediction. Additionally, it can accommodate various types of time series data, including hourly, daily, or weekly observations and irregularly spaced intervals. The link to the code is https://gist.github.com/lovesingla001/d9893289840db40ee7a53196961b373d.

10.4 RESULTS

It was seen that there will be a lot of variations in air pollution in the coming times. Here, ten different cities are taken into consideration; we are going to discuss city-wise.

10.4.1 AHMEDABAD

From the data interpretation via the machine learning model, it was seen that there would be a dip in the concentration of particulate matter having 2.5 microns ($PM_{2.5}$) in

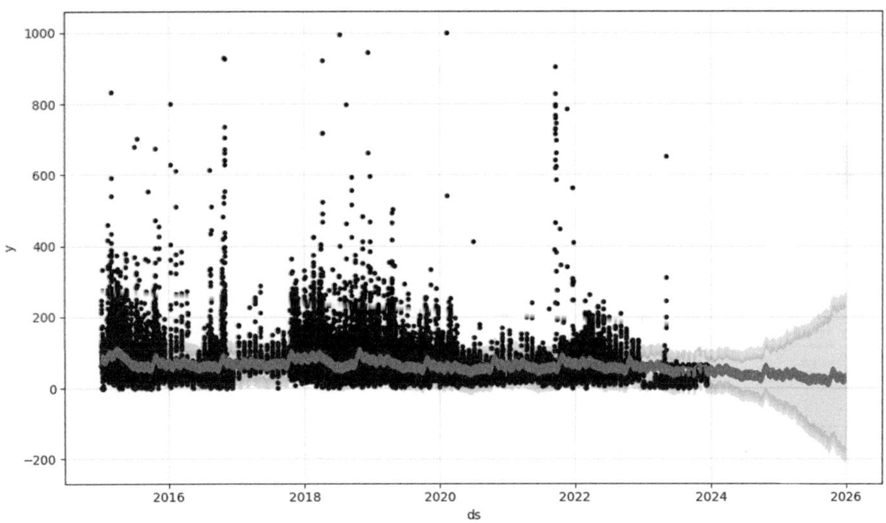

FIGURE 10.1 Trend of $PM_{2.5}$ across the years (2013–2023) with future prediction in coming years (2024–2026).

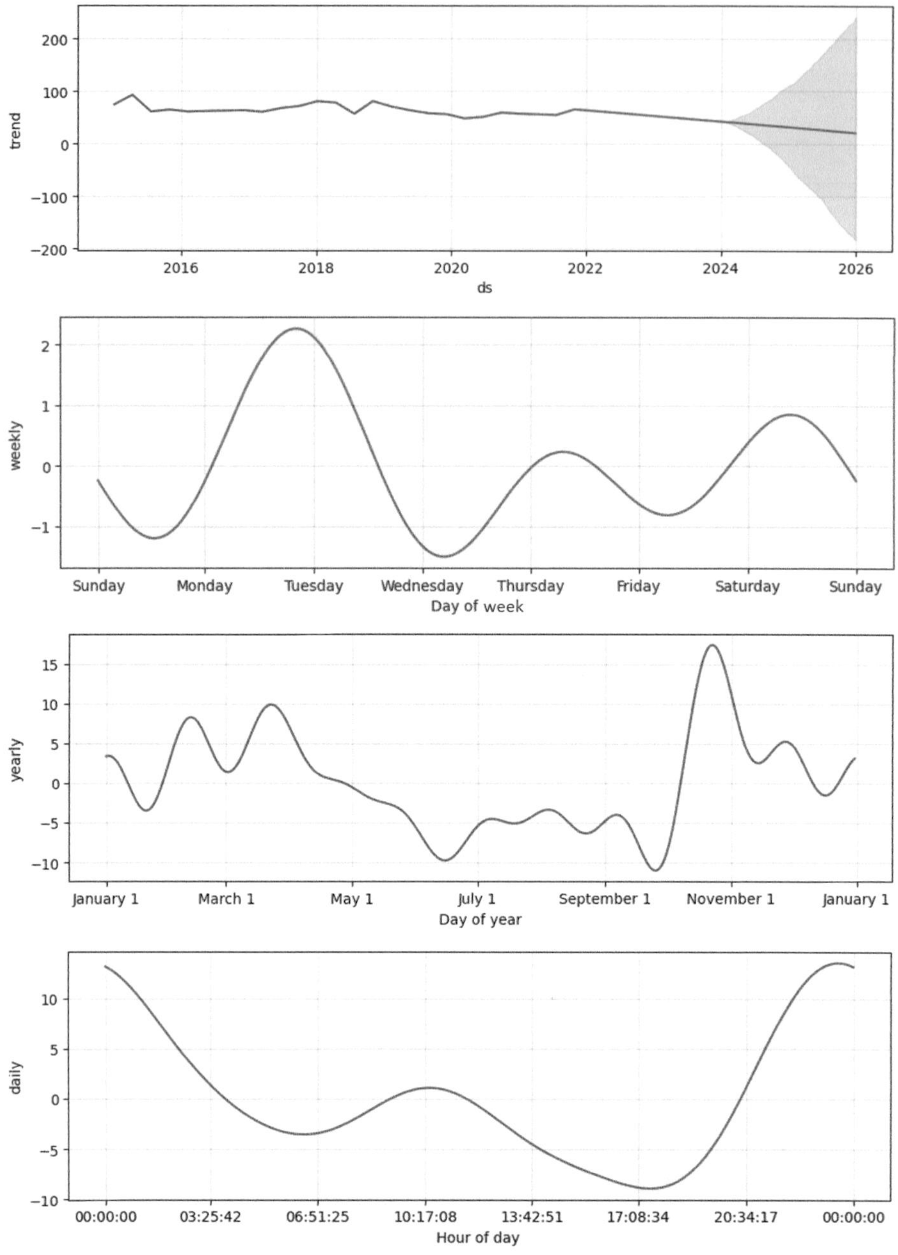

FIGURE 10.2 Trend in the concentration of PM 2.5 in Ahmedabad city on a yearly, daily, and weekly basis.

the coming two years (Figure 10.1). Figure 10.2 shows that maximum concentration was recorded across the year at the end of October – the start of November. According to the weekly report, maximum concentration was reported on Monday–Tuesday, with the least on Wednesday, day-off time. The highest concentration is recorded at midnight.

10.4.2 Bengaluru

From the data interpretation via the machine learning model, it was seen that there would be constant growth in the concentration of particulate matter having 2.5 microns ($PM_{2.5}$) in the coming two years (Figure 10.3). Figure 10.4 shows that across the year, maximum concentration was recorded at the end of December – the start of the year. The weekly report reported maximum concentration on Wednesday–Friday, with the least on Thursday, day-off time. The highest concentration is recorded at 21:00.

10.4.3 Chandigarh

From the data interpretation via the machine learning model, it was seen that there would be a variable growth in the concentration of particulate matter having 2.5 microns ($PM_{2.5}$) in the coming two years (Figure 10.5). Figure 10.6 shows that across the year, maximum concentration was recorded at the end of December – the start of the year. The weekly report shows maximum concentration on Friday–Saturday, with the least on Sunday, day-off time. The highest concentration is recorded at 22:00.

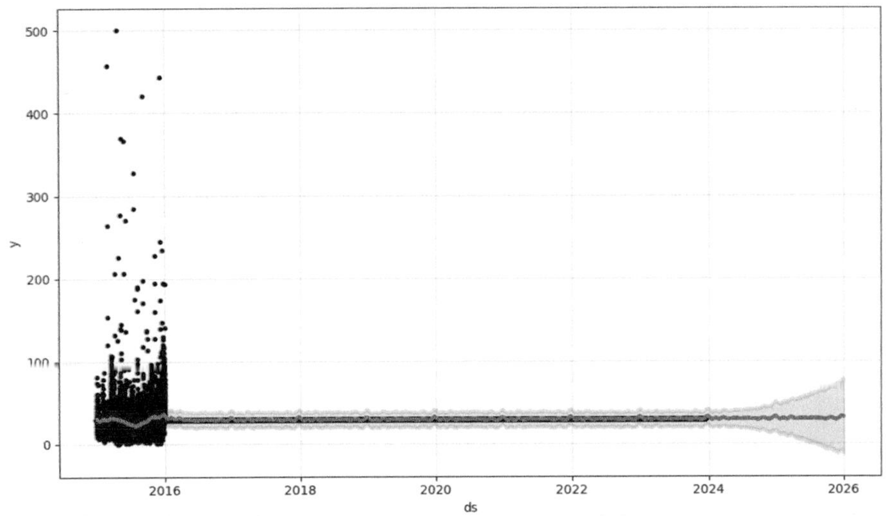

FIGURE 10.3 Trend of $PM_{2.5}$ across the years (2013–2023) with future prediction in coming years (2024–2026).

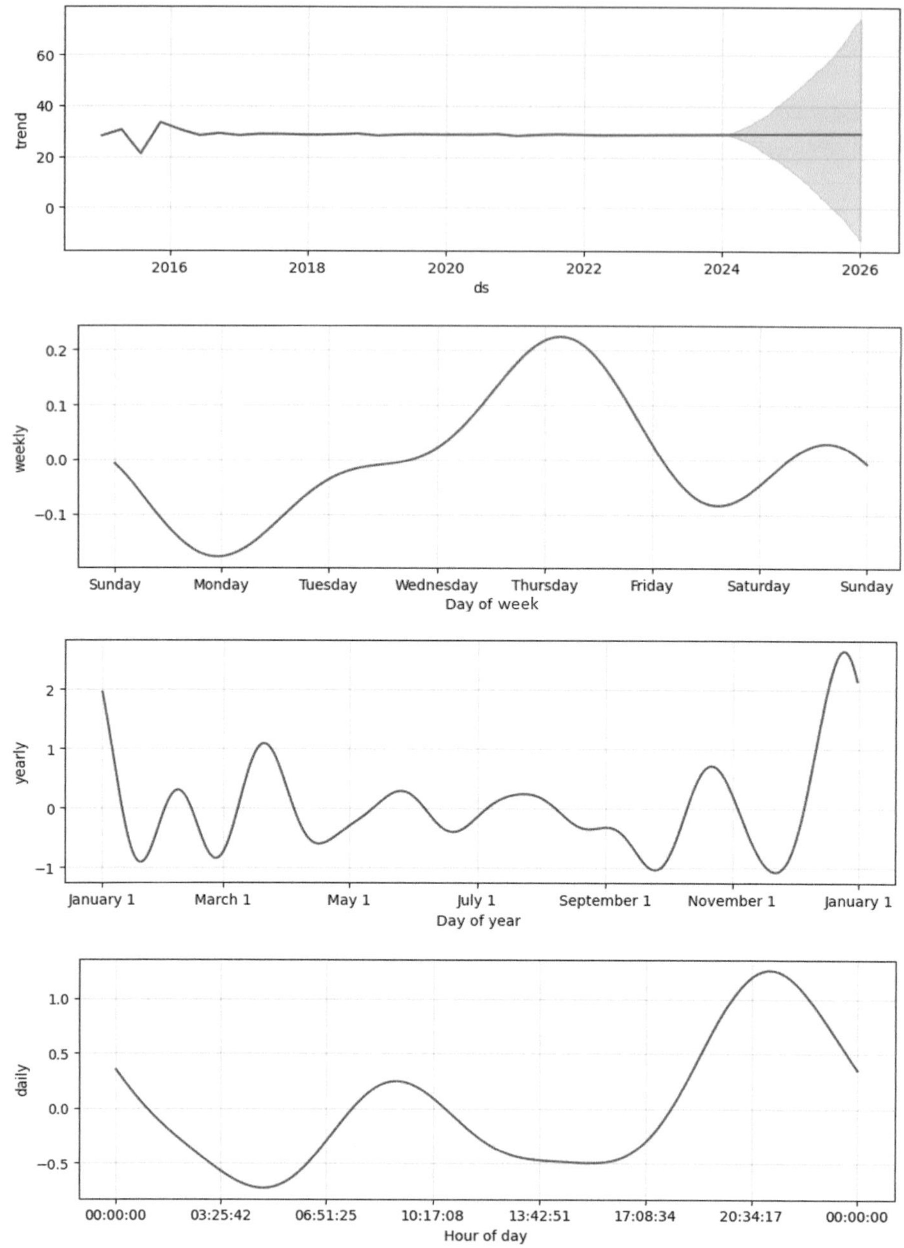

FIGURE 10.4 Trend in the concentration of $PM_{2.5}$ in Bengaluru city on a yearly, daily, and weekly basis.

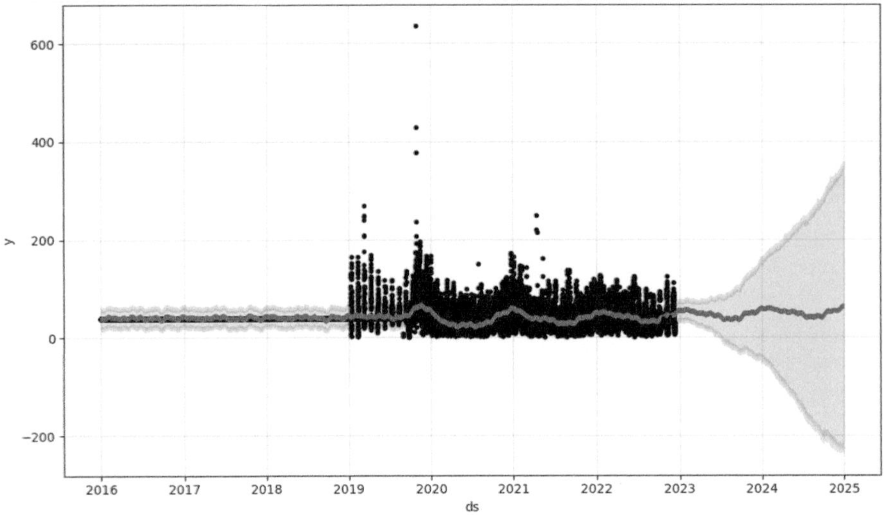

FIGURE 10.5 Trend of $PM_{2.5}$ across the years (2013–2023) with future prediction in coming years (2024–2026).

10.4.4 Chennai

From the data interpretation via the machine learning model, it was seen that there would be a slight dip in the concentration of particulate matter having 2.5 microns ($PM_{2.5}$) in the coming two years (Figure 10.7). Figure 10.8 shows that, across the year, maximum concentration was recorded in mid-October. The weekly report shows maximum concentration on different days, such as Tuesday and Sunday, with the least on Monday. The highest concentration is recorded at 09:00.

10.4.5 Delhi

From the data interpretation via the machine learning model, it was seen that there is a particular pattern in the levels of particulate matter having 2.5 microns ($PM_{2.5}$) around these years. It is predicted that there will be a slight dip with the increase, then a decrease in the concentration of particulate matter having 2.5 microns ($PM_{2.5}$) in the coming two years (Figure 10.9). The overall trend shows a drop in the concentration of particulate matter with 2.5 microns ($PM_{2.5}$) until 2026 (Figure 10.10). Figure 10.10 shows that maximum concentration was recorded across the year at the end of December, with an increase starting from October. The weekly report shows maximum concentration mid-Friday, with the least on Tuesday. The highest concentration is recorded at midnight.

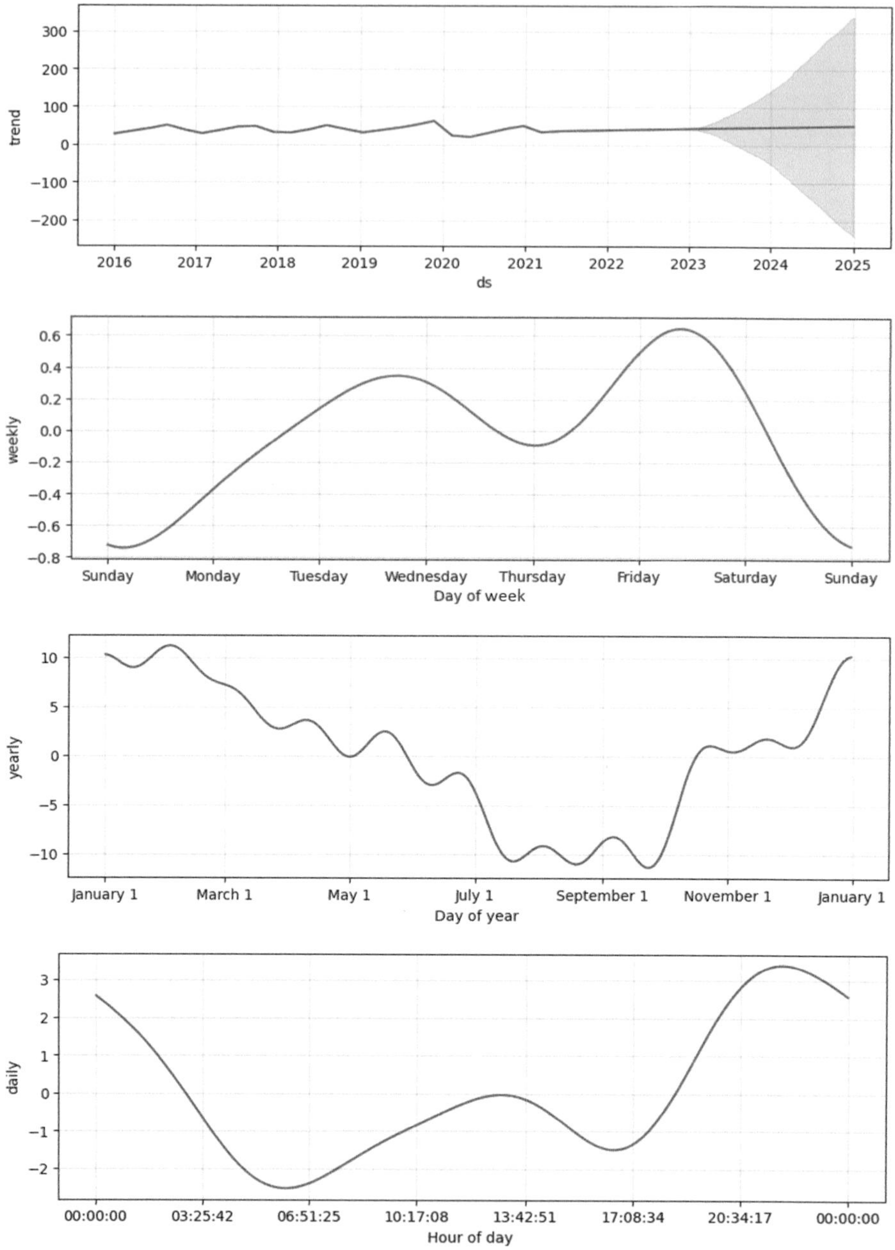

FIGURE 10.6 Trend in the concentration of $PM_{2.5}$ in Chandigarh city on a yearly, daily, and weekly basis.

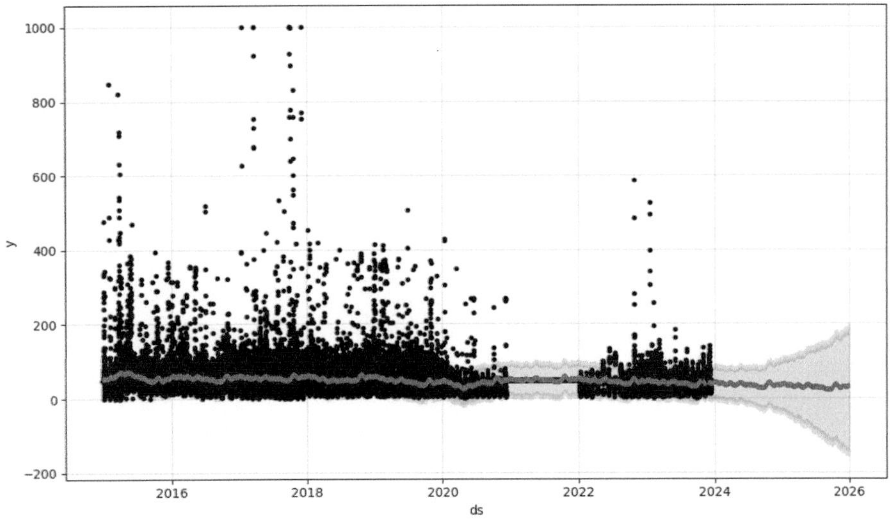

FIGURE 10.7 Trend of $PM_{2.5}$ across the years (2013–2023) with future prediction in coming years (2024–2026).

10.4.6 Gurugram

From the data interpretation via the machine learning model, it was seen that there is a particulate pattern in the levels of particulate matter having 2.5 microns ($PM_{2.5}$) around these years. It is predicted that there will be a slight dip with the increase, then a decrease in the concentration of particulate matter having 2.5 microns ($PM_{2.5}$) in the coming two years (Figure 10.11). The overall trend showed a minute dip in the concentration of particulate matter with 2.5 microns ($PM_{2.5}$) until the start of 2026 (Figure 10.12). Figure 10.12 shows that maximum concentration was recorded across the year at the end of December, with an increase starting from October. The weekly report shows that maximum concentration mid-Friday increased from Thursday, with the least at the dawn of Saturday. The highest concentration is recorded at 22:30.

10.4.7 Hyderabad

From the data interpretation via the machine learning model, it was seen that there is a particular pattern in the levels of particulate matter having 2.5 microns ($PM_{2.5}$) around these years. It is predicted that there will be a slight dip with the increase, then a decrease in the concentration of particulate matter having 2.5 microns ($PM_{2.5}$) in the coming two years (Figure 10.13). The overall trend shows a minute with the constant increase in the concentration of particulate matter having 2.5 microns ($PM_{2.5}$) till the start of 2026 (Figure 10.14). Figure 10.14 shows that the maximum concentration was recorded between October and December. The weekly report shows that the maximum concentration is at dawn on Wednesday and mid-Friday. The highest concentration is recorded at 21:30.

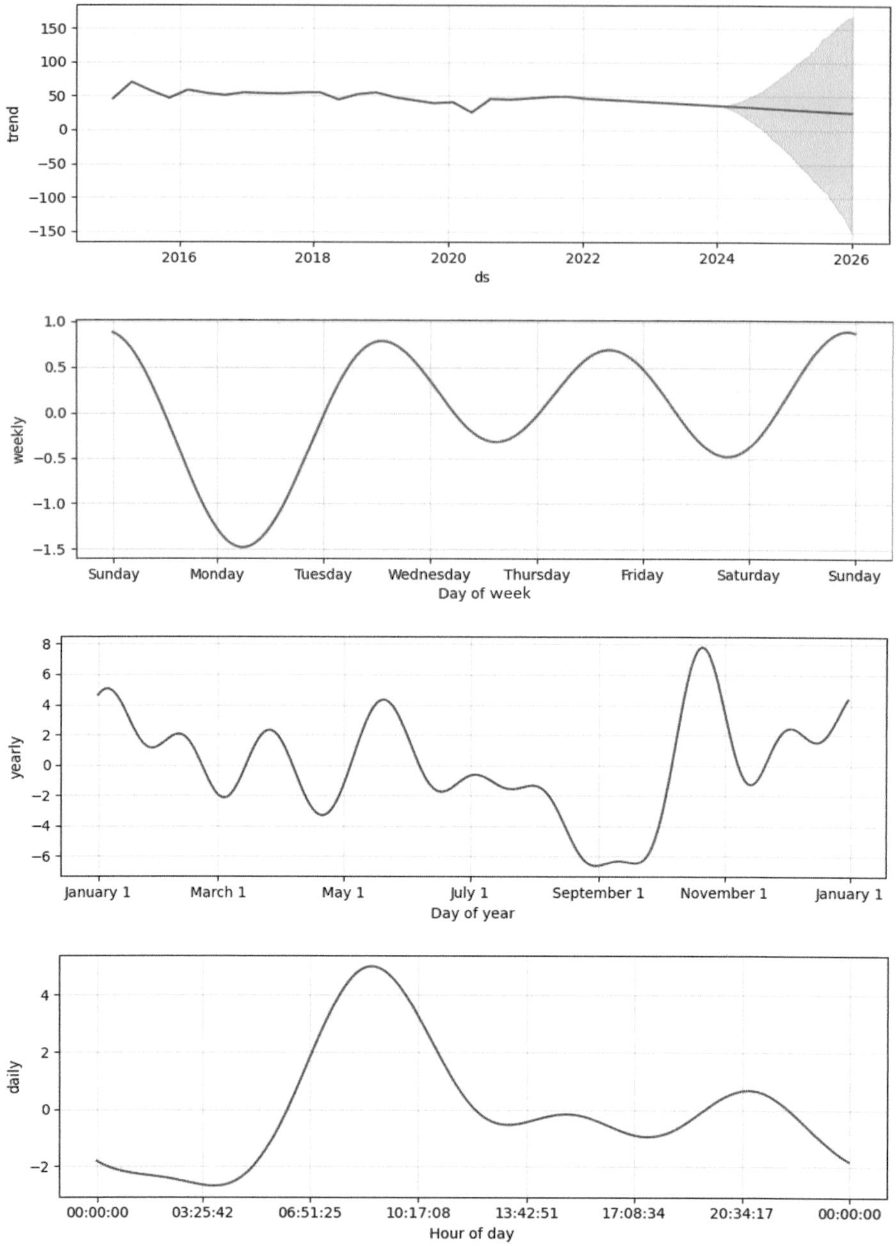

FIGURE 10.8 Trend in the concentration of $PM_{2.5}$ in Chennai city on a yearly, daily, and weekly basis.

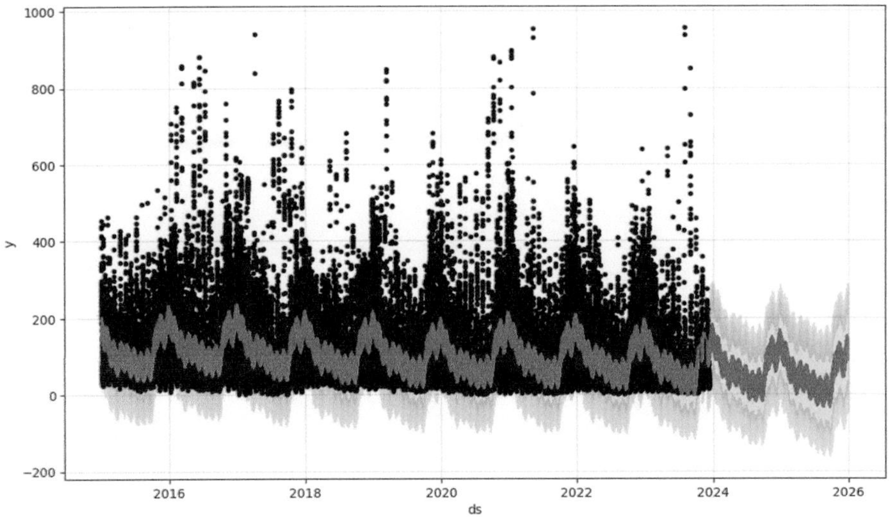

FIGURE 10.9 Trend of $PM_{2.5}$ across the years (2013–2023) with future prediction in coming years (2024–2026).

10.4.8 Lucknow

From the data interpretation via the machine learning model, it was seen that there is a particular pattern in the levels of particulate matter having 2.5 microns ($PM_{2.5}$) around these years. It is predicted that there will be a dip with the sudden increase, then a decrease in the concentration of particulate matter having 2.5 microns ($PM_{2.5}$) in the coming two years (Figure 10.15). The overall trend shows constant levels of particulate matter having 2.5 microns ($PM_{2.5}$) until 2026 (Figure 10.16). Figure 10.16 shows that maximum concentration was recorded in the middle of November and end of December. The weekly report shows that the maximum concentration is at Sunday midnight. The highest concentration of particulate matter having 2.5 microns ($PM_{2.5}$) was recorded at 22:00.

10.4.9 Mumbai

From the data interpretation via the machine learning model, it was seen that there is a particular pattern in the levels of particulate matter having 2.5 microns ($PM_{2.5}$) around these years. It is predicted that there will be a dip with the gradual increase, then a decrease in the concentration of particulate matter having 2.5 microns ($PM_{2.5}$) in the coming two years (Figure 10.17). The overall trend shows a constant reduction in particulate matter having 2.5 microns ($PM_{2.5}$) levels until 2026 (Figure 10.18). Figure 10.18 shows that the maximum concentration was recorded at the end of December. The weekly report shows that the maximum concentration is on Thursday. The highest concentration of particulate matter having 2.5 microns ($PM_{2.5}$) was recorded at 10:00 in the day.

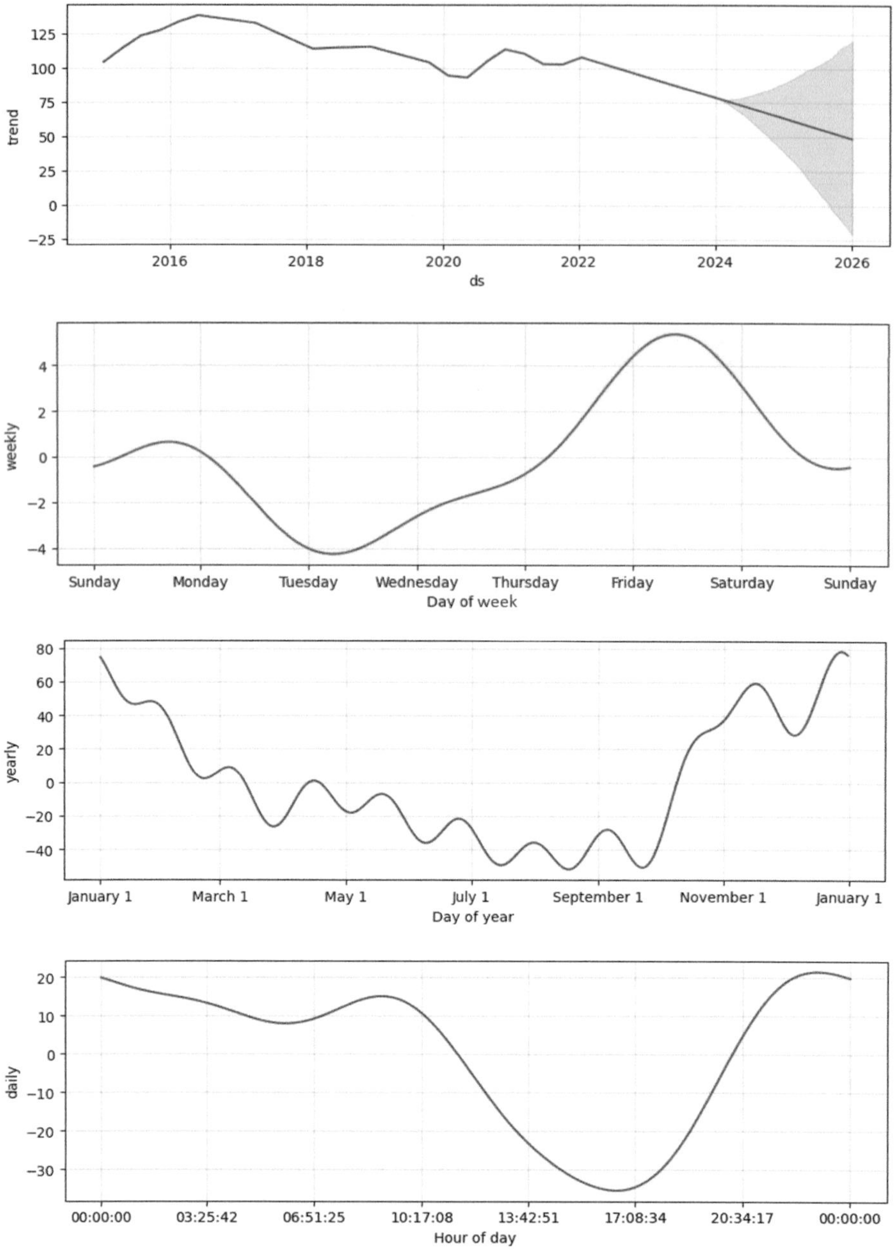

FIGURE 10.10 Trend in the concentration of $PM_{2.5}$ in Delhi city on a yearly, daily, and weekly basis.

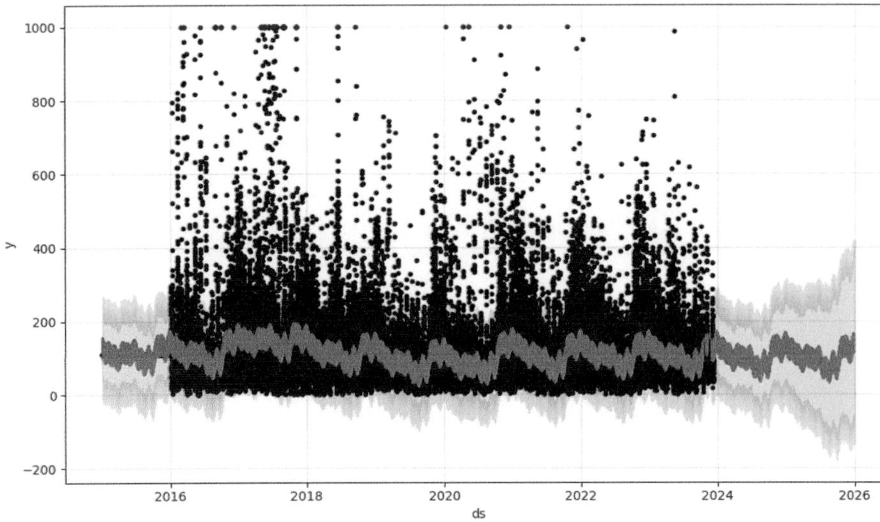

FIGURE 10.11 Trend of $PM_{2.5}$ across the years (2013–2023) with future prediction in coming years (2024–2026).

10.4.10 Patna

From the data interpretation via the machine learning model, it was seen that there is a particular pattern in the levels of particulate matter having 2.5 microns ($PM_{2.5}$) around these years. It is predicted that there will be a dip with the gradual increase, then a decrease in the concentration of particulate matter having 2.5 microns ($PM_{2.5}$) in the coming two years (Figure 10.19). The overall trend shows a constant but minute reduction in particulate matter having 2.5 microns ($PM_{2.5}$) levels until 2026 (Figure 10.20). Figure 10.20 shows that the maximum concentration was recorded at the end of December. The weekly report shows that the maximum concentration is on Thursday morning and Tuesday early morning. The highest concentration of particulate matter having 2.5 microns ($PM_{2.5}$) was recorded at 22:00.

10.5 DISCUSSION

The significance of a reliable AQI resides in its capacity to furnish valuable data regarding the caliber of the air we inhale. The significance of a reliable Air Quality Index (AQI) lies in its ability to provide crucial data about the quality of the air we breathe. Maintaining a healthy AQI is vital for public health, as regular monitoring aids particularly vulnerable demographics. Additionally, it supports awareness and education efforts by enhancing public consciousness about air pollution. Moreover, it informs policy and regulation by enabling authorities to track air quality patterns, identify pollution sources, and develop strategies to reduce emissions and improve overall air quality. Furthermore, a reliable AQI is essential for understanding the environmental impact of air pollution, such as its contribution to acid rain, agricultural yield

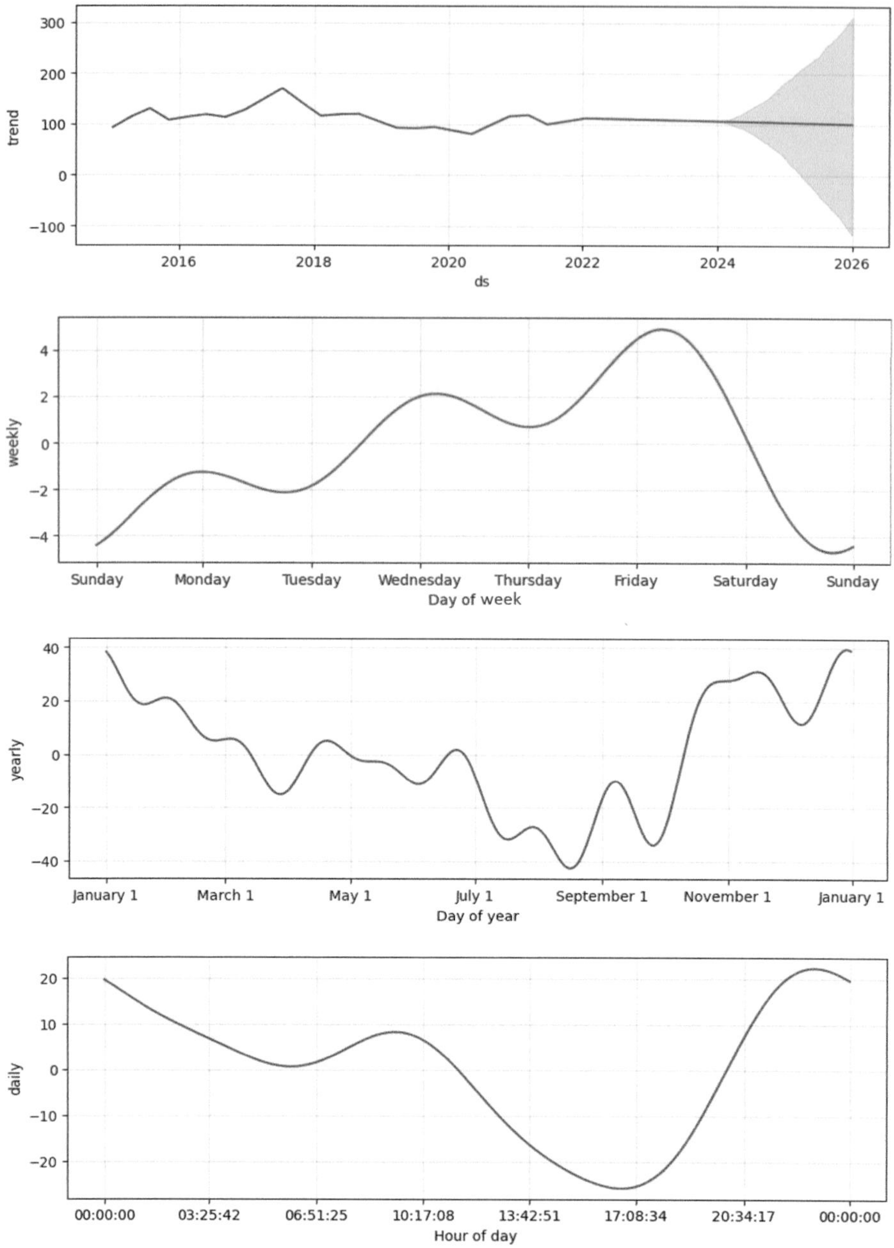

FIGURE 10.12 Trend in the concentration of $PM_{2.5}$ in Gurugram city on a yearly, daily, and weekly basis.

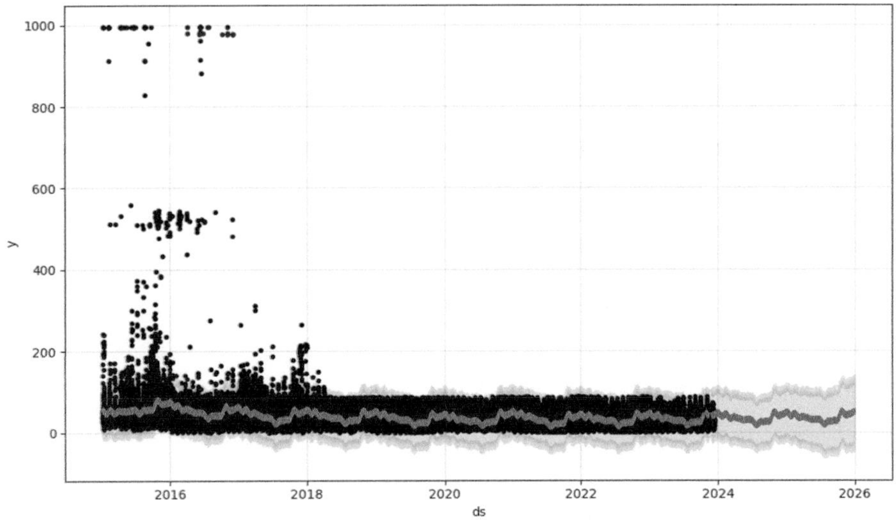

FIGURE 10.13 Trend of $PM_{2.5}$ across the years (2013–2023) with future prediction in coming years (2024–2026).

impairments, and negative effects on aquatic ecosystems. Additionally, it highlights economic implications, including healthcare costs related to pollution-induced illnesses, decreased worker productivity, and damage to crops and structures. Lastly, it facilitates research and development. To enhance the future AQI, we have used FB Prophet as a machine learning model to predict air quality levels across the Indian subcontinent, considering some Indian cities. To indicate air quality levels, we used Particulate matter ($PM_{2.5}$) as the basis of our study.

10.6 CONCLUSION

The predictive analysis showed that nine out of ten cities show a defined pattern of particulate matter having 2.5 microns ($PM_{2.5}$) levels over the years (January 2013 to December 2023).

- Some cities such as Ahmedabad, Chennai, and Delhi show a decrease in particulate matter having 2.5 microns ($PM_{2.5}$) levels with chances of better AQI in the coming times (2024 – 2026).
- Cities like Bengaluru, Gurugram, Lucknow, Mumbai, and Patna have shown constant levels over the past few years (2013–2023) that might also be constant in the coming years.
- The cities most concerned with a steady growth in air pollutants are Chandigarh and Hyderabad, which have shown a slow but steady increase in particulate matter having 2.5 microns ($PM_{2.5}$) levels that will be dangerous with lots of consequences.

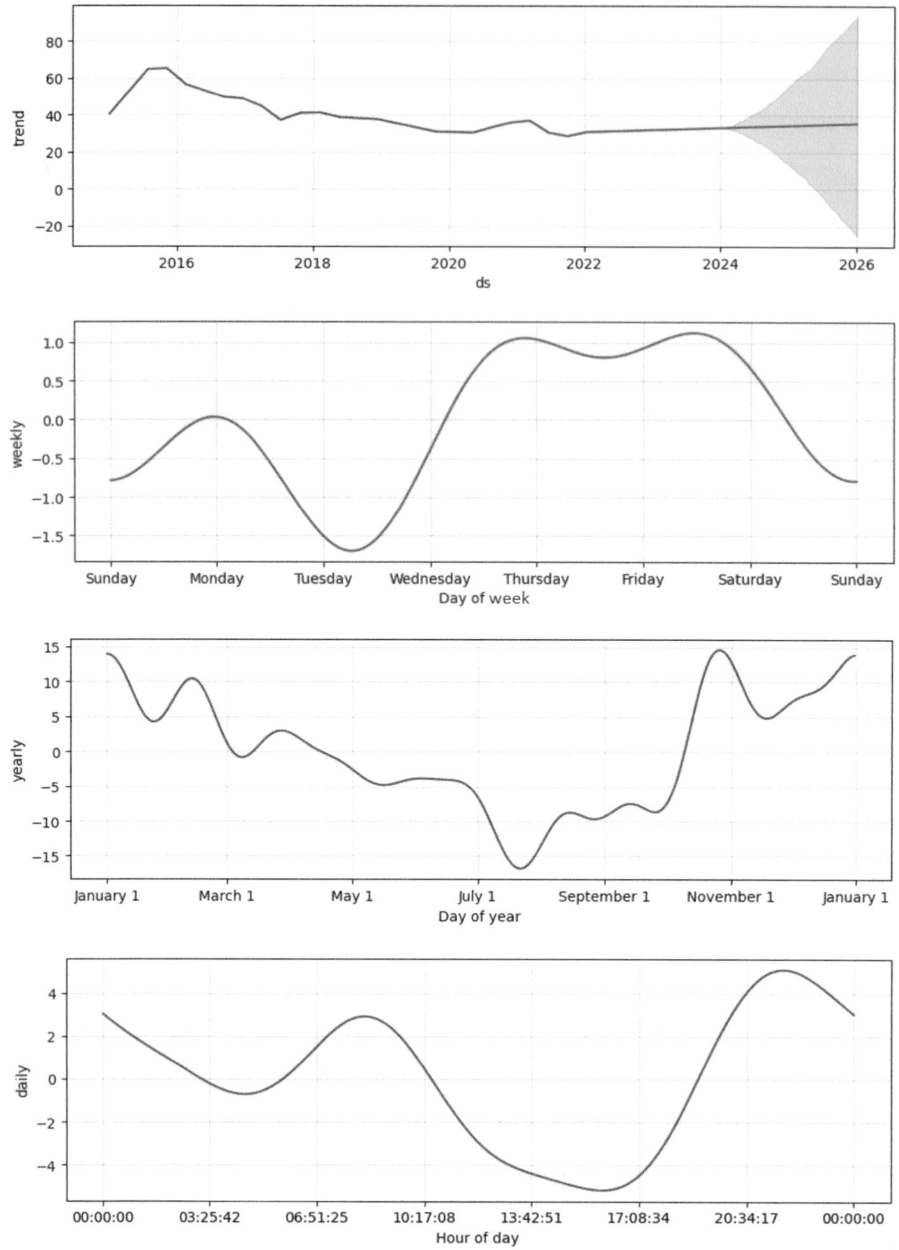

FIGURE 10.14 Trend in the concentration of $PM_{2.5}$ in Hyderabad city on a yearly, daily, and weekly basis.

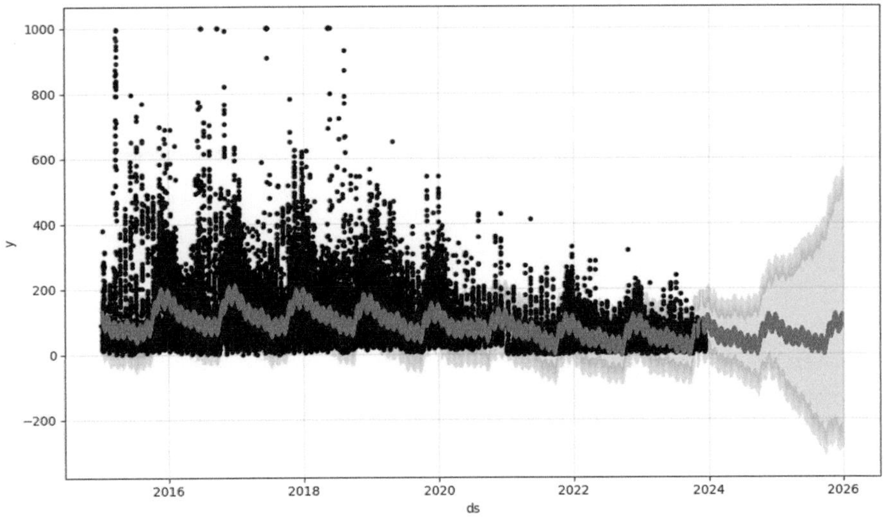

FIGURE 10.15 Trend of $PM_{2.5}$ across the years (2013–2023) with future prediction in coming years (2024–2026).

These predictive analyses can indicate preventive measures that have to be taken to decrease the levels of particulate matter having 2.5 microns ($PM_{2.5}$) to bring the AQI to good criteria.

REFERENCES

Agarwal, R., Awasthi, A., Mital, S. K., Singh, N., & Gupta, P. K. (2014). Statistical model to study the effect of agriculture crop residue burning on healthy subjects. *Mapan*, *29*, 57–65.

Anand, G., Nippani, S. K., Kuchhal, P., Vasishth, A., & Sarah, P. (2017). Dielectric, ferroelectric and piezoelectric properties of Ca0. 5Sr0. 5Bi4Ti4O15 prepared by solid state technique. *Ferroelectrics*, *516*(1), 36–43.

Bishop, C. M., & Nasrabadi, N. M. (2006). *Pattern Recognition and Machine Learning* (Vol. 4, Issue 4). Springer.

CCR. (n.d.). Retrieved December 26, 2023, from https://airquality.cpcb.gov.in/ccr/#/caaqm-dashboard-all/caaqm-landing

Central Pollution Control Board. (2023). *National Ambient Air Quality Status & Trends 2019*. https://cpcb.nic.in/upload/NAAQS_2019.pdf

Chen, M., Mao, S., & Liu, Y. (2014). Big data: A survey. *Mobile Networks and Applications*, *19*(2), 171–209. https://doi.org/10.1007/S11036-013-0489-0/METRICS

CPCB | Central Pollution Control Board. (n.d.). Retrieved December 26, 2023, from https://cpcb.nic.in/

George, P. E., Thakkar, N., Yasobant, S., Saxena, D., & Shah, J. (2024). Impact of ambient air pollution and socio-environmental factors on the health of children younger than 5 years in India: a population-based analysis. *The Lancet Regional Health - Southeast Asia*, *20*, 100328. https://doi.org/10.1016/J.LANSEA.2023.100328

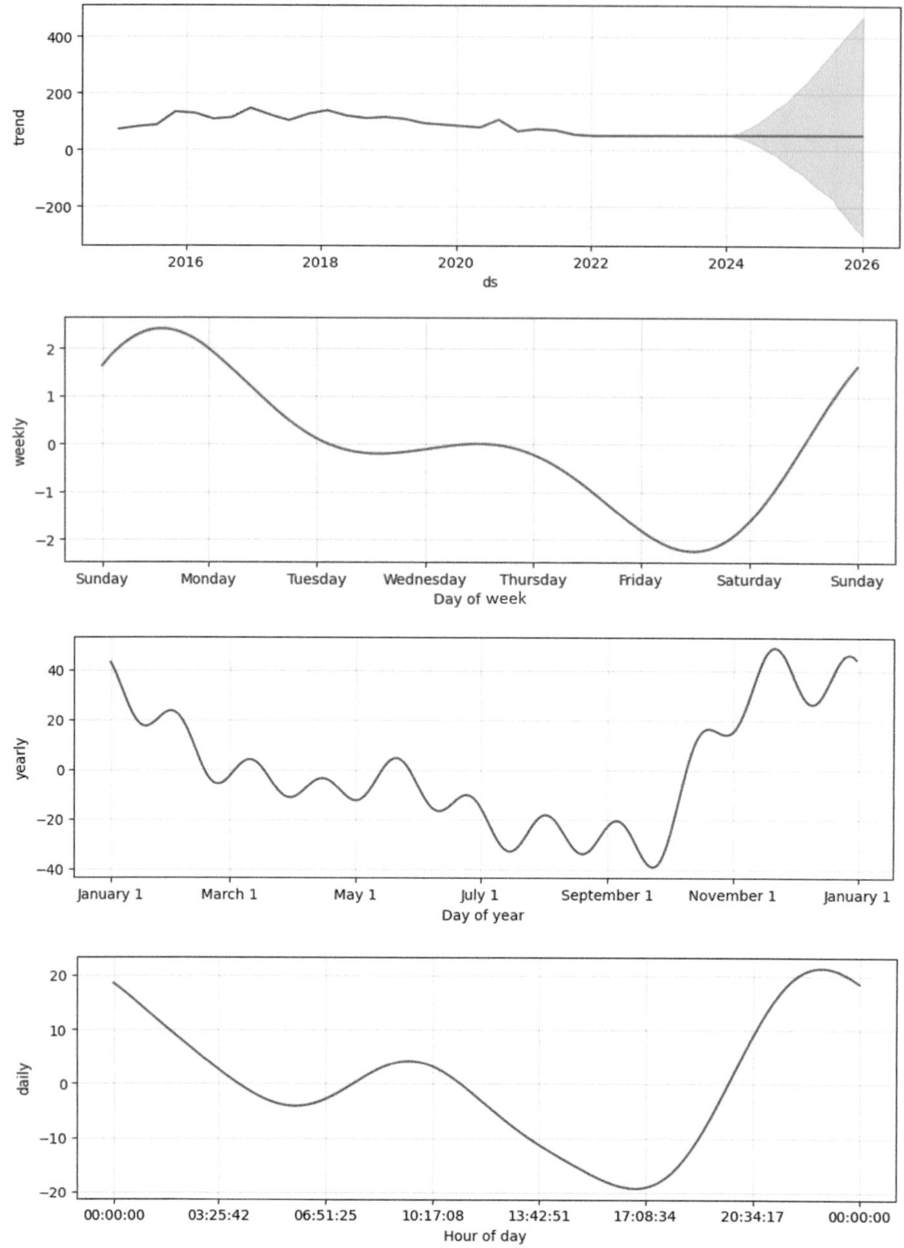

FIGURE 10.16 Trend in the concentration of $PM_{2.5}$ in Lucknow city on a yearly, daily, and weekly basis.

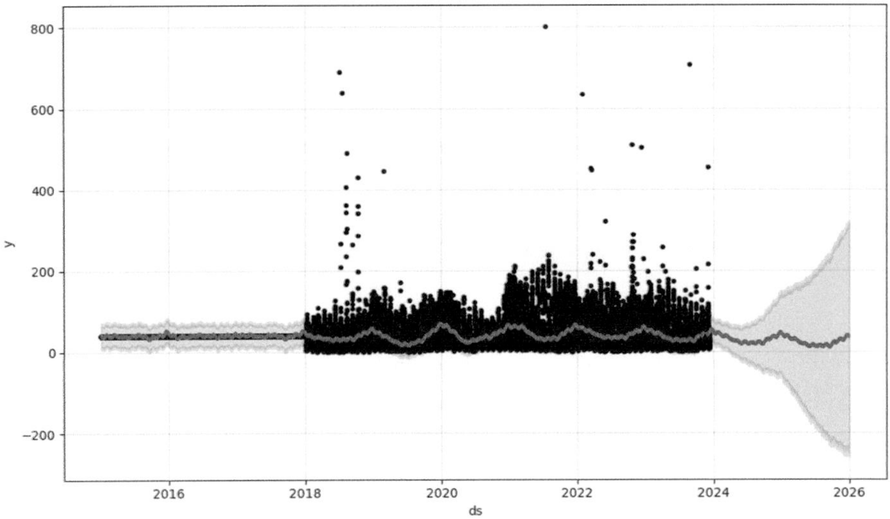

FIGURE 10.17 Trend of $PM_{2.5}$ across the years (2013–2023) with future prediction in coming years (2024–2026).

Glencross, D. A., Ho, T. R., Camiña, N., Hawrylowicz, C. M., & Pfeffer, P. E. (2020). Air pollution and its effects on the immune system. *Free Radical Biology and Medicine*, *151*, 56–68. https://doi.org/10.1016/J.FREERADBIOMED.2020.01.179

Liang, Y. C., Maimury, Y., Chen, A. H. L., & Juarez, J. R. C. (2020). Machine learning-based prediction of air quality. *Applied Sciences 2020*, *10*(24), 9151. https://doi.org/10.3390/APP10249151

Neo, E. X., Hasikin, K., Mokhtar, M. I., Lai, K. W., Azizan, M. M., Razak, S. A., & Hizaddin, H. F. (2022). Towards integrated air pollution monitoring and health impact assessment using federated learning: A systematic review. *Frontiers in Public Health*, *10*. https://doi.org/10.3389/FPUBH.2022.851553/FULL

Northey, S. A., Mudd, G. M., & Werner, T. T. (2018). Unresolved complexity in assessments of mineral resource depletion and availability. *Natural Resources Research*, *27*(2), 241–255. https://doi.org/10.1007/S11053-017-9352-5

Pandey, A., Brauer, M., Cropper, M. L., Balakrishnan, K., Mathur, P., Dey, S., Turkgulu, B., Kumar, G. A., Khare, M., Beig, G., Gupta, T., Krishnankutty, R. P., Causey, K., Cohen, A. J., Bhargava, S., Aggarwal, A. N., Agrawal, A., Awasthi, S., Bennitt, F., … Dandona, L. (2021). Health and economic impact of air pollution in the states of India: The Global Burden of Disease Study 2019. *The Lancet Planetary Health*, *5*(1), e25–e38. https://doi.org/10.1016/S2542-5196(20)30298-9

Pant, A., Sharma, S., & Pant, K. (2023). Evaluation of machine learning algorithms for Air Quality Index (AQI) prediction. *Journal of Reliability and Statistical Studies*, 229-242–229-242. https://doi.org/10.13052/JRSS0974-8024.1621

Pozzer, A., Dominici, F., Haines, A., Witt, C., Münzel, T., & Lelieveld, J. (2020). Regional and global contributions of air pollution to risk of death from COVID-19. *Cardiovascular Research*, *116*(14), 2247–2253. https://doi.org/10.1093/CVR/CVAA288

Ravikiran, U., Sarah, P., Anand, G., & Zacharias, E. (2019). Influence of Na, Sm substitution on dielectric properties of SBT ceramics. *Ceramics International*, 45(15), 19242–19246.

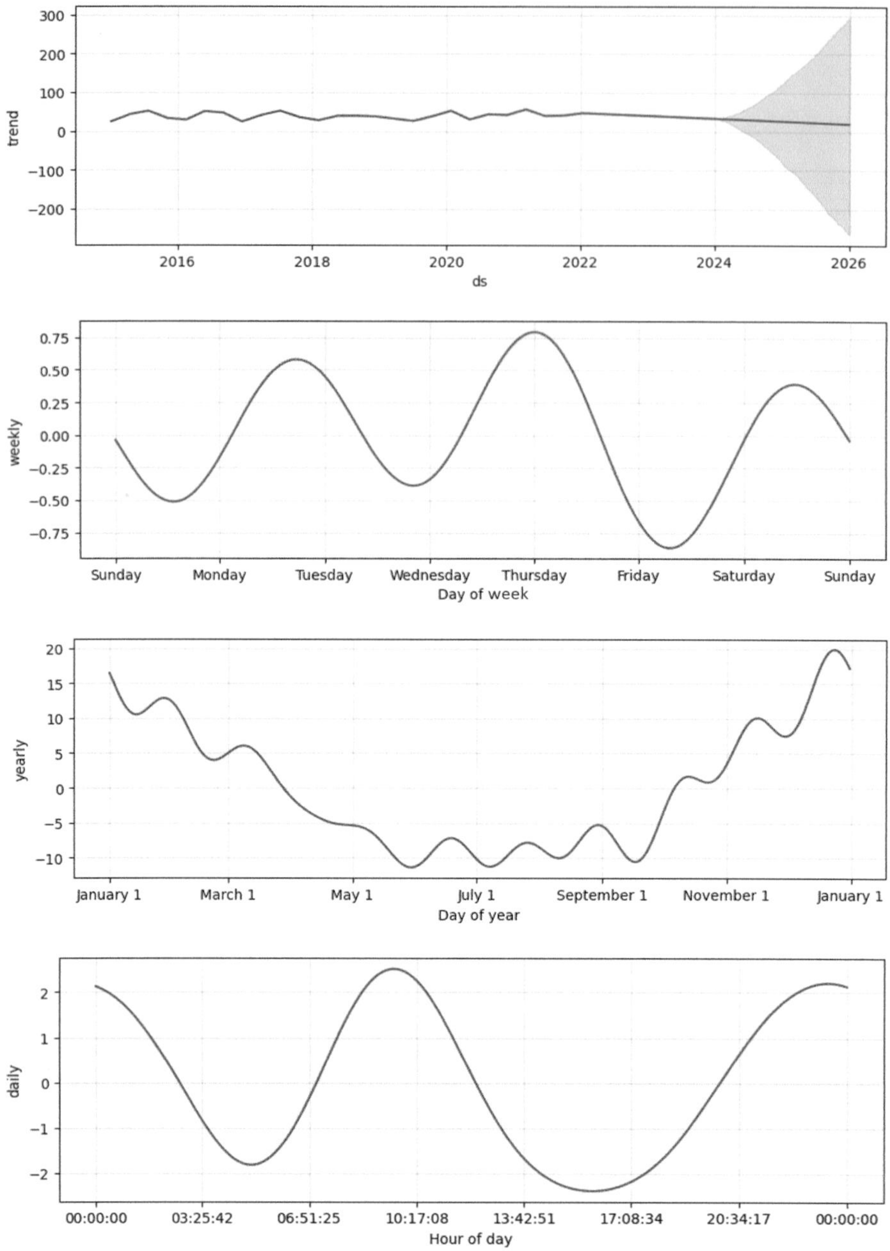

FIGURE 10.18 Trend in the concentration of $PM_{2.5}$ in Mumbai city on a yearly, daily, and weekly basis.

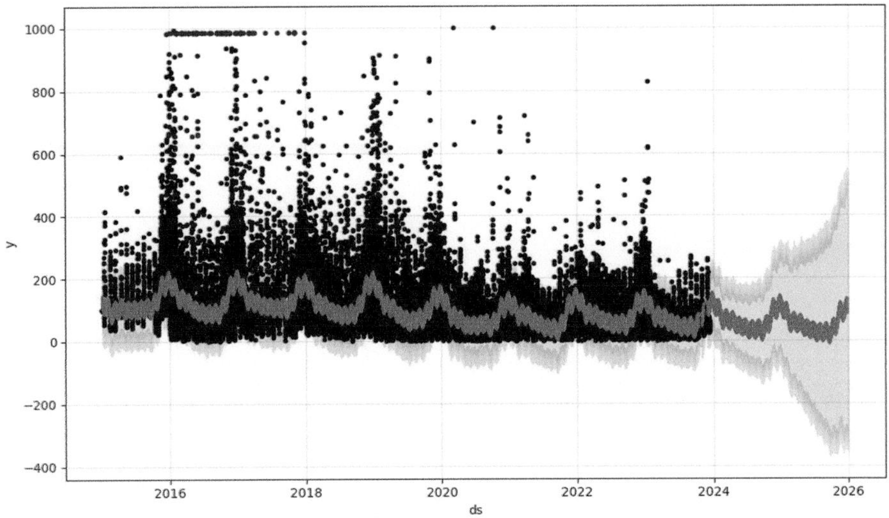

FIGURE 10.19 Trend of $PM_{2.5}$ across the years (2013–2023) with future prediction in coming years (2024–2026).

Sarkar, N., Gupta, R., Keserwani, P. K., & Govil, M. C. (2022). Air Quality Index prediction using an effective hybrid deep learning model. *Environmental Pollution*, *315*, 120404. https://doi.org/10.1016/J.ENVPOL.2022.120404

Singh, N., Agarwal, R., Awasthi, A., Gupta, P. K., & Mittal, S. K. (2010). Characterization of atmospheric aerosols for organic tarry matter and combustible matter during crop residue burning and non-crop residue burning months in Northwestern region of India. *Atmospheric Environment*, 44(10), 1292–1300.

Taylor, S. J., & Letham, B. (2017). *Forecasting at scale*. https://doi.org/10.7287/PEERJ.PREPRINTS.3190V2

The Global Health Cost of PM2.5 Air Pollution: A Case for Action Beyond 2021. (2022). *The Global Health Cost of PM2.5 Air Pollution: A Case for Action Beyond 2021*. https://doi.org/10.1596/978-1-4648-1816-5

Understanding FB Prophet: A Time Series Forecasting Algorithm | by Pratyush Khare | ILLUMINATION | Medium. (n.d.). Retrieved December 26, 2023, from https://medium.com/illumination/understanding-fb-prophet-a-time-series-forecasting-algorithm-c998bc52ca10

Vito, S. De, Piga, M., Martinotto, L., & Di Franca, G. (2009). CO, NO2 and NOx urban pollution monitoring with on-field calibrated electronic nose by automatic bayesian regularization. *Elsevier*, *143*(1), 182–191. https://doi.org/10.1016/j.snb.2009.08.041

Vitousek, P. M. (1994). Beyond global warming: Ecology and global change. *Ecology*, *75*(7), 1861–1876. https://doi.org/10.2307/1941591

When to use Facebook Prophet - Crunching the Data. (n.d.). Retrieved December 26, 2023, from https://crunchingthedata.com/when-to-use-facebook-prophet/

Yadav, N., Rajendra, K., Awasthi, A., Singh, C., & Bhushan, B. (2023). Systematic exploration of heat wave impact on mortality and urban heat island: A review from 2000 to 2022. *Urban Climate*, *51*, 101622.

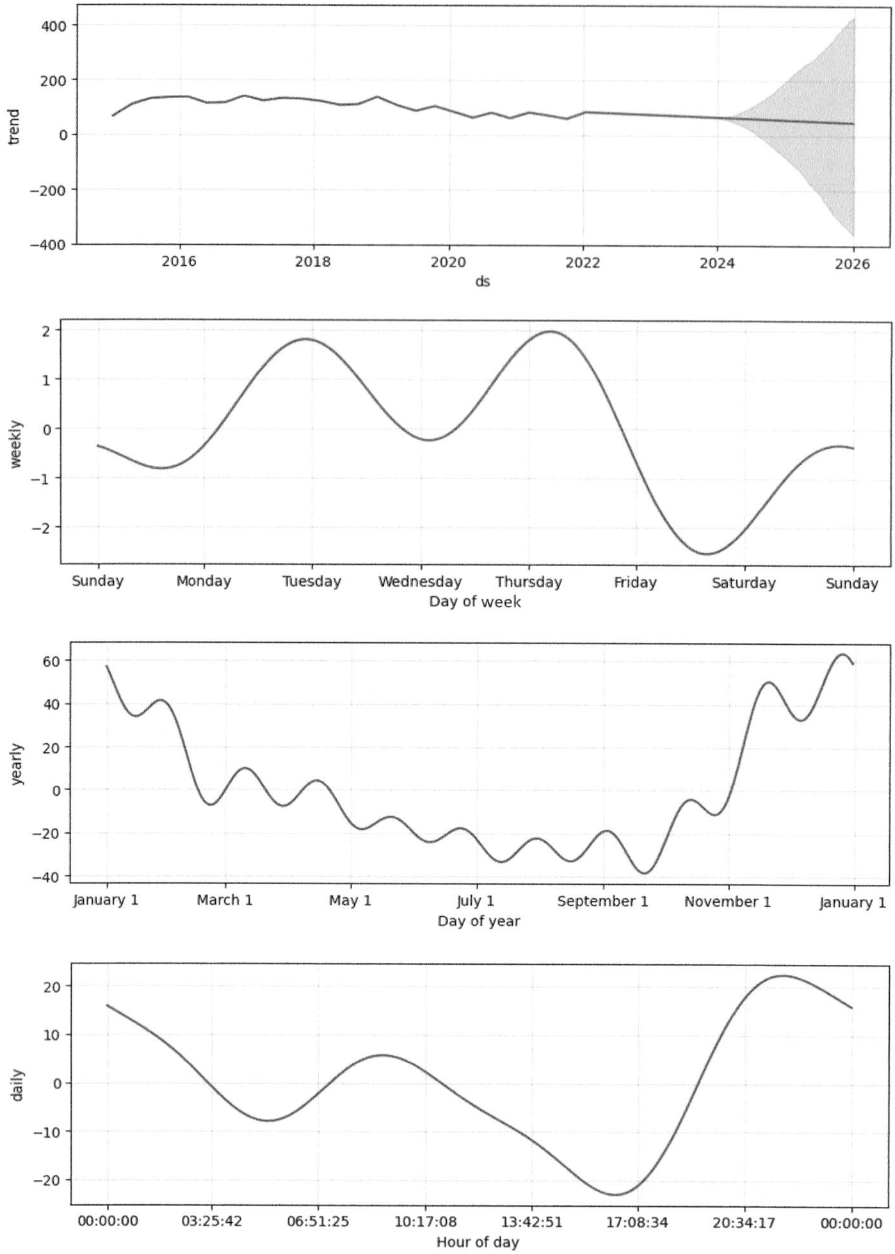

FIGURE 10.20 Trend in the concentration of $PM_{2.5}$ in Patna city on a yearly, daily, and weekly basis.

Yilmaz, O., Kara, B. Y., & Yetis, U. (2017). Hazardous waste management system design under population and environmental impact considerations. *Journal of Environmental Management*, *203*(Pt 2), 720–731. https://doi.org/10.1016/J.JENVMAN.2016.06.015

Zhang, H., Wang, S., Hao, J., Wang, X., Wang, S., Chai, F., & Li, M. (2016). Air pollution and control action in Beijing. *Journal of Cleaner Production*, *112*, 1519–1527. https://doi.org/10.1016/J.JCLEPRO.2015.04.092

Zhang, Q., Jiang, X., Tong, D., Davis, S. J., Zhao, H., Geng, G., Feng, T., Zheng, B., Lu, Z., Streets, D. G., Ni, R., Brauer, M., Van Donkelaar, A., Martin, R. V., Huo, H., Liu, Z., Pan, D., Kan, H., Yan, Y., … Guan, D. (2017). Transboundary health impacts of transported global air pollution and international trade. *Nature*, *543*(7647), 705–709. https://doi.org/10.1038/NATURE21712

Zhang, X., Zhang, H., Wang, Y., Bai, P., Zhang, L., Wei, Y., & Tang, N. (2023). Characteristics and determinants of personal exposure to typical air pollutants: A pilot study in Beijing and Baoding, China. *Environmental Research*, *218*, 114976. https://doi.org/10.1016/J.ENVRES.2022.114976

Zhu, S., Lian, X., Liu, H., Hu, J., Wang, Y., & Che, J. (2017). Daily air quality index forecasting with hybrid models: A case in China. Environmental *P*ollution, 231, 1232–1244.

Zou, B., You, J., Lin, Y., Duan, X., Zhao, X., Fang, X., Campen, M. J., & Li, S. (2019). Air pollution intervention and life-saving effect in China. *Environment International*, *125*, 529–541. https://doi.org/10.1016/J.ENVINT.2018.10.045

SUPPLEMENTARY DATA

Central Pollution Control Board

List of CAAQMS Stations Date: 26-12-2023

S. No.	State	City	Station Name	Controlling Authority	Link to the authorized sites
1.	Gujarat	Ahmedabad	SVPI Airport Hansol, Ahmedabad	Indian Institute of Tropical Meteorology, Ministry of Earth Sciences, Government of India	https://ews.tropmet.res.in/ahmedabad/index.php
2.	Karnataka	Bengaluru	City Railway Station, Bengaluru	Karnataka State Pollution Control Board	https://kspcb.karnataka.gov.in/
3.	Chandigarh	Chandigarh	Sector 22, Chandigarh	Chandigarh Pollution Control Committee	https://cpcc.chd.gov.in/
4.	Tamil Nadu	Chennai	Alandur Bus Depot, Chennai	Central Pollution Control Board	https://cpcb.nic.in/chennai/

(*continued*)

Central Pollution Control Board					
List of CAAQMS Stations Date: 26-12-2023					
S. No.	**State**	**City**	**Station Name**	**Controlling Authority**	**Link to the authorized sites**
5.	Delhi	Delhi	DTU, Delhi	Delhi Pollution Control Committee	www.dpcc.delhigovt.nic.in/#gsc.tab=0
6.	Haryana	Gurugram	Sector-51, Gurugram	Haryana State Pollution Control Board	www.hspcb.org.in/
7.	Telangana	Hyderabad	Ramachandrapuram, Hyderabad	Telangana State Pollution Control Board	https://tspcb.cgg.gov.in/default.aspx
8.	Uttar Pradesh	Lucknow	B R Ambedkar University, Lucknow	Uttar Pradesh Pollution Control Board	www.uppcb.com/
9.	Maharashtra	Mumbai	Chhatrapati Shivaji Intl. Airport (T2), Mumbai	Maharashtra Pollution Control Board	https://mpcb.gov.in/node
10.	Bihar	Patna	IGSC Planetarium Complex, Patna	Bihar State Pollution Control Board	https://bspcb.bihar.gov.in/

11 Air Quality Forecast Using Machine Learning Algorithms

Saurabh Kumar

11.1 INTRODUCTION

In the twenty-first century, air pollution has grown to be a major environmental issue that is receiving the attention of researchers worldwide. Worsening of the air quality standards raises alarms worldwide about their substantial effects on human health and the natural world (WHO, 2016). Most urban regions have seen a sharp rise in air pollution levels due to infrastructural expansion, industrial activity, and higher population density (Yadav et al., 2023; Ma et al., 2019). Global recognition of air pollution problems compels many countries, especially in the Asian, American, and European regions to make the primary goal of their national environmental prediction and monitoring programs (Chen et al., 2018). In order to mitigate air pollution, many methods have been discovered by the scientific community worldwide. A cost-effective method for identifying trends in the state of air pollution both at local and global scale has emerged among these: air quality prediction. Both environmental research and living quality improvement depend on accurate air quality forecasting (Lee et al., 2020). Researchers have focused their emphasis on advanced methodologies for air quality prediction in environmental management plans. On the other hand, without a feasible and precise forecasting model, it is impossible to forecast air quality effectively. The World Health Organization (WHO) refers to air pollution as a "silent killer," causing the early death of approximately 7 million people each year, 600,000 of whom are children.

Because air pollution has had several potentially hazardous effects for individuals, it should be measured regularly to ensure it is efficiently regulated. An effective way to manage the issue of air pollution is to understand its origin, forcing factors, and source. The introduction of harmful or excessive amounts of particular compounds into the atmosphere causes air pollution. Solid particles, liquid droplets, and gaseous substances are examples of such things. There are two types of air pollutants: primary and secondary. The pollutants classified as primary are those that are released into the atmosphere straight from their source. Primary air pollutants can be caused by natural phenomena such as volcanic eruptions, sand storms, and so on, or by anthropogenic activity such as the combustion of fossil fuels, unsupervised incineration of agricultural waste, the leakage of gaseous substances from appliances, waste disposal

DOI: 10.1201/9781032683805-11

burning and so on. Carbon monoxide (CO), nitrogen oxides (NO_X), particulate matters (PMs) and sulfur dioxide (SO_2) are examples of primary pollutants (Singh et al., 2010). Secondary pollutants in the atmosphere are created due to interaction of different primary pollutants. Usually, these interactions entail physical or chemical reactions. Examples of secondary pollutants are secondary organic aerosols (SOA) and photochemical oxidants etc.

One essential indicator for assessing air quality is the air quality index (AQI). The AQI is typically used to display the degree of contaminants in air quality. The AQI is a tool used to estimate the rate at which the current level of air pollution could damage humans. It is essential for evaluating the quality of the air since it measures different pollutants and their possible effects on human health. A thorough understanding of air quality levels is obtained by the computation of AQI, facilitating the adoption of well-informed choices, precautions and initiatives to protect the environment and public health. We can shield individuals from the harmful effects of pollution by alerting them if the AQI rises above the maximum amount. The AQI alerts individuals to changes in the air quality in their area, provides information about which populations may be adversely affected by the air, and suggests preventative measures. Policymakers can take action to lessen the impact of pollution by using air pollution forecasts, which provide information about pollution levels. The following six pollutants are usually taken into account in the calculation of the AQI: carbon monoxide (CO), ozone (O_3), nitrogen dioxides (NO_2), sulfur dioxide (SO_2), particulate matters ($PM_{2.5}$ and PM_{10}).

A piecewise linear function is used to transform pollutant concentrations to calculate the AQI.

The AQI sub-index, denoted as $I_{P,S}$ is calculated for each criteria pollutant and monitoring station using equation (1), considering concentration breakpoints (C_{High} and C_{Low}), pollutant concentration (CP), and corresponding breakpoint indices (I_{High} and I_{Low}).

$$I_{P,S} = I_{High} - I_{Low}\ C_{High} - C_{Low} (C_P - C_{Low}) + I_{Low} \quad \text{......... ....} \quad (11.1)$$

Further using equation 2, the AQI for each station is calculated as the highest value among the sub-indices.

$$AQI_S = Max(I_{O3}, I_{PM10}, I_{PM2.5,} I_{CO}, I_{SO2}, I_{NO2}) \quad \text{..................} \quad (11.2)$$

On a standardized scale, AQIs range from 0 to 500. Higher AQI scores are associated with severe levels of air pollution and higher health risks. Numbers below 50 on the AQI signify "good" air quality, whilst numbers above 300 are classified as hazardous. Table 11.1 illustrates six color-coded categories and corresponding warning messages that make AQI understanding easier for the general audience.

To prevent periods of excessive pollution, for instance, traffic restrictions can be put in place. Many megacities worldwide have adopted traffic limitations such as odd-even, light-duty vehicles and CNG motors.

There are two types of non-deep learning approaches for predicting air quality: statistical approaches and deterministic approaches. To mimic the process of air pollution release, alteration, diffusion, emission and elimination according to air chemical

TABLE 11.1
Classification of AQI

AQI Class	AQI ranges	State	Color code
1	AQI<50	Good	
2	50<AQI<100	Moderate	
3	100<AQI<150	Unhealthy for a particular class	
4	150<AQI<200	Unhealthy	
5	200<AQI<300	Very Unhealthy	
6	300<AQI	Hazardous	

and physical interactions, deterministic approaches use mathematical equations and meteorological concepts. Meteorological parameters can be predicted using the weather research forecast (WRF) model. Furthermore, WRF-Chem, CMAQ, CAMx, Polyphemus, and other CTMs are utilized to model air pollution concentrations. These state-of-the-art CTMs take into account a broad range of variables related to the discharge of air pollutants, including meteorological, anthropogenic, fire, and biogenic emissions, etc. Although these deterministic models offer insightful information on air pollutant prediction, they are limited in terms of data availability and usage restrictions. The nonlinearity between several parameters that are essential for the generation of air pollution is not taken into account. To overcome these shortcomings, scientists have begun using statistical techniques. Statistical techniques, as opposed to deterministic models, reveal nonlinear correlations between several variables that could influence the concentrations of air pollutants. Machine learning (ML) techniques have a significant benefit over conventional statistical methods in that they can handle non-linear information, which leads to improved prediction accuracy. Furthermore, a thorough grasp of the chemical and physical interactions among the air contaminants along with others relevant atmospheric factors are not necessary for the application.

The most widely used approaches for predicting air quality are ML techniques. There are several investigations reported in the scientific literature that suggest application of various models to achieve the highest level of accuracy when predicting pollutant concentrations or the AQI.

The key components of the chapter are as follows:

- Machine learning and regression
- Deep learning algorithm
- Limitations
- Conclusion and future work

ML is a subfield of artificial intelligence that seeks to empower machine to pick up specific skills on their own without needing human input. The foundation of this method lies in the creation of data-driven learning prototypes and generate predictions or judgments in the presence of new data. Deep learning (DL) is an advancement in ML that makes use of an Artificial Neural Network (ANN), which is a multi-layered structure. Because features are automatically extracted, DL algorithms require less human intervention. DL requires a lot of data to function successfully, which sets it apart from other ML techniques.

11.2 MACHINE LEARNING AND REGRESSION

Numerous ML techniques are available for use in addressing various scientific issues. This section discusses ML algorithms that are used to forecast air quality using the technique of regression analysis.

11.2.1 Multiple Linear Regression (MLR)

A group of independent variables and a dependent variable are related to one another through regression analysis (Agarwal et al., 2014). An estimation of the dependent variable is derived from this relationship and the values of the independent variables. Let us consider y is a dependent and x_i where i = 1,…… ..k are the independent variables, respectively. MLR equation can generally be expressed as

$$y = \sum_{i=1}^{k} a_i x_i + \varepsilon \tag{11.3}$$

where $a_1, a_2 \ldots \ldots\ldots, a_k$ are linear regression parameters and ε is an estimated error term. The regression modelling objective is to estimate parameters $a_1, a_2, \ldots \ldots, a_k$. This can be accomplished by applying the minimum square error technique. Regression parameters are estimated with the goal of minimizing mean square error.

$$\mathrm{Min}\left\{\left(y - f\left(x_1, \ldots\ldots, x_k\right)\right)^2\right\}.$$

If k =1, simple linear regression is the name given to the aforementioned regression method. .

11.2.2 Auto-Regressive Integrated Moving Average (ARIMA)

An ARIMA model is a type of statistical techniques for evaluating and projecting time series data. With the inclusion of integration, ARIMA is a generalization of the more straightforward ARIMA. Using ML techniques, the ARIMA model is used to forecast air quality. There are three components in the ARIMA model. Autoregression refers to a technique which utilizes the reliant relation between a recorded and an earlier measure of lagged monitored data. In other words, the number of delays incorporated into the model is known to be referred to as the auto-regressive component. For instance, an auto-regressive value of five indicates that the current value can only be explained by the five values that came before it. The degree of differentiation needed to turn the given time-series into a stationary one is represented by the integrated constituent. For example, the time-series is made stationary by deducting an observation from an observation at the preceding time step. Further moving average is a technique that employs a moving average model's residual error to delayed observations, utilizing the dependence between the observations. To swiftly identify the specific ARIMA model in use, the conventional notation ARIMA (p, d, q) is employed, where integer values are substituted for the parameters.

The parameters of the ARIMA model are defined as follows:

p: The number of delay observations incorporated in the model, also known as the lag order.
d: The number of times that the raw observations are differenced, termed as the degree of differencing.
q: The moving average window size, also called the order of moving average.

11.2.3 Support Vector Regression (SVR)

A support vector machine model used for regression is called support vector regression (SVR), and it was initially introduced by Drucker et al. (1996). SVR, which makes use of Support Vector Machines (SVMs) for regression analysis, is a widespread method for time series prediction. Using SVR, a ML technique, a model can determine the significance of a given variable in describing the relationship between input and output variables. The structured risk minimization principle serves as the foundation for the support vector regression technique. SVR seeks to reduce an upper bound on the generalization error rather than identify empirical shortcomings.

Let us assume S = $\{(x_i, y_i)$: i= 1, 2,……,n} is the collection of n samples, where x_i and y_i are the independent and dependent variables respectively. Then a regression function f(x) can be roughly expressed to represent the non-linear relation between the independent and dependent variables. The SVR model's function can be expressed as

$$f(x) = w.\varphi(x) + b \tag{11.4}$$

Where $x$is treated as the model input vector, $w\, and\, b$ are the coefficient parameters of the function. $\varphi(x)$is the non-linear function. The objective of the SVR algorithm is to find the optimum $w$and$b$along with some arguments of $\varphi(x)$. The optimized solution of the SVR algorithm can be achieved by the solution of the subsequent optimization problem:

$$\min_{w,b,\varepsilon,\varepsilon^*} (1/2)w^2 + c\sum_{i=1}^{n}(\varepsilon_i + \varepsilon_i^*) \tag{11.5}$$

$$subject\ to\ constraints: \begin{cases} y_i - \left(w.\varphi(x_i) + b\right) \le (\varepsilon_i + \varepsilon_i^*) \\ \left(w.\varphi(x_i) + b\right) - y_i \le (\varepsilon_i + \varepsilon_i^*) \\ \varepsilon_i, \varepsilon_i^* \ge 0, i = 1, 2, \ldots\ldots\ldots, n \end{cases} \tag{11.6}$$

Where ε_i and ε_i^* are called as slack variables that measure the training error below and above the tube. ε and c is a positive constant penalty coefficient factor that

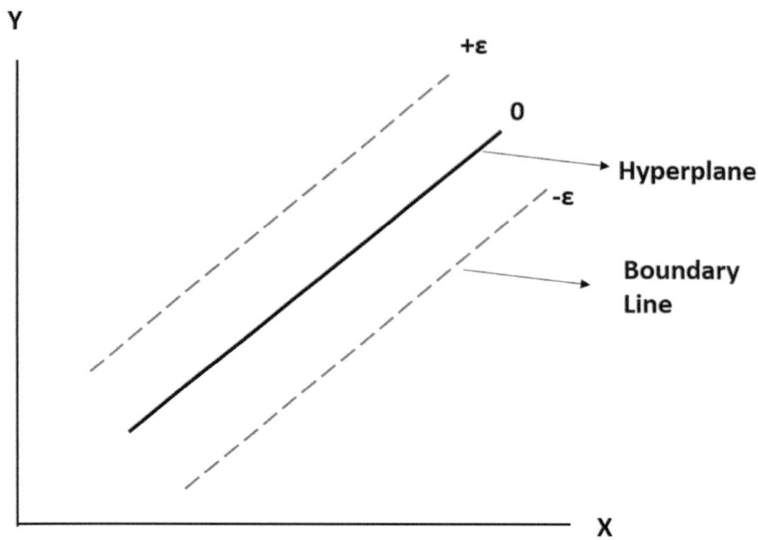

FIGURE 11.1 The parameters for the support vector regression.

determines the degree of penalized loss during training stage. Above, equation (6) represents the constraints equation, where left hand side terms represent the generalization of the model and right-side terms show the risk based on empirical data. The minimization of these two parameters is the aim of SVR.

More precisely, the formulation of SVR as an optimization problem involves identifying the flattest tube containing the majority of the training cases and establishing a convex ε-insensitive loss function to be minimized (Figure 11.1).

SVR has the following substantial benefits over other ML techniques:

I. Because SVR reduces both empirical risk error (the accuracy of fitting the training dataset) and generalization error (the ability to fit new dataset), it has a high capacity for generalization and is flexible enough to handle a variety of regression and classification tasks. An equilibrium between the trained model complexity and the quality of the approximation is sought after by setting up the targets.
II. Since SVR is unperturbed by the dimensions of the input vector space, it can be trained using a limited number of training datasets.
III. Global optimization algorithms can be used to address the SVR regression problem since it can be formulated as a convex problem or a linear function. Consequently, SVR can identify the global optimal solution.

11.2.4 Decision Trees (DT)

Decision trees are included in the category of Supervised Learning. Both non-linear and non-continuous characteristics characterize the decision tree regression. It is an illustration of a function that accepts an attribute values vector as input and outputs

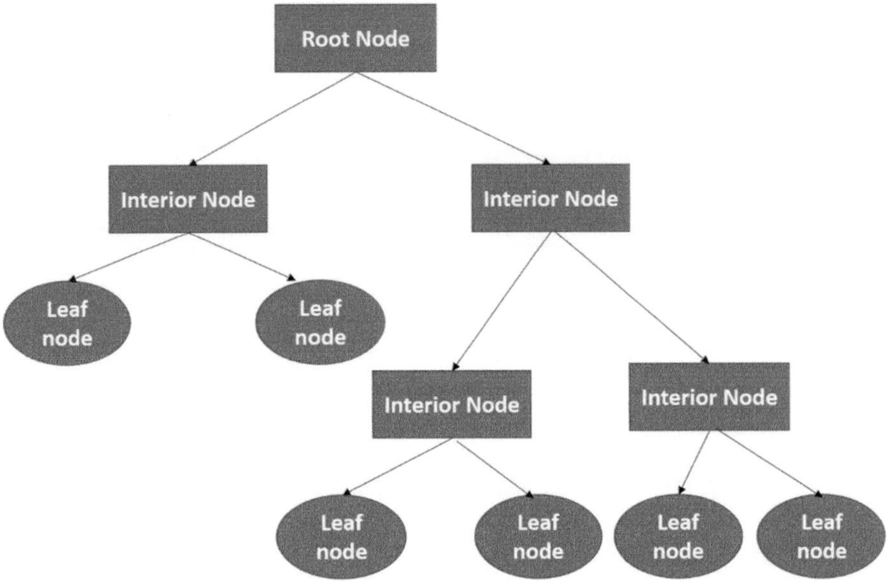

FIGURE 11.2(a) Decision tree structure.

a choice. A decision tree makes a decision by carrying out a number of actions. Decision nodes and leaves make up a tree's structure and recursive division serves as the algorithm's base. A tree is made up of leaves and decision nodes. In order to decide how to divide a node into two or more branches, DT regression typically takes the standard deviation reduction into account. The root node is the first node to be split according to the most important independent variable. This will help to decide the initial choice. The lowest sum of the squared estimate of errors (SSE) for a parameter is taken into consideration as the decision node, and nodes are divided yet another time. The chosen variable's values are used to partition the dataset. The procedure comes to an end when a predetermined threshold for termination is met. The dependent variable prediction is provided by the final nodes, also referred to as departure nodes (Figure 11.2(a)).

11.2.5 Random Forest (RF)

This is one of the most widely used ML methods. A decision tree based integrated prediction model known as a "random forest" combines several decision trees, each of which depends in some way on values from an independently selected random vector. In the random forest, every decision tree has the same distribution. The number of features and the number of trees are the two most significant RF factors. The total number of features denotes the number of characteristics that were chosen arbitrarily for each decision tree and the quantity of decision trees shows the quantity of trees in the forest. This method creates data storage groups by using supervised ensembled algorithms. The classification of a random forest with n number of trees is depicted in Figure 11.2(b).

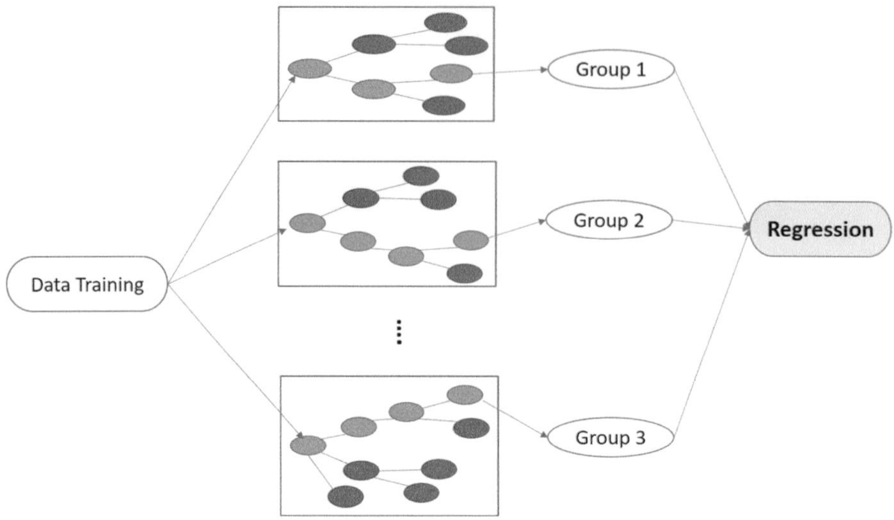

FIGURE 11.2(b) Random forest structure.

11.2.6 K-Nearest Neighbors Regression (KNN)

One ML technique that is useful for the classification is K-nearest neighbor (KNN). While it can be used for regression-related problems as well, the k-nearest neighbors approach is typically employed in classification problems. Its core concept is to use distance measurements between various sample types to predict correlations among samples. The Manhattan distance, Markov distance, Euclidean distance, and other methods were frequently employed to determine the distance. The weighted k-nearest-neighbors (WKNN) algorithm is an advancement of this algorithm. In this instance, the prediction's computation takes a weighted arithmetic mean into account. The general structure of a KNN model is depicted graphically in Figure 11.2(c).

11.3 DEEP LEARNING ALGORITHM

Deep learning (DL) methods have been widely used in big data analysis to solve different types of classification and regression problem, such as computer vision, image classification, speech recognition, time series forecasting, and specifically in the field of air quality prediction (Zhang et al., 2022), due to the rapid development of artificial intelligence and big data techniques. Furthermore, several studies have demonstrated that the DL method can identify the fundamental causes of seasonal changes, regional geographical implications, chemical processes, and other factors that may affect the concentrations of air pollutants (Bai et al., 2019; Zhang et al., 2021). ANNs are used in DL algorithms. This section will provide a brief overview of the many ANN types that have been utilized in the literature to predict air quality.

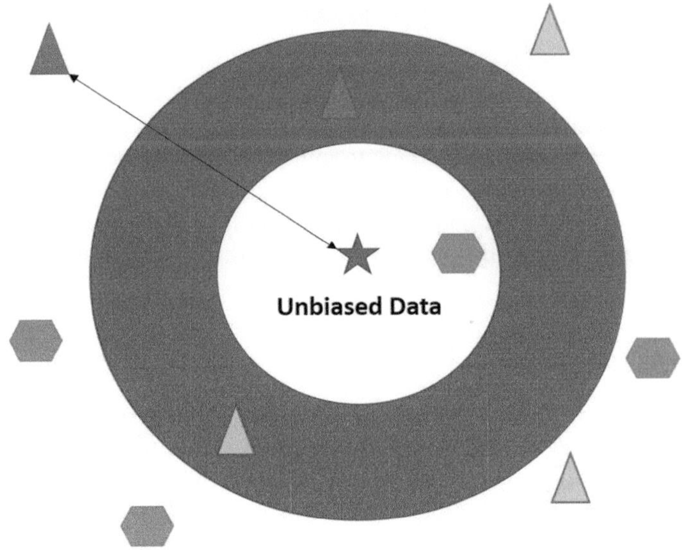

FIGURE 11.2(c) KNN structure.

11.3.1 Multi-Layer Perceptron Neural Networks (MLP)

This is one of the simplest feed-forward networks. MLPs have one or more hidden layers in a typical neural network structure. There are three different types of layers in it: input, hidden and output layers. The signal that has to be processed is received at the input stage. The output layer completes essential functions such as prediction and classification. An arbitrary number of hidden layers placed between the input and output layers contain the MLP's actual calculation. The MLP neurons are trained using the back propagation learning algorithm.

11.3.2 Convolutional Neural Networks (CNN)

CNNs are typically used on visual data. They are an extension of MLP with alternating convolutional and pooling layer architectures.

The purpose of convolutional layers is to collect attributes from the input imagery. A matrix known as a kernel or filter performs a convolutional operation, producing convoluted features as the result. The pooling layers' input consists of these features. The primary objective of pooling layer's is to minimize the dimensionality of the complex characteristics in order to lower the amount of processing power needed to handle the image. CNNs have been used by numerous academics to extract the spatial correlation from data on air pollution concentrations (Wen et al., 2019). Multiscale shift invariant properties of data can be analyzed by CNN. Between two successive convolutional layers, a subsampling process is carried out. Max pooling and mean pooling are two frequently utilized subsampling techniques.

11.3.3 Recurrent Neural Networks (RNN)

The recurrent neural network is a specific kind of ANN which employs successive time-series data. Similar to feedforward and CNNs, RNNs acquire information through data collected for training purpose. RNNs stand out because of their capacity to apply knowledge from earlier inputs to influence the current input and outcome. RNNs possess intrinsic storage in which a given neuron may provide feedback on its own via taking in its own output as input. This makes it possible for the neural network to develop temporary memory, a feature that is necessary for time-series forecasting.

11.3.4 Long-Short Term Memory Neural Networks (LSTM)

The most prevalent problem with normal and deep RNNs involves disappearing or inflating gradients. The loss function's gradient in an RNN with activation functions gets closer to zero while more layers are introduced. If further layers of activation mechanisms are included, the gradient associated with the loss function can reach 0 (vanish), maintaining each function in their original form. This puts a halt to additional training and prematurely concludes the process. As a result, parameters only record immediate dependencies; data from previous time steps can be lost. As a result, the model settles on an untenable result. On the other side exploding gradients may occur because erroneous gradients might be unstable. Hochreiterand and Schmidhuber (1997) introduced LSTM networks, which solve the problem of disintegrating and surging gradients. An expansion of RNNs is an LSTM. Their ability to manage long-term dependency stems from their larger memory. Information can be retained by LSTMs for any length of time. The cell state, which contains information and is processed with the data, is the fundamental part. These gates stop errors from vanishing or blowing up and allow for extra adjustments to accommodate for non-linearity. The new data that will be utilized to update the information is determined by the gate of input. The forget gate determines which data from earlier stages should be discarded. The input gate generates a vector candidate using a hyperbolic-tangent function. The output gate, which is the final gate, selects which portion of the modified cell state will be sent out using a tanh function of activation. A more straightforward form of LSTMs is known as a gated recurrent unit (GRU). GRU integrates the input and forget gates. They produce outcomes that are comparable to those of LSTMs.

11.3.5 Encoder–Decoder Neural Networks (EDNN)

The Encoder–Decoder neural network (EDNN) is a kind of RNN which is utilized primarily for forecasting challenges that vary from one sequence to another. EDNN is made up of three parts: an encoder, an intermediate vector, and a decoder. Standard RNNs, LSTMs, GRUs, and CNN can all be utilized as encoders. Again, a common strategy is to employ an LSTM or GRU-based RNN to initialize the decoder's first hidden state. To improve the precision of the decoder forecasting, each encoder unit evaluates a segment of the input pattern and attempts to capture the related memory within the transitive vector. The context that the encoder provides and any items that the decoder generates for previous states serve as guidelines for this incremental process. There are several studies where an encoder-decoder network structure is used within the RNN-based forecast of air quality. Mathematically the encoder–decoder system can be expressed as:

$$h = w\left(\rho_t ; \beta_1\right) \tag{11.7}$$

$$X_{t+1;t+s} = f\left(h; \beta_2\right) \tag{11.8}$$

Where w is the encoder and f is the decoder; ρ_t is the input data at the given t timestep; h represents the semantic vector output from the encoder; $X_{t+1;t+s}$ is the s-timestep forward forecast data; β_1 *and* β_2 are trained parameter.

Apart from the machine and DL techniques outlined above, numerous other algorithms of this type have been applied worldwide to predict air quality. Recently, Ravindiran et al (2023) utilized several ML models, including Adaptive boosting (Adaboost), Catboost, eXtreme Gradient Boosting (XGBoost), Light Gradient Boosting Machine (LightGBM) and random forest (RF) for the AQI forecasting over the Indian coastal city Visakhapatnam. Based on the results, the Catboost model performed better than the other models. Ayus et al (2023) utilized five distinct DL methods such as RNNs, Bidirectional Gated Recurrent unit (Bi-GRU), Bidirectional Long Short-Term Memory (BiLSTM), Convolutional Neural Network BiLSTM (CNN-BiLSTM), and Convolutional BiLSTM (Conv1D-BiLSTM) to predict the AQI for ten major cities of China. Comparisons have been made between these models' performance and that of a ML model XGBoost. The findings imply that XGBoost, performs better than the DL models. In Tianjin, China, Pan (2018) employed XGBoost to forecast $PM_{2.5}$. The MLR models XGBoost, Random Forest, MLR, DTR, and SVR were compared. XGBoost fared better than the other models with comparatively better statistical metrics. A variety of statistical indices, including Correlation Coefficient (R), Index of Agreement (IoA), Mean Square Errors (MSE), Mean Absolute Error (MAE) and Root Mean Square Error (RMSE) are generally adopted to assess the validation of various models.

11.4 LIMITATIONS

11.4.1 Data Availability

ML-based models need a large set of data to get good forecast outcomes, but there isn't always enough data to anticipate air quality. For instance, extreme circumstances, such as bad weather conditions and equipment maintenance can disrupt data collection, especially in high pollution episodes. In many places, air-related statistics are only occasionally gathered because of inadequate air-quality monitoring systems. The effectiveness of ML models has been undermined by a lack of training data.

11.4.2 Computational Cost

DL-based models typically require GPUs or TPUs rather than CPUs to perform computation. Additionally, they require a lot more memory due to the increasing number of parameters. It would be worthwhile to investigate more ways to guarantee that DL models do not use excessive amounts of processing power while training and testing outcomes.

11.4.3 Multi-Dimension Data

The majority of the raw data used by existing ML and DL algorithms to quantify the air quality are in the form of one dimension, and additional effect parameters such as population density, land usage, forest fire, and the rate of forest cover have not been given enough weightage. Air pollution prediction can be improved using multi-dimensional data that fuses additional factors, but the costs are very expensive. The "curse of dimensionality" and high computing costs are unaffordable for current algorithms, and the issue of how to select the right number of acceptable factors needs to be resolved.

11.4.4 Long-Range Forecasting and Comprehension Issue

The majority of ML and DL techniques concentrate on short- or medium-term air quality forecasting. Long-term forecasting necessitates more intricate and adaptable spatiotemporal relationships, as well as more unidentified components. Its methodology is likewise different from that of short or medium-range forecasting, and it may include several, as-yet-uncertain attributes and may lessen the influence of the previous dataset. On the other hand, ML and DL models have been plagued by the "black box" issue since their inception. Their intricate structure and numerous characteristics make their operations difficult to understand. The problem of building an ML and DL model with great comprehensibility needs to be addressed immediately.

11.5 CONCLUSION AND FUTURE WORK

Relying on ML methods, the scientific community prefers DL algorithms over regression techniques. Moreover, a few hybrid algorithms combine the two categories of algorithms into a single model. The most often used DL algorithms are MLP and LSTM. CNN, RNN, GRU, and auto-encoders are other, less commonly utilized DL techniques. RF and SVR are particularly often utilized regression techniques. Less commonly used alternative regression algorithms include DT, ARIMA, KNN, and Boosting. Recently, Balogun et al. (2021) have presented a thorough analysis of the relationship across the air pollution and climate change to create ML methods and algorithms which improve the alert systems and enable an efficient countermeasure to air pollution brought on by climate change, ultimately promoting ecologically sound cities and societies as a whole. In recent times there are various ML models and algorithms that have been developed by scientists across the globe to improve air quality forecasting. Recently the Deep Transfer Network has been used to forecast O_3 levels in Madrid (Méndez et al., 2022) and $PM_{2.5}$ prediction in China (Zhang et al., 2022). Graph Neural Networks are another intriguing class of models that have recently gained favor in air quality forecasting. Li et al. (2021) have conducted a study whereby they utilized Graph Neural Networks to forecast $PM_{2.5}$ concentration. The primary feature of this kind of network is the data gathered from the dynamic interactions across the various streets, neighborhoods, or cities, which are weighted based on distance. Most recently, $PM_{2.5}$ levels have been forecasted using temporal convolutional networks (TCNs) algorithm (Ni et al., 2023). Academics working in the area of forecasting air quality may find this chapter to be a helpful resource since it offers an overview of the topic and research developments in the area.

To continuously forecast the pollution levels of the following day, it would be extremely beneficial for future work in this area if these models were applied to online data generation. In this scenario, data daily readings would be fed into the model. Data on air quality is primarily sourced from national monitoring stations. It is insufficient, therefore scientists have suggested creating new data collection tools, like low-cost air pollution sensors and smart mobile pollution monitoring devices. In addition, the diverse dataset can be combined to create a substantial database of information for additional purposes. Given that different approaches may be available for a given prediction task, a set of widely recognized selection criteria is still necessary. It's still unclear how to select the best option for a certain assignment in order to get the best feasible forecasting accuracy. Sometimes environmental factors or device malfunctions cause raw data on air quality that monitoring stations record to differ from actual values. The final prediction made by ML techniques may be impacted by these noise-contaminated data. To minimize this drawback, the handling of data and forecasting models are now carried out through two distinct processes. It is of the utmost importance to develop a strong model that might handle this issue efficiently. One interesting pathway for further study on ML based air quality forecasting is the integration of a function for predicting air quality into the modern urban framework. Furthermore, given the potential of these algorithms, creating a real-time AQI forecasting tool that can make use of such models would be worthwhile. These technologies have the potential to provide policymakers with current and precise data, empowering them to execute prompt and efficient actions aimed at enhancing air quality. More research and exploration are necessary to determine the scalability and applicability of these algorithms in different contexts.

REFERENCES

Agarwal, R., Awasthi, A., Mital, S.K., Singh, N., Gupta, P.K. (2014). Statistical model to study the effect of agriculture crop residue burning on healthy subjects. *Mapan,* 29, 57–65.

Ayus, I., Natarajan, N., Gupta, D. (2023). Comparison of machine learning and deep learning techniques for the prediction of air pollution: A case study from China. *Asian J. Atmos. Environ.* 17, 4. https://doi.org/10.1007/s44273-023-00005-w

Bai, Y., Li, Y., Zeng, B., Li, C., Zhang, J. (2019). Hourly pm2. 5 concentration forecast using stacked autoencoder model with emphasis on seasonality. *J. Clean. Prod.* 224, 739–750.

Balogun, A-L., Tella, A., Baloo, L., Adebisi, N. (2021). A review of the inter-correlation of climate change, air pollution and urban sustainability using novel machine learning algorithms and spatial information science. *Urban Climate*, 40, 100989.

Chen, L.J., Ho, Y.H., Hsieh, H.H., Huang, S.T., Lee, H.C., Mahajan, S. (2018). ADF: An anomaly detection framework for large-scale PM2.5 sensing systems. *IEEE Internet Things Journal*, 5(2), 559–570.

Drucker, H., Burges, C.J.C., Kaufman, L., Smola, A.J., Vapnik, V.N. (1996). Support Vector Regression Machines. In *Proceedings of the NIPS on Advances in Neural Information Processing Systems*, Denver, CO, USA, pp. 155–161, 3–5 December 1996.

Hochreiterand, S., Schmidhuber, J. (1997). Long Short-Term memory. *Neural Comput*. 9(8).

Lee, M., Lin, L., Chen, C.Y., Tsao, Y., Yao, T.H., Fei, M.H., Fang, S.H. (2020). Forecasting air quality in Taiwan by using machine learning. *Scientific Reports*, 10(1), 4153.

Li, D., Yu, H., Geng, Y. A., Li, X., & Li, Q. (2021, December). Ddgnet: A dual-stage dynamic spatio-temporal graph network for pm 2.5 forecasting. In *2021 IEEE International Conference on Big Data (Big Data)* (pp. 1679–1685). IEEE.

Ma, J., Cheng, J.C.P., Lin, C., Tan, Y., Zhang, J. (2019). Improving air quality prediction accuracy at larger temporal resolutions using deep learning and transfer learning techniques. *Atmos Environ*. 214, 116885.

Méndez, M., Montero, C., Núñez, M. (2022). Using deep transformer based models to predict ozone levels. In *Intelligent Information and Database Systems—14th Asian Conference, ACIIDS 2022, Part I. Lecture Notes in Computer Science*, vol. 13757, pp. 169–182. Springer.

Ni, S., Jia, P., Xu, Y., Zeng, L., Li, X., Xu, M. (2023). Prediction of CO concentration in different conditions based on gaussian-TCN. *Sens Actuators B*, 376, 33010.

Pan, B. (2018). *Application of XGBoost Algorithm in hourly PM2.5 Concentration Prediction in IOP Conference Series Earth and Environmental Science*, Harbin, China, vol. 113, pp. 12127–12135. IOP.

Ravindiran, G., Hayder, G., Kanagarathinam, K., Alagumalai, A., Sonne, C. (2023). Air quality prediction by machine learning models: A predictive study on the indian coastal city of Visakhapatnam. *Chemosphere,* 338, art. no. 139518., doi: 10.1016/j.chemosphere.2023.139518

Singh, N., Agarwal, R., Awasthi, A., Gupta, P.K., Mittal, S.K. (2010). Characterization of atmospheric aerosols for organic tarry matter and combustible matter during crop residue burning and non-crop residue burning months in Northwestern region of India. *Atmospheric Environ*. 44(10), 1292–1300.

Wen, C., Liu, S., Yao, X., Peng, L., Li, X., Hu, Y., Chi, T. (2019). A novel spatiotemporal convolutional long short-term neural network for air pollution prediction. *Sci. Total Environ.* 654, 1091–1099.

Yadav, N., Rajendra, K., Awasthi, A., Singh, C., Bhushan, B. (2023). Systematic exploration of heat wave impact on mortality and urban heat island: A review from 2000 to 2022. *Urban Climate,* 51, 101622.

Zhang, B., Zou, G., Qin, D., Lu, Y., Jin, Y., Wang, H. (2021). A novel encoder-decoder model based on read-first lstm for air pollutant prediction. *Sci. Total Environ*. 765, 144507.

Zhang, C., Liu, C., Li, B., Zhao, F., Zhao, C. (2022). Spatiotemporal neural network for estimating surface NO2 concentrations over north China and their human health impact. *Environ. Pollut*. 119510.

12 Deep Learning Approaches in Air Quality Prediction

Prabhjot Kaur, Soni Chaurasia, Mamta Bansal, Tanupriya Choudhury, and Ayan Sar

12.1 INTRODUCTION

Deep learning (DL), is an AI machine learning subset, which is recognized as a broader AI field. It thrives in the creation of networks, namely neural networks, with the intention of doing tasks without directly programming them. The term DL was coined for using a certain deep architecture for neural networks as they include many different layers that are interconnected with each other and formed of artificial neurons (or nodes) (Goodfellow et al., 2016). DL is a subset of Machine Learning (ML), which is a subset of AI, as presented in Figure 12.1. The ML technique has predominantly been employed in the computation/prediction of various metrological parameters (Tandon et al., 2023). However, its computational efficiency degrades dramatically with the usage of larger datasets. However, the domain of DL overcomes this limitation. Using DL, air quality monitoring and forecasting can be refined taking into account the ability of DL to handle the infrastructure of complex and nonlinear relationships among environmental data. There are several ways in which DL can be utilized in this context such as: sensor data analysis, forecasting, e.g., temporal and spatial pattern analysis, time series prediction, early warning systems, and predictive modelling, which have been included in the literature review (Z. Zhang et al., 2022). For example, through the processing of information and with the analyzing of the data collected by different sensors of air quality, the DL algorithms might be monitored for obtaining different kinds of data on all components of air pollution that would certainly include Particulate Matter (PM), nitrogen dioxide, ozone, sulfur dioxide and lastly carbon monoxide. These models are believed to be more capable of detecting and making the proper use of temporal and spatial patterns which might emerge from air pollution data and ultimately capturing the interconnectedness and dynamics of all the features exhibited over the time and space. More or less, the predictions related to the time series are also made with the use of RNNs (Recurrent Neural Networks) or gradually LSTMs (Long-short-term-Memory), which efficiently allowed the forecasting of concentrations of different pollutant concentrations at every specified time intervals (Xu & Yoneda, 2021). Also, DL models can be used to predict air quality with the use of multivariate data of different environmental

DOI: 10.1201/9781032683805-12

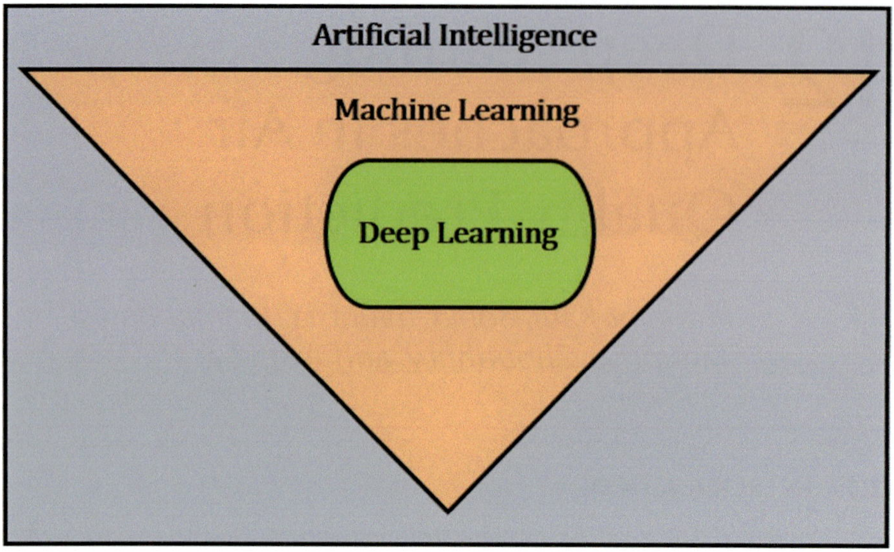

FIGURE 12.1 AI, ML and DL context.

Source: Adapted from Goodfellow et al., 2016.

factors obtained from environmental monitoring devices. The Convolutional Neural Networks (CNNs) are trained to process the satellite imagery to detect the air quality distance wide. These PoMs can trace the patterns inevitable from pollution sources and changes in wild species that may in the end influence air quality. DL models also can perform unsupervised data training that enables them to detect the presence of outliers and uncommon patterns in the air quality data. Via this, one will be in a position to spot all pollution events and abnormalities which require immediate action.

Providing air quality data along with weather prediction using integrated models will help to improve the accuracy and suitability of predictions. The DL model can successfully combine these multiple data sets and encapsulate intricate effects of its numerous contributors influencing air quality. DL is able to utilize mobile and IoT devices with air quality sensors (Bekkar et al., 2021). It can process data in a similar manner to how network traffic is collected from the web (Kaur et al., 2018). It allows for providing timely data and further develops the system of monitoring by introducing dynamics to it. The DL models can be incorporated into the early warnings system to give alerts when the levels of air quality are beyond the prescribed limits and hence enable timely response.

12.1.1 Motivation

The motivation is the absence of a reliable site survey on the path of these two disciplines – deep learning and air quality research – as outlined by the author. This chapter aims to supplement the research on the topic by delivering a summary, analysis, and description of the current state of the field through which researchers and practitioners are able to gain deeper insights into the relevant research.

12.1.2 Contribution

This review brings forth the following contributions:

a) From cultivating the strengths as well as providing reliable summing up of multiple DL approaches used in research on air quality.
b) The datasets used for simulation in this domain are put fourth in a separate section.
c) To execute performance analysis of the current state-of-the-art techniques, here we will put our efforts.

The subsequent segments of this chapter encompass various aspects: Section 2 presents a survey of existing DL models and techniques, Section 3 delves into the performance analysis of these models, Section 4 outlines the details of datasets used for simulation in this domain, Section 5 encompasses a discussion on challenges faced in employing DL models for monitoring and predicting air quality, and finally, Section 6 concludes the chapter while also outlining future directions.

12.2 STATE-OF-THE-ART DEEP LEARNING (DL) MODELS FOR AIR QUALITY PREDICTION

In this section, we delve into the realm of state-of-the-art DL models designed specifically for air quality prediction. Leveraging advanced DL architectures and sophisticated algorithms, these models showcase remarkable performance in forecasting pollutant levels, offering significant advancements in environmental monitoring and mitigation strategies. The diagrammatic representation of various categories of DL techniques is shown in Figure 12.2. The DL techniques are not limited to 21 as shown in figure; instead there are continuous developments in this field and we can see many more techniques in future. Table 12.1 consists of the acronyms and corresponding expansion that are used in this chapter.

12.2.1 Long Short-Term Memory Networks (LSTMs)

LSTMs are a type of recurrent neural network (RNN) designed to capture long-term dependencies in sequential data. They are well suited for time series prediction, making them effective for forecasting air quality over time. This was developed in the 1980s (Werbos, 1988). There are several advancements made in the LSTM networks; one such is transfer learning based stacked bidirectional long short-term memory (TLS-BLSTM) network for air quality prediction. In the TLS-BLSTM model, existing knowledge acquired using a pre-trained model is utilized to predict air quality for new stations and this model achieved a significant ~35% lower RMSE in comparison to recent models in this area (Ma et al., 2020). Another study is conducted

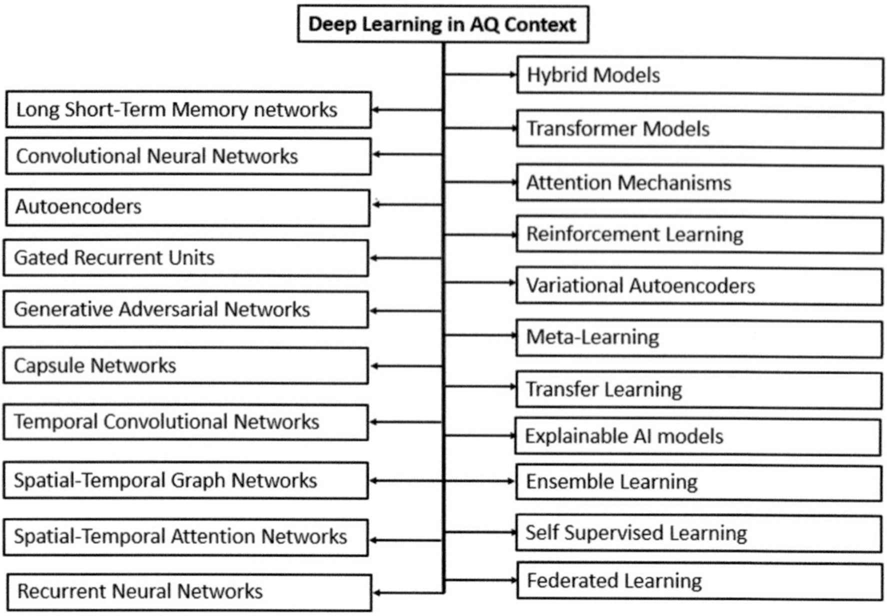

FIGURE 12.2 Deep learning (DL) techniques AQ context.

TABLE 12.1
Acronym and Corresponding Expansion

Acronym	Expansion
AI	Artificial Intelligence
AQ	Air Quality
AQI	Air Quality Index
CapsNet	Capsule Network
CNN	Convolutional Neural Networks
CO	Carbon Monoxide
CR-LSTM	Convolutional Recursive Long Short-Term Memory
DL	Deep Learning
R2	Coefficient Of Determination
DAE	Deep Autoencoder
DAEDN	Denoising Autoencoder Deep Network
EvaNet	Extreme Value Attention Network
GAN	Generative Adversarial Networks
GNN	Graph Neural Network
GRU	Gated Recurrent Unit
ILSTM	Improved Long Short-Term Memory
IoT	Internet Of Things
LSTM	Long Short-Term Memory

TABLE 12.1 (Continued)
Acronym and Corresponding Expansion

Acronym	Expansion
LSTM-FC	Long Short-Term Memory – Fully Connected
MAE	Mean Absolute Error
MAPE	Mean Absolute Percentage Error
NO_2	Nitrogen Dioxide
O_3	Ozone
PM	Particulate Matter
RL	Reinforcement Learning
RMSE	Root Mean Square Error
RNN	Recurrent Neural Network
SO_2	Sulphur Dioxide
STA-LSTM	Spatiotemporal Attention Mechanism
SAE	Stacked Autoencoder
SMAPE	Symmetric Mean Absolute Percentage Error
TCN	Temporal Convolutional Network
TLS-BLSTM	Transfer Learning Based Stacked Bidirectional Long Short-Term Memory
UCI	University Of California Irvine
VAE	Variational Autoencoder
XAI	Explainable AI

over Delhi region in India to predict the O_3, $PM_{2.5}$, NOx, and CO concentrations using LSTM network (Krishan et al., 2019). Another prediction is carried out in Seoul, Korea with 10-day PM_{10} and $PM_{2.5}$ concentrations to obtain RMSE of 15.03 and 15.43 respectively using LSTM batch size of 64 (Xayasouk et al., 2020). Similarly, in Australia's Queensland, the concentration level of particulate matter is measured and the corresponding RMSE value close to ~13% is obtained using Convolutional Neural Network based LSTM network (Sharma et al., 2020). Another study in China, showed the results using LSTM-FC concept (Zhao et al., 2019).

12.2.2 Convolutional Neural Networks (CNNs)

CNNs are particularly useful for processing spatial data, such as satellite imagery. They can be applied to analyze images related to air quality, identifying patterns associated with pollution sources or changes in environmental conditions. Zhang et al. used the CNN model in conjunction with LSTM and obtained MAE, RMSE and R^2 values of 3.17, 5.46 and 8.45 percentage respectively (J. Zhang & Li, 2022). Another study conducted by Duan et al. using the CNN-LSTM model resulted in RMSE, MAE and R2 values of 7.5, 5.285, and 0.989 respectively (Duan et al., 2023). Wang et al. used an improved LSTM with CNN to obtain better results of 8.4, 202.19 and 0.96 for MAE, MSE, and R2 performance metric respectively (J. Wang, Li, Jin, et al., 2022). Another CNN-BiLSTM based model has outperformed existing basic

CNN models with hourly forecast of air quality index, whereby MAE, RMSE and R2 values are obtained as 5.9, 9.2 and 0.974 respectively (J. Wang, Li, Wang, et al., 2022).

12.2.3 Recurrent Neural Networks

Chang-Hoi et. al predicted the value of $PM_{2.5}$ based on RNN and obtained an accuracy value of ~81% (Chang-Hoi et al., 2021)(Tsai et al., 2018). Belavadi et al. conducted a study in India for air quality parameters such as CO, NH3, NO_2, OZONE, PM_{10}, $PM_{2.5}$ and SO_2 and their corresponding RMSE value obtained is 3.14, 0.23, 1.02, 5.11, 4.72, 4.96, and 1.04 respectively (Belavadi et al., 2020).

12.2.4 Autoencoders

Autoencoders can be used for feature extraction and dimensionality reduction. By compressing the input data into a lower-dimensional representation and then reconstructing it, autoencoders can learn meaningful features that contribute to air quality predictions. Autoencoders can also be used with other DL models and once such technique is proposed with the combination of autoencoders with LSTM networks (Xu & Yoneda, 2021). Xu and Yoneda presented performance of their proposed model using RMSE, MAPE, and SMAPE with predicted values of 14.52, 8.22, and 29.37 respectively (Zhao et al., 2019). Cai et. al proposed a model named denoising autoencoder deep network (DAEDN) based on autoencoders and LSTM networks (Cai et al., 2020).

12.2.5 Gated Recurrent Units (GRUs)

GRUs – the counterparts of LSTMs but featuring a little more simple architecture – are another type of RNN. They are able to make the duration dependence explicit in sequential data via sequential prediction tasks. They are, thus, particularly useful for detecting the air quality. In China the value of RMSE is calculated as approximately ~18% of GRU and LSTM by the help of the forecasters that predict the CH 4 and even $PM_{2.5}$ concentration (B. Wang et al., 2019). For instance, Rahim and colleagues modeled the amounts of PM~2.5~ and received an overall RMSE of 0.20345 (Rahim et al., 2023).

12.2.6 Hybrid Models

The connection of CNNs with LSTMs/GRUs models in a hybrid model is of great benefit because it helps to study spatial and temporal features of air quality in a comprehensive manner. By the use of models the linkages between the geography of pollution and its time-dependency can be cloned. By means of a hybrid model dubbed CRLSTM launched in Beijing, China, a handful of studies were able to estimate the concentration levels of $PM_{2.5}$ will result in R_2 and RMSE values of 0.74 and 18.96, respectively (W. Wang et al. 2021). The other kind of the model is based on LSTM as well as deep auto-encoders, having the capacity to predict the concentration levels of $PM_{0.5}$ and PM_{10} (Xayasouk et al., 2020).

12.2.7 Transformer Models

Transformer models, such as the transformer architecture used in natural language processing, can be adapted for sequential data in air quality monitoring. These models excel at capturing long-range dependencies and can be applied to time series prediction tasks. A recent study conducted by Zhang et al. on two datasets *viz.* Beijing $PM_{2.5}$ and Taizhou $PM_{2.5}$ showcases the performance of their transfer models with best values obtained on latter dataset as 0.924, 5.79, 3.76 for R_2, RMSE and MAE respectively (Z. Zhang & Zhang, 2023). A similar work was also carried by Zhang et. al using two datasets (Z. Zhang et al., 2022). Another study aimed at prediction of hourly $PM_{2.5}$ concentrations in wildfire-prone areas using a spatio-temporal transformer model (Yu et al., 2023).

12.2.8 Attention Mechanisms

Attention mechanisms can be integrated into various DL architectures to allow the model to focus on relevant parts of the input data. This can be beneficial for capturing critical features in air quality monitoring datasets. Zhao et al. proposed an STA-LSTM based attention mechanism for prediction of $PM_{2.5}$ (Zou et al., 2021). Another work carried out by Dairi et al. performs predictive analysis on NO_2, CO, O_3 and SO_2 parameters and the computed values of RMSE for corresponding environmental parameters is obtained is 12.37, 0.013, 1.087, and 0.241 respectively (Dairi et al., 2021). Chen et al. showcased the results of their proposed multi-graph spatial-temporal attention network model for predicting air quality on two datasets for Beijing and Tainjin cities (X. Chen et al., 2024). Chen et al. proposed a model named EvaNet based on Extreme Value Attention Network for predicting air quality (Z. Chen et al., 2020).

12.2.9 Generative Adversarial Networks (GANs)

In the context of air pollutants, GANs can increase the size of the available data or generate synthetic air quality data. Through learning a GAN with such data, it becomes capable of creating extra samples which are used to bolster the data set and thus to provide better representation. The model complied by Cheng was a GAN based model which was built on two cities in the Jing-jin-ji province of China (Cheng et al., 2022). The solution was patching this air quality data gap by using a specific algorithm known as the GAN that was mastered by Zhou et al. (2021).

12.2.10 Reinforcement Learning (RL)

RL can be contemplated among other control technologies to achieve better air quality standards. With the RL models we can search for the optimal policies of how to modify premises of environmental control systems for the purpose of increasing air quality. Chang yet al. refined the statistical forecasting model, drawing influence from RL in the process (Chang et al., 2020).

12.2.11 Spatial-Temporal Graph Networks

Example of applying the GNNs could be mapping the spatial relationships between different monitoring stations of different kind or simply between any geographical points. These models will be able to explain and capture the different intricate dependencies that feature in the precise details of the air quality data that will flow through the neural network (Ouyang et al., 2021; Dun et al., 2022).

12.2.12 Capsule Networks (CapsNets)

Capsule networks were proposed to tackle the deficiencies of neural networks with regard to 2D spatial interactions according to the work of Liu team of researchers published in 2023. Traditionally, these CapsNets typically have a distinct advantage over other models, especially those that demand in learning the arrangement of feature such as this one showed by W. Li and team in 2024 that was mentioned as a study on pollution patterns.

12.2.13 Variational Autoencoders (VAEs)

On the one hand Variational Autoencoders (VAEs) as per the research paper by Bonet et al. in 2022 are the models that provide a means of discovering the dominant distribution patterns of air quality data. These are the patterns that are known to exist but only the models have the capability of seeing them. Their worth is in discovering the data hidden in a selected dataset, only visible by having the full representation of the data available through modeling thus generating realistic samples – a scientific exploration domain which has been widely studied in air quality levels.

12.2.14 Meta-Learning

Meta learning capabilities will be exploited to train models which could then adjust to locations or change in environmental conditions very swiftly as proposed by Yao and others in 2019. This versatility is very essential, especially in situations where the need arise to do more air quality monitoring than conducted by Dr. Wu and her group in 2023.

12.2.15 Transfer Learning

In the process of the transfer finding you will teach a model how to learn on the big data set and then edit it for a job of air monitoring using a smaller dataset (Fong et al., 2020). This way may also be significant if there exists a situation where no data are available.

12.2.16 Explainable AI (XAI) Models

The future of learning shall be illuminated further by XAI models providing the transparency and the comprehension needed through the learning forecasts (Tasioulis &

Karatzas 2024). Techniques, for example SHapley Additive exPlanations (SHAP) and layer wise Relevance Propagation (LRP), help with understanding how features affect the model's predictions regarding air quality change (Yang et al., 2022).

12.2.17 Temporal Convolutional Networks (TCNs)

TCNs are brought into existence to have sequence-to-sequence learning at ease, which finally increase the performance by reusing the computational resources repeatedly, as Wang et al. (2021) proposed. In the given case, these networks serve the purpose of prediction of air quality, not just in isolated situations but in terms of trends which according to Samal and his colleagues (2021) is a further essential factor to consider when making forecasts.

12.2.18 Ensemble Learning

Ensemble methods involve combining predictions from multiple models to improve overall performance and robustness (Lin et al., 2021). Stacking or bagging is an example of method which is applied to a machine learning model to ascertain the air quality (Singh et al., 2013).

12.2.19 Self-Supervised Learning

In the case of self-supervised learning, models are trained on unlabeled data in order to identify the inherent structures and features present in the data and thereby represent them in a meaningful way (Hoffmann & Lessig, 2023). This can be helpful when there is a lack of labelled data on air quality because the model can still extract pertinent features from the information that is available (Han and al., 2003).

12.2.20 Spatial-Temporal Attention Networks

Air quality models illustrated in this chapter make use of the attention mechanism in the forecasting networks, which helps in location and time (Wang et al., 2022). It gives the model the ability to choose as per the requirement to pick different locations and moments in the chain of time and thus expose the hidden aspects (Zhao et al., 2020).

12.2.21 Federated Learning

Federated learning would always allow for the models to be trained efficiently across the different forms of decentralized devices or any monitoring stations without any sharing (Liu et al., 2021). This could be very much helpful and useful in different scenarios where the concern majorly lies in the privacy part related to data (Le et al., 2022).

12.3 PERFORMANCE ANALYSIS

This section includes all the details of all the recent state-of-the-art methods, which includes the major performance evaluations in different datasets. Table 12.2 showcases

TABLE 12.2
Performance Analysis of DL Techniques in the Context of AQ

Year	Country	City	DL Model	Parameters	Dataset	Performance metric	Performance	Reference
2021	Malaysia	Kuala Lumpur	Auto encoder	$PM_{2.5}$	N/A	RMSE, MAE, SMAPE	14.52, 8.22, 29.37	(Xu & Yoneda, 2021)
2021	China	Beijing	Bi-GRU	$PM_{2.5}$	UCI repository	MAE, RMSE, R_2	9.692, 16.712, 0.978	(Bekkar et al., 2021)
2021	China	Beijing	Bi-LSTM	$PM_{2.5}$	UCI repository	MAE, RMSE, R_2	8.868, 15.597, 0.982	(Bekkar et al., 2021)
2020	China	Beijing	CLSTM	$PM_{2.5}$	Queensland, Australia.	RMSE	13.97	(Sharma et al., 2020)
2021	China	Beijing	CNN	$PM_{2.5}$	UCI repository	MAE, RMSE, R_2	9.591, 16.981, 0.979	(Bekkar et al., 2021)
2021	China	Beijing	CNN-GRU	$PM_{2.5}$	UCI repository	MAE, RMSE, R_2	9.459, 17.7, 0.977	(Bekkar et al., 2021)
2021	China	Beijing	CNN-LSTM	$PM_{2.5}$	UCI repository	MAE, RMSE, R_2	6.742, 12.921, 0.989	(Bekkar et al., 2021)
2021	China	Beijing	CRLSTM	$PM_{2.5}$	Beijing Municipal Ecology and Environment Bureau	R_2, RMSE	0.74, 18.96	(W. Wang et al., 2021)
2020	China	Beijing	DAEDN	$PM_{2.5}$	Beijing Municipal Environmental Monitoring Center (www.bjmemc.com.cn/)	RMSE, MAE	15.504, 6.789	(Cai et al., 2020)

2019	South Korea	Seoul	GRU-LSTM	$PM_{2.5}$	China National Environmental Monitoring Centre	RMSE	18	(B. Wang et al., 2019)
2023	Beijing, China	Beijing	GRU	$PM_{2.5}$	Primary: Collected from United State Embassy and Consulate building in Kuala Lumpur	RMSE	0.20345	(Rahim et al., 2023)
2021	China	Beijing	GRU	$PM_{2.5}$	UCI repository	MAE, RMSE, R_2	9.541, 16.609, 0.978	(Bekkar et al., 2021)
2021	South Korea	Seoul	LSTM	$PM_{2.5}$	UCI repository	MAE, RMSE, R_2	9.503, 16.217, 0.98	(Bekkar et al., 2021)
2019	Australia.	Queensland	LSTM	O_3, $PM_{2.5}$, NO_x, and CO concentrations	Hourly concentration	N/A	N/A	(Krishan et al., 2019)
2020	Beijing, China	Beijing	LSTM, DAE	PM_{10}, $PM_{2.5}$	Korea Meteorological Agency website	RMSE	15.03, 15.43	(Xayasouk et al., 2020)
2019	China	Beijing	LSTM-FC	$PM_{2.5}$	N/A	N/A	N/A	(Zhao et al., 2019)
2021	China	Beijing	RNN	$PM_{2.5}$	National Institute of Environmental Research, the Ministry of Environment	Accuracy	81	(Chang-Hoi et al., 2021)
2023	India	Delhi	SpatioTemporal Transformer	N/A	N/A	RMSE	6.92	(Yu et al., 2023)

(continued)

TABLE 12.2 (Continued)
Performance Analysis of DL Techniques in the Context of AQ

Year	Country	City	DL Model	Parameters	Dataset	Performance metric	Performance	Reference
2016	China	Beijing	Stacked Autoencoder	$PM_{2.5}$	Ministry of Environmental Protection of China	RMSE, MAE, MAPE	15.15, 9.11, 22.95	(Li et al., 2016)
2020	Beijing, China	Beijing	TLS-BLSTM	$PM_{2.5}$	China National Environmental Monitoring Centre	RMSE	35.21	(Ma et al., 2020)
2023	USA	Los Angeles	Transformer	$PM_{2.5}$	Beijing $PM_{2.5}$ dataset, Taizhou $PM_{2.5}$ dataset	R_2, RMSE, MAE	0.937, 19.04, 11.13 and 0.924, 5.79, 3.76	(Z. Zhang & Zhang, 2023)
2022	China	Beijing	Transformer	$PM_{2.5}$	Beijing $PM_{2.5}$ dataset, Taizhou $PM_{2.5}$ dataset	R_2, RMSE, MAE	18.51, 11.06, 22.91 and 5.7,3.66, 20.23	(Z. Zhang et al., 2022)
2022	China	Beijing	CNN-LSTM	$PM_{2.5}$	Beijing air quality	MAE, RMSE, R_2	3.17, 5.46, 8.45	(J. Zhang & Li, 2022)
2023	China	Beijing	CNN-LSTM	$PM_{2.5}$	Resource and Environmental Science and Data Center of the Chinese Academy of Sciences.	RMSE, MAE, R_2	7.5, 5.285, 0.989	(Duan et al., 2023)

2022	China	Shijiazhuang	CNN-ILSTM	$PM_{2.5}$	Air quality data from 00:00 on January 1, 2017, to 23:00 on June 30, 2021	MAE, MSE, R_2	8.4134, 202.1923, 0.9601	(J. Wang, Li, Jin, et al., 2022)
2022	China	Shijiazhuang	CNN-BiLSTM	$PM_{2.5}$	Hourly AQI from 00:00 on January 1, 2020, to 23:00 on September 30, 2020	MAE, RMSE, R_2	5.98, 9.23, 0.974	(J. Wang, Li, Wang, et al., 2022)
2018	Taiwan	Taipei	RNN	$PM_{2.5}$	Environmental Protection Administration of Taiwan	N/A	N/A	(Tsai et al., 2018)
2020	India	Amaravati	RNN-LSTM	CO, NH_3, NO_2, O_3, PM_{10}, $PM_{2.5}$, SO_2	Primary: Amaravati dataset	RMSE	3.14, 0.23, 1.02, 5.11, 4.72, 4.96, 1.04	(Belavadi et al., 2020)
2021	China	Beijing	Attention mechanism: STA-LSTM	$PM_{2.5}$	Chinese weather website from January 1, 2018, to December 31, 2018	RMSE (1-hourly)	12.23	(Zou et al., 2021)
2021	USA	California	Attention mechanism	NO_2, CO, O_3, SO_2	United States Environmental Protection Agency (2000–2016)	RMSE	12.37, 0.013, 1.087, 0.241	(Dairi et al., 2021)

(continued)

TABLE 12.2 (Continued)
Performance Analysis of DL Techniques in the Context of AQ

Year	Country	City	DL Model	Parameters	Dataset	Performance metric	Performance	Reference
2024	China	Tianjin	Multi-graph spatial-temporal attention network	$PM_{2.5}$	Beijing-2018 dataset and Tianjin-2014 dataset	MAE, RMSE	8.39, 12.96; 18.53, 27.15	(X. Chen et al., 2024)
2020	China	Beijing	EvaNet	$PM_{2.5}$	Beijing Shunyi District Data	RMSE, MAE	3.26, 2.18	(Z. Chen et al., 2020)
2022	China	Beijing, Tianjin	GAN	$PM_{2.5}$	Air Quality data	RMSE	43.34, 56.93	(Cheng et al., 2022)
2020	Taiwan	N/A	RL	$PM_{2.5}$	Taiwan LASS environmental sensing network system	RMSE	49.95	(Chang et al., 2020)

the details such as year, country, city, DL model used, parameters predicted (e.g. PM_{10}, $PM_{2.5}$, CO, NOx, etc.), dataset used for simulation, performance metric (e.g. RMSE, MAE, MAPE, R_2 etc.), and performance value. The authors have tried to include works from all 21 categories, as shown in Figure 12.2. It has been observed from Table 12.2 that the RMSE performance metric is used by most of the DL models, thus making it a standard metric for prediction evaluation in the context of AQ. The mathematical representation of RMSE, MAE and MAPE are given in eq (12.1) to (12.3) respectively.

$$\mathrm{RMSE} = \sqrt{\frac{1}{n}\sum_{k=1}^{n}\left(Q_k - R_k\right)^2} \qquad \text{eq (12.1)}$$

$$\mathrm{MAE} = \sqrt{\frac{1}{n}\sum_{k=1}^{n}\left|Q_k - R_k\right|} \qquad \text{eq (12.2)}$$

$$MAPE = \sqrt{\frac{1}{n}\sum_{k=1}^{n}\frac{\left|Q_k - R_k\right|}{Q_k}} \qquad \text{eq (12.3)}$$

Here, 'n' is the number of samples, Q_k is the instance of observed air quality, R_i is the instance of predicted air quality. These metrics are used in the state-of-the-art methods as shown in Table 12.2. Consider a study conducted in Malaysia in 2021 by Xu et al. (2021) also used these metrics to compute the performance of their model based on auto encoder deep learning framework. Similarly, other studies conducted by researchers use these performance evaluation metrices. Most of the work is carried in detection of $PM_{2.5}$ and less work is available on other particulate matters of other sizes, thus there is a scope of work for prospective researcher.

12.4 DATASET DESCRIPTION

This section briefly includes details pertaining to datasets used to simulate air quality prediction models. Table 12.3 consists of the details of datasets descriptions consisting of dataset name and corresponding web link to access it. All of the links given in Table 12.3 were last accessed in February 2024. Different countries maintain meteorological data which can be accessed from the websites for research and development purposes. For example, the first data row of Table 12.3 consists of the website details of Korea Meteorological Agency, whereby it provides data of different meteorological parameters. The details of respective websites for corresponding country are mentioned in Table 12.3.

12.5 DISCUSSION

In the discussion of DL approaches in air quality prediction, several key aspects merit consideration. First and foremost, the effectiveness of these models heavily relies on the quality and representativeness of the input data. It is essential to obtain pertinent

TABLE 12.3
Dataset Description

S. No.	Dataset	Link to dataset
1	Korea Meteorological Agency website	https://data.kma.go.kr/cmmn/main.do
2	China National Environmental Monitoring Centre	www.cnemc.cn
3	Beijing Municipal Ecology and Environment Bureau	http://sthjj.beijing.gov.cn/
4	Beijing Municipal Environmental Monitoring Center	www.bjmemc.com.cn/
5	Ministry of Environmental Protection of China	http://datacenter.mep.gov.cn
6	Beijing $PM_{2.5}$	www.kaggle.com/djhavera/beijing-pm25-data-data-set
7	EPA USA dataset	www.kaggle.com/epa/epa-historical-air-quality
8	AirNet Dataset	https://openreview.net/forum?id=SkymMAxAb
9	Air Quality Dataset	www.kaggle.com/datasets/fedesoriano/air-quality-data-set
10	Dutch-emission-data-1990-2015	https://data.world/alexandra/dutch-emission-data-1990-2015
11	Air Quality Data in India 2017–2022	www.kaggle.com/datasets/fedesoriano/air-quality-data-in-india
12	AQI – Air Quality Index	www.kaggle.com/datasets/azminetoushikwasi/aqi-air-quality-index-scheduled-daily-update
13	Air Quality	https://archive.ics.uci.edu/dataset/360/air+quality
14	Resource and Environment Science and Data Center of the Chinese Academy of Sciences	www.resdc.cn/Default.aspx
15	Environmental Protection Administration of Taiwan (EPA)	https://airtw.epa.gov.tw/ENG/default.aspx
16	United States Environmental Protection Agency	www.epa.gov/outdoor-air-qualitydata/download-daily-data
17	Madrid Open data	https://datos.madrid.es/

pollutants databases and relevant meteorological elements for a high-end accuracy. Furthermore, overcoming data gaps issues and outliers errors which can create of biases from the sensor stations is compulsory for model application. The DL framework choice is another for this purpose. Whilst RNNs and LSTMs dual in the process of capturing temporal sequences, the latter can easily pinpoint spatial behavior peculiarities. Choosing an appropriate model architecture is subject to being adapted to the air quality data's special characteristics as well as the prediction's purpose.

Interpretability remains a challenge in DL models, especially in complex domains like air quality. It is important to know the models' method to make forecasts, which will help in building trust and applying these models in decision-making processes. The future of DL will largely depend on the investigations aimed at providing clear interpretability of the DL model; this will inevitably facilitate the acceptance and utilization of these models. Besides, it has to be established that the situation with air pollution is in permanent change and is caused by such influential factors as climate change, urbanization, and industrial production. The continual process of model evaluation and updating to new and changing patterns must always be sought to enable a successful forecast over a long time. The collaboration between scientists, policy makers as well as industrial sectors is the bedrocks for the integration of already existing data and emerging information for the design of smart strategies towards good air quality management.

12.6 CONCLUSION AND FUTURE SCOPE

The most indicational of DL approaches in the field of air quality monitoring and forecasting is that they show great promise in enabling these capabilities to be significantly enhanced. Further than this, in the light on the technology progress, tackling challenges through technological breakthroughs would help to create the robust systems of monitoring and predicting air quality that will end in health improvement and the environment protection.

First, the immense rise in populization of DL methods in air quality prediction carry with it the unique promise of being both an efficient and effective solution to the complex intricacies that both atmospheric dynamics and pollutant concentrations bear. Using state-of-the-art neural network architectures such as Recurrent Neural Networks (RNNs) and Long Short-Term Memory networks (LSTMs), these models exhibit the exquisite ability to handle complex temporal and spatial attributes contained in air quality data. One of the most obvious qualities of geospatial data analysis is that in real-world applications, it offers great accuracy and efficiency in such results as urban planning and public health interventions. Nevertheless, more investigations and partnerships are needed to increase the accuracy of the models and find out how to improve the models functioning and integration with the data regarding with environmental change.

The development of DL in air quality prediction holds in having even more sophisticated models developed, data quality improved, and the interpretability enhanced. Pure DL models were examined in order to reduce the contamination that traditional statistical methods make to the establishment of air quality dynamics. Furthermore, the interdisciplinary team made up of researchers, environmentalists, and data engineers can add to the capacity in building models that are well-thought-out and place-specific. Moreover, among the real-time sensor data, satellite observations, and initiating devices from Internet of Things (IoT), these predictive models have been richly supplemented. Overall, the development of DL approaches to air quality prediction will lead us on the path towards the creation of environments that are more clean and healthy. We will be better placed to make informed and sustainable decisions through such an approach for the good of not just our single communities but community globally.

REFERENCES

Bekkar, A., Hssina, B., Douzi, S., & Douzi, K. (2021). Air-pollution prediction in smart city, deep learning approach. *Journal of Big Data*, *8*(1), 161. https://doi.org/10.1186/s40537-021-00548-1

Belavadi, S. V, Rajagopal, S., Ranjani, R., & Mohan, R. (2020). Air quality forecasting using LSTM RNN and wireless sensor networks. *Procedia Computer Science*, *170*, 241–248. https://doi.org/10.1016/j.procs.2020.03.036

Bonet, E. R., Do, T. H., Qin, X., Hofman, J., La Manna, V. P., Philips, W., & Deligiannis, N. (2022, August). Conditional variational graph autoencoder for air quality forecasting. In *2022 30th European Signal Processing Conference (EUSIPCO)* (pp. 1442–1446). IEEE.

Cai, J., Dai, X., Hong, L., Gao, Z., & Qiu, Z. (2020). An air quality prediction model based on a noise reduction self-coding deep network. *Mathematical Problems in Engineering*, *2020*, 1–12. https://doi.org/10.1155/2020/3507197

Chang, S.-W., Chang, C.-L., Li, L.-T., & Liao, S.-W. (2020). Reinforcement learning for improving the accuracy of PM2.5 pollution forecast under the neural network framework. *IEEE Access*, *8*, 9864–9874. https://doi.org/10.1109/ACCESS.2019.2932413

Chang-Hoi, H., Park, I., Oh, H.-R., Gim, H.-J., Hur, S.-K., Kim, J., & Choi, D.-R. (2021). Development of a PM2.5 prediction model using a recurrent neural network algorithm for the Seoul metropolitan area, Republic of Korea. *Atmospheric Environment*, *245*, 118021. https://doi.org/10.1016/j.atmosenv.2020.118021

Chen, X., Hu, Y., Dong, F., Chen, K., & Xia, H. (2024). A multi-graph spatial-temporal attention network for air-quality prediction. *Process Safety and Environmental Protection*, *181*, 442–451. https://doi.org/10.1016/j.psep.2023.11.040

Chen, Z., Yu, H., Geng, Y., Li, Q., & Zhang, Y. (2020). EvaNet: An extreme value attention network for long-term air quality prediction. *2020 IEEE International Conference on Big Data (Big Data)*, 4545–4552. https://doi.org/10.1109/BigData50022.2020.9378094

Cheng, M., Fang, F., Navon, I. M., Zheng, J., Tang, X., Zhu, J., & Pain, C. (2022). Spatio-temporal hourly and daily ozone forecasting in China using a hybrid machine learning model: Autoencoder and generative adversarial networks. *Journal of Advances in Modeling Earth Systems*, *14*(3). https://doi.org/10.1029/2021MS002806

Dairi, A., Harrou, F., Khadraoui, S., & Sun, Y. (2021). Integrated multiple directed attention-based deep learning for improved air pollution forecasting. *IEEE Transactions on Instrumentation and Measurement*, *70*, 1–15. https://doi.org/10.1109/TIM.2021.3091511

Duan, J., Gong, Y., Luo, J., & Zhao, Z. (2023). Air-quality prediction based on the ARIMA-CNN-LSTM combination model optimized by dung beetle optimizer. *Scientific Reports*, *13*(1), 12127. https://doi.org/10.1038/s41598-023-36620-4

Dun, A., Yang, Y., & Lei, F. (2022). Dynamic graph convolution neural network based on spatial-temporal correlation for air quality prediction. Ecological Informatics, 70, 101736.

Fong, I. H., Li, T., Fong, S., Wong, R. K., & Tallón-Ballesteros, A. J. (2020). Predicting concentration levels of air pollutants by transfer learning and recurrent neural network. Knowledge-Based Systems, 192, 105622. https://doi.org/10.1016/j.knosys.2020.105622

Goodfellow, I., Bengio, Y., & Courville, A. (2016). *Deep Learning*. MIT Press.

Han, J., Liu, H., Xiong, H., & Yang, J. (2023). Semi-supervised air quality forecasting via self-supervised hierarchical graph neural network. IEEE Transactions on Knowledge and Data Engineering, 35(5), 5230–5243. https://doi.org/10.1109/TKDE.2022.3149815

Hoffmann, S., & Lessig, C. (2023). AtmoDist: Self-supervised representation learning for atmospheric dynamics. Environmental Data Science, 2, e6. https://doi.org/10.1017/eds.2023.1

Kaur, P., Chaudhary, P., Bijalwan, A., & Awasthi, A. (2018). Network traffic classification using multiclass classifier. In *Advances in Computing and Data Sciences: Second International Conference, ICACDS 2018, Dehradun, India, April 20–21, 2018, Revised Selected Papers, Part I* 2 (pp. 208–217). Springer Singapore.

Krishan, M., Jha, S., Das, J., Singh, A., Goyal, M. K., & Sekar, C. (2019). Air quality modelling using long short-term memory (LSTM) over NCT-Delhi, India. *Air Quality, Atmosphere & Health, 12*(8), 899–908. https://doi.org/10.1007/s11869-019-00696-7

Le, D.-D., Tran, A.-K., Dao, M.-S., Nguyen-Ly, K.-C., Le, H.-S., Nguyen-Thi, X.-D., Pham, T.-Q., Nguyen, V.-L., & Nguyen-Thi, B.-Y. (2022). Insights into multi-model federated learning: An advanced approach for air quality index forecasting. Algorithms, 15(11), 434. https://doi.org/10.3390/a15110434

Li, X., Peng, L., Hu, Y., Shao, J., & Chi, T. (2016). Deep learning architecture for air quality predictions. *Environmental Science and Pollution Research, 23*(22), 22408–22417. https://doi.org/10.1007/s11356-016-7812-9

Lin, C.-Y., Chang, Y.-S., & Abimannan, S. (2021). Ensemble multifeatured deep learning models for air quality forecasting. Atmospheric Pollution Research, 12(5), 101045. https://doi.org/10.1016/j.apr.2021.03.008

Liu, Y., Nie, J., Li, X., Ahmed, S. H., Lim, W. Y. B., & Miao, C. (2021). Federated learning in the sky: Aerial-ground air quality sensing framework with UAV swarms. IEEE Internet of Things Journal, 8(12), 9827–9837.

Ma, J., Li, Z., Cheng, J. C. P., Ding, Y., Lin, C., & Xu, Z. (2020). Air quality prediction at new stations using spatially transferred bi-directional long short-term memory network. *Science of The Total Environment, 705*, 135771.

Ouyang, X., Yang, Y., Zhang, Y., & Zhou, W. (2021, July). Spatial-temporal dynamic graph convolution neural network for air quality prediction. In *2021 International Joint Conference on Neural Networks (IJCNN)* (pp. 1–8). IEEE.

Rahim, M. S. A., Yakub, F., Omar, M., Ghani, R. A., Salim, S. A. Z. S., Masuda, S., & Dhamanti, I. (2023). Prediction of indoor air quality using long short-term memory with adaptive gated recurrent unit. *E3S Web of Conferences, 396*, 01095. https://doi.org/10.1051/e3sconf/202339601095

Samal, K. K. R., Babu, K. S., & Das, S. K. (2021). Temporal convolutional denoising autoencoder network for air pollution prediction with missing values. Urban Climate, 38, 100872. https://doi.org/10.1016/j.uclim.2021.100872

Sharma, E., Deo, R. C., Prasad, R., Parisi, A. V., & Raj, N. (2020). Deep air quality forecasts: Suspended particulate matter modeling with convolutional neural and long short-term memory networks. *IEEE Access, 8*, 209503–209516. https://doi.org/10.1109/ACCESS.2020.3039002

Singh, K. P., Gupta, S., & Rai, P. (2013). Identifying pollution sources and predicting urban air quality using ensemble learning methods. Atmospheric Environment, 80, 426–437.

Tandon, A., Awasthi, A., & Pattnayak, K. C. (2023). Comparison of different machine learning methods on precipitation dataset for Uttarakhand," *2023 2nd International Conference on Ambient Intelligence in Health Care (ICAIHC)*, Bhubaneswar, India, 2023, pp. 1–6. doi: 10.1109/ICAIHC59020.2023.10431402

Tasioulis, T., & Karatzas, K. (2024). Reviewing Explainable Artificial Intelligence Towards Better Air Quality Modelling. In: Wohlgemuth, V., Kranzlmüller, D., Höb, M. (eds), *Advances and New Trends in Environmental Informatics 2023*. ENVIROINFO 2023. Progress in IS. Springer, Cham. (pp. 3–19). https://doi.org/10.1007/978-3-031-46902-2_1

Tsai, Y.-T., Zeng, Y.-R., & Chang, Y.-S. (2018). Air pollution forecasting using RNN with LSTM. *2018 IEEE 16th Intl Conf on Dependable, Autonomic and Secure Computing, 16th Intl Conf on Pervasive Intelligence and Computing, 4th Intl Conf on Big Data Intelligence and Computing and Cyber Science and Technology Congress(DASC/PiCom/DataCom/CyberSciTech)*, 1074–1079. https://doi.org/10.1109/DASC/PiCom/DataCom/CyberSciTec.2018.00178

Wang, B., Kong, W., Guan, H., & Xiong, N. N. (2019). Air quality forecasting based on gated recurrent long short term memory model in Internet of Things. *IEEE Access*, *7*, 69524–69534. https://doi.org/10.1109/ACCESS.2019.2917277

Wang, J., Li, J., Wang, X., Wang, T., & Sun, Q. (2022). An air quality prediction model based on CNN-BiNLSTM-attention. *Environment, Development and Sustainability*. https://doi.org/10.1007/s10668-021-02102-8

Wang, J., Li, X., Jin, L., Li, J., Sun, Q., & Wang, H. (2022). An air quality index prediction model based on CNN-ILSTM. *Scientific Reports*, *12*(1), 8373. https://doi.org/10.1038/s41598-022-12355-6

Wang, W., Mao, W., Tong, X., & Xu, G. (2021). A novel recursive model based on a convolutional long short-term memory neural network for air pollution prediction. *Remote Sensing*, *13*(7), 1284. https://doi.org/10.3390/rs13071284

Werbos, P. J. (1988). Generalization of backpropagation with application to a recurrent gas market model. *Neural Netw.*, *1*(4), 339–356.

Wu, Z., Liu, N., Li, G., Liu, X., Wang, Y., & Zhang, L. (2023). Meta-learning-based spatial-temporal adaption for coldstart air pollution prediction. International Journal of Intelligent Systems, 2023, 1–22. https://doi.org/10.1155/2023/3734557

Xayasouk, T., Lee, H., & Lee, G. (2020). Air pollution prediction using long Short-Term Memory (LSTM) and Deep Autoencoder (DAE) Models. *Sustainability*, *12*(6), 2570. https://doi.org/10.3390/su12062570

Xu, X., & Yoneda, M. (2021). Multitask air-quality prediction based on LSTM-Autoencoder model. *IEEE Transactions on Cybernetics*, *51*(5), 2577–2586. https://doi.org/10.1109/TCYB.2019.2945999

Yang, Y., Mei, G., & Izzo, S. (2022). Revealing influence of meteorological conditions on air quality prediction using explainable deep learning. IEEE Access, 10, 50755–50773. https://doi.org/10.1109/ACCESS.2022.3173734

Yao, H., Liu, Y., Wei, Y., Tang, X., & Li, Z. (2019). Learning from multiple cities: A meta-learning approach for spatial-temporal prediction. The World Wide Web Conference, 2181–2191. https://doi.org/10.1145/3308558.3313577

Yu, M., Masrur, A., & Blaszczak-Boxe, C. (2023). Predicting hourly PM2.5 concentrations in wildfire-prone areas using a spatio temporal transformer model. *Science of The Total Environment*, *860*, 160446. https://doi.org/10.1016/j.scitotenv.2022.160446

Zhang, J., & Li, S. (2022). Air quality index forecast in Beijing based on CNN-LSTM multi-model. *Chemosphere*, *308*, 136180. https://doi.org/10.1016/j.chemosphere.2022.136180

Zhang, Z., & Zhang, S. (2023). Modeling air quality PM2.5 forecasting using deep sparse attention-based transformer networks. *International Journal of Environmental Science and Technology*, *20*(12), 13535–13550. https://doi.org/10.1007/s13762-023-04900-1

Zhang, Z., Zhang, S., Zhao, X., Chen, L., & Yao, J. (2022). Temporal difference-based graph transformer networks for air quality PM2.5 prediction: A case study in China. *Frontiers in Environmental Science*, *10*. https://doi.org/10.3389/fenvs.2022.924986

Zhao, J., Deng, F., Cai, Y., & Chen, J. (2019). Long short-term memory: Fully connected (LSTM-FC) neural network for PM2.5 concentration prediction. *Chemosphere*, *220*, 486–492. https://doi.org/10.1016/j.chemosphere.2018.12.128

Zhao, P., & Zettsu, K. (2020). MASTGN: Multi-attention spatio-temporal graph networks for air pollution prediction. *2020 IEEE International Conference on Big Data (Big Data)* (pp. 1442–1448). https://doi.org/10.1109/BigData50022.2020.9378156

Zou, X., Zhao, J., Zhao, D., Sun, B., He, Y., & Fuentes, S. (2021). Air quality prediction based on a spatiotemporal attention mechanism. *Mobile Information Systems*, *2021*, 1–12. https://doi.org/10.1155/2021/6630944

13 Incorporation of AI with Conventional Monitoring Systems

Tania Ghatak (Chakraborty), Rashi Jain, and Abhijit Sarkar

13.1 INTRODUCTION

One of the few factors which have helped life to evolve and be sustainable on this planet was air. Pure air sustains life and growth of both flora and fauna and helps them to flourish. But with the gradual advent of the human race and its ever-increasing progress, civilization and industrialization have impacted the whole environment like never before. Rapid industrialization, automobile explosion, population boom, and overall imprudent utilization of fossil fuels and other natural resources for lifestyle enhancements have led to severe degradation of the environment. In today's world, thus pollution of nature has emerged as the crucial facet that can lead to the utter destruction of life on this planet. Air pollution is responsible not only for climate change but also creates poverty and disease outbreaks throughout the world. It is recommended in the report that around 92% of the world population is affected by extremely high levels of outdoor air pollution (COP26 Special Report on Climate Change and Health, n.d.), which is responsible for the burning of fossil fuels for expanding degrees of transportation, urban area expansion, and overall growth of industrial sectors which does not always follow the safety protocols. The World Health Organization (2005) first reported deaths of close to 7 million people each year due to air pollution. One project also highlighted globally that polluted air contributed to the early mortality of 5 million people per annum (GBD 2019 Risk Factors Collaborators, 2020). This same incident has also been reported by WHO (2018) showing the high mortality rate by polluted air in both urban and rural areas. Numerous reports also stated that a number of life-threatening diseases like lung cancer and respiratory disorders, heart diseases, even heart attacks, and other complications such as gynecological problems, pregnancy problems, and birth-related instances are directly linked to poor quality of surrounding air (Neo et al., 2022). Air pollution is considered a significant environmental and health concern (Nagpure et al. 2016), and is considered as one of the top five global risk factors in terms of mortality (Murray et al., 2020). Therefore, the ever-increasing threat in the form of air pollution has forced WHO to initiate the "let us breathe clean air" plan in 2021 (WHO, 2021).

DOI: 10.1201/9781032683805-13

13.1.1 Indian Scenario

As a consequence of ever-increasing population and consequent lack of proper employment opportunities, the major cities of India welcome a surge of migrants from different rural and underdeveloped areas of the country every day. Surging urbanization, increased automobile combustion, unsustainable and ignorant anthropogenic activities like burning fossil fuels, and abolishing water bodies and plantations results in air pollutant emissions and poor air quality. According to Laurance (2010), by 2030, around 50% of the global population is expected to transfer to urban areas. It is also very well assessed that more than 80% of the population in urban areas is exposed to emissions that exceed the standards set by WHO (WHO, 2016). PM pollution was the third significant reason responsible for death and this level is the maximum observed in India with 1.1 million premature deaths, in which outdoor $PM_{2.5}$ contributed 56% and 44% of household pollution (Joshi et al., 2022). According to WHO (2016), around 3 million deaths were because of outdoor air pollution (Gurjar et al., 2021). Not only Indian megacities like Delhi, Mumbai, and Kolkata but many other densely populated cities also harbor a grave air quality index. India is among the top five most polluted countries in terms of $PM_{2.5}$ by WHO (2019). The Indian cities, on average, exceeded the WHO standard by 500% (Gurjar, Apr 2021). Jat, Gurjar, and Lowe (2021) examined air pollutants by using Weather Research and Forecasting (WRF) coupled with chemistry and reported the very high concentration of $PM_{2.5}$, oxides of nitrogen and sulfur, organic and black carbons, and non-methane volatile organic carbons in the winter months. Along with vehicle emissions, power plants and household activities, agricultural sources of air pollution are also a crucial source of emissions in India. A country which primarily depends on agriculture to feed its citizens and to boost its economy faces a severe challenge to maintain a sustainable way to promote agriculture especially in rural areas. Enteric fermentation, animal manure, wetlands, and fertilizer are also important to produce different types of air pollutants (Gurjar et al., 2004). Open burning of the residue of the crop is responsible for the deterioration of air quality (Agarwal et al., 2014; Singh et al., 2010). All these are reasons from neighboring Northern states contribute to the agricultural pollution load in the capital city that make the air totally unbreathable and tremendously fatal nowadays (Nagpure et al., 2016).

13.2 MONITORING OF AIR POLLUTION

It is evident from recent reports from WHO that air pollution contributes significantly to posing health issues globally. To prevent these high mortality rates and other adverse effects, countries must take joint actions not only to curb these pollution but also to monitor pollution extent, source and pattern for better understanding. During 2021, the globally recognized health monitoring agency WHO declared that nations should maintain a balance between their national interests with available resources and also implement some measures to mitigate their air quality problems. But to start, it is crucial to know the problem thoroughly, so before managing air quality, the recognition regarding the pros and cons of quality management, cost, output effects on socio-economic scenarios etc. are to be understood systematically.

The data generated by cutting-edge techniques often empowered by artificial intelligence (AI) can generate sample data that could help policy-makers and government officials to evaluate their country's general air quality levels and to draw a route map for air quality monitoring and the proper management of air pollution. This data could be used for protecting public health and thus the adverse effects and diseases can be averted. Not only the developed countries but also the developing and underdeveloped countries can also take help of these methods and try to pinpoint their weak spot to stop the main cause of air pollution, to lessen the population's exposure to air pollution, and other necessary steps. Countries should employ a combination of approaches with measurements and modeling methods for the betterment of their local air quality issues on one hand and the other adjust their national priorities and resource availability. Therefore, it can be said that various techniques should work in a synchronized way for an all-inclusive air quality management knowledge which will lead to a complete route map to control this universal problem of air pollution.

13.3 OVERVIEW OF ARTIFICIAL INTELLIGENCE IN RELATION TO AIR POLLUTION

Forecasting and prediction models can be categorized into three types: statistical, physical, and AI models; the physical forecasting model generally analyzes weather phenomena and meteorological conditions. Chemical Transport Models (CTM) consider chemical and meteorological changes in the atmosphere, and also take note of the release, spread, and mixing of air pollutants (Liu H et al., 2021). It is very much understandable about that with physical methods, one needs to collect an enormous number of computations, and a large amount of information regarding pollutants and their sources. As the characteristics of the model vary with the amount of data, it is a major drawback that affects the efficiency of the physical model (Feng et al., 2015). Whereas, statistical models are more complex to utilize (Liu H, 2021).

In this very context, a pressing requisite arises for an ace analytical system for smart air quality monitoring. To bridge this gap, AI models are now becoming a method of preference over physical and statistical forecasting methods as they imbibe human learning, reasoning, and vision (Chen et al, 2020). AI is nothing but a novel technology that first collects information and then develops methods, assumptions, models, and claim systems for understanding human intelligence and the comparison of different AI models is interesting to understand the problem (Tandon et al., 2023; Wang N. et al., 2018). An Artificial Neural Network (ANN) is a replication by nonlinear series by simulating the nervous system and human brain (Aayush et al., 2020). It can perform tedious and complex tasks that would generally need human aptitude and most importantly enormous amount of time and labor. AI can not only save essential time and expenditure of work-hour but also extract data with utmost efficacies and convenience which also provide long-term sustainability (Chowdhury & Sadek, 2012; Borana, 2016). For this reason, it's evident that almost all developed countries now are keen to adopt AI techniques and systems in their policies. Hence, research on AI and its applications has increased manyfold (Belinger et al., 2017).

Relating to this time-saving aspect, Gu et al. (2022) explained the role of AI-assisted Bibliometrics. Bibliometric study deals with mathematical, statistical, and supplementary analyzing techniques to assess the allocation structure, change arrangements, measurable associations, and mathematical executives of literature and intelligence. Bibliometry generally searches the literature system for particular structures, properties, and other scientific and technological aspects of the object of study. Use of computers for bibliometric work has been practiced for many years now. But nowadays, some advanced approaches and tools create a graphic of the patterns, distribution, and structure of scientific knowledge. According to Chen C (2017), CiteSpace software inspects co-citations in the literature, searches the citation space for information clustering and circulation, as well as offers analysis of co-occurrence between other knowledge resources, which improves presentation and helps to comprehend the development of scientific understanding.

Today, AI has been proven helpful for a varied number of requirements like weather prediction, expert creation, disease estimation, environmental monitoring, and pollutant forecasting (Hino et al., 2018; Hashimoto, 2020; Masood et al., 2021; May et al., 2020). It is a kind of technological boon which has incredible potential to impede climate change and help us to conceive new adaptation methods in fields like energy, land use, and disaster response. Though there are many obstacles and hold-ups present in socio economic and political aspects that thwart AI from achieving its peak position (Stein, 2020).

The AI-aided system plays an important character for experts to analyze air quality, preserve its excellence, and predict its influence regarding the achievement of less polluted regions. AI has become an invincible tool to fight air pollution as recent reports emphasized that not only can it assess but also manage and avert damaging consequences of diverse air contaminants which has tremendous benefits in the various medical and atmospheric fields (Mo et al., 2019). AI techniques are now being used as medically conclusive confirm methods to detect, sustain, and cure different aero-related diseases. A recent report by Heuvelmans et al. (2021) discussed deep learning, which can forecast lung cancer based on the CT scans of patients. The cutting-edge model has proved to be a super performer as it can diagnose benign pulmonary tumors at its early onset though the standardized screening procedures should be further upgraded for better accuracy. Nilashi et al. (2020) reported a new methodology using fuzzy logic and support vector machines (SVM) for detecting heart diseases. The result from the 573 examples of historical heart diseases showed that the collective model displayed precise observation for the categorization of heart-related symptoms and also greatly helped in the early diagnosis of these problems through computation models.

Not only that, AI has been proven to be an essential tool to aid regional environmental protection agencies by providing them with air pollution-reducing measures which will help to make necessary decisions to decrease public experience danger (Jerrett et al., 2005). Compared to other models, AI is capable of understanding non-linear and complicated connections that often stand between air quality parameters and which enables to predict better pollution incidents (Fernandez and Vico, 2013). AI can effortlessly identify the current pollution baseline and accurately and quickly predict pollution hot spots. AI based techniques also create analytical algorithms of different quantities from the meteorological data and air, which is very much helpful in predicting the occurrences of visibility, haze, smog, and other meteorological interference. Big data analytical systems have imparted AI as the number one choice for researchers developing accurate and advanced forecasting systems (Bai et al., 2018; Masood and Ahmad, 2020; Rahimi, 2017).

AI-empowered models have been thoroughly tested earlier with more common techniques for pollution. McKendry (2002) matched ANN with other procedures that use various explanatory variables to calculate PM_{10} and $PM_{2.5}$. In recent times, Hu et al. developed an RF model, including meteorological fields, land use parameters, and aerosol optical depth to estimate the $PM_{2.5}$ over the US (Hu et al., 2017). At the same time, Li et al. also evaluated air pollutant concentration by inventing a short-term memory NN extended model (Li et al, 2017). Huang and Kuo also used a deep neural network model to apply the $PM_{2.5}$ forecasting system in smart cities (Huang et al., 2018).

All such experiments validate that the AI approach has proved to be the strongest one in air pollution forecasting. As discussed earlier, AI-assisted models have been verified to be advantageous for having extreme accuracy, rapid processing speed, and deep error tolerance when compared with the traditional measurement technique. Furthermore, deep learning models have proved to be very efficient in evaluating parameters with several layers. Nowadays researchers are keen on studying and exploring numerous forms of smart prediction models to better analyze and control air quality but to choose which one is most suitable depending upon the different variables present. Hybrid models of AI not only use different algorithms but also the amalgamation of various properties of other factors, which ultimately provides improved output (Gu et al., 2022).

Therefore nowadays, scientists monitor air quality using sensors at monitoring stations. The sensors measure pollutant levels which are then remotely transferred to observe if they are well below the WHO and United States Environmental Protection Agency (USEPA) standards. AI-assisted models here become an integral part as they process all air pollution monitoring networks, deduce the sensor devices' measurement signals, and combine data from all present monitoring stations, which ensures no monitoring gaps.

Summing up the topic, it is evident from different research that present-day understanding and forecasting of scenarios of air pollution demands the implementation of intricate numerical models that can simulate weather conditions and chemistry of air pollution. In today's world, AI with hybrid models empowered with cost-effective air pollution sensors open a new avenue for exhaustive analysis and detailed understanding of air pollution maps, which in turn would definitely enhance better and more effective control measures that are currently available. Even now it is possible to deliver the exact quantification of pollutants inhaled using AI-based pollution monitoring coupled with various physiological sensors and medical information systems. This in-person prediction technology will tremendously benefit vulnerable individuals to organize outdoor activities and restrict them to be present near harmful environments (Neo et al., 2022).

13.4 OUTLINE OF AI TECHNIQUES

After the general discussion about the significance and potential uses of AI technology, let us search for comprehensive knowledge about the most commonly used AI functions in the field of air pollution forecasting.

13.4.1 Artificial Neural Networks (ANN)

ANNs are technique that understands information using algorithms that imitate the functions of a human nervous system (Olivieri and Allegrini, 2019; Sapra et al., 2015). The new age AI technique functions like the biological brain as it mimics the functioning of the neural network of the brain. Just like the brain's functional unit neurons, they also have interconnected nonlinear processing units (Chatterjee and Pandya, 2016; Ghritlahre and Prasad, 2018). These neuron-like units work simultaneously and can be used to analyze a plethora of problems, pattern grouping, function estimation, prediction, clustering, retrieval by content, control and optimization (Masood, 2021). Figure 13.1 shows a schematic of an ANN.

The ANN model typically consists of three layered structures namely training, design, and validation. The "design" deals with the suitable layering or structuring of the neurons, algorithm and input, output data (Subramanyam, 2008). The training stage includes the process of minimizing the difference between the observed and predicted data, whereas in the validation stage, the overall performance of the ANN is

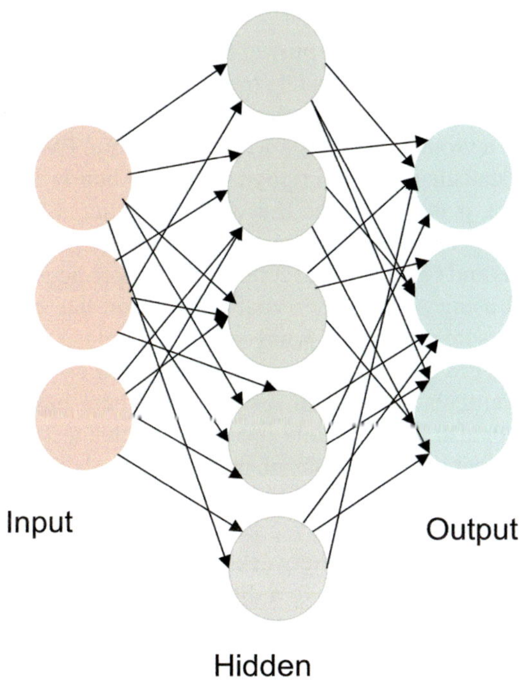

FIGURE 13.1 Schematic diagram of an artificial neural network.

tested for unknown data. After passing these tests, only the ANN system can be used for forecasting purposes (Masppd and Ahmad, 2021; Hasanuzzaman and Rahim, 2019). The multilayer perceptron (MLP) has been the most commonly used ANN that is used to forecast different pollution events (Agirre et al., 2006).

13.4.1.1 Applications

ANN in recent years has been proved to be immensely helpful in numerous ways to predict different parameters related to air pollution. An array of research studies has used ANN as a most effective and accurate tool to forecast occurrences of different pollutants even well in advance.

a. Gaseous Pollutants Forecasting

The very first study which used ANN aided pollution forecasting was done by Mlakar et al. (1994). In this study concentration levels of SO_2 were measured; relative humidity, etc. were given as input and SO_2 concentrations were predicted at the thermal power plant in Sostanj. It shows the system can forecast almost exact SO_2 levels almost 30 minutes before. Another study utilized the MLP model of ANN to predict concentration of SO_2 in an industrial area (Arena et al., 1995). Similar studies were carried out for other gaseous pollutants using ANN models viz. SO_2, CH_4, CO, nitric oxide (NO), NO_2, total hydrocarbons (THC), and ozone (O_3) (Sohn, S.H.et al; 1999). Most studies showed ANN forecasting is beneficial, but there is scope for better performance by providing other weather parameters as inputs (Masood, 2021).

b. Particulate Matter Forecasting

Artificial neural networking has also proved to be very much essential in forecasting of Particulate matter. Generally particulate matters in air are measured as $PM_{2.5}$ and 10 which indicates severe air pollution. A three-layered MLP (multiple layered perceptron) was created by a group of scientists which can predict PM_{10} accurately based on the meteorological, statistical and descriptive data in the hourly mode (Fernando et al., 2012). From the data, it was observed that ANN was successful in predicting PM_{10} concentration almost perfectly but could be made more efficient by improving the input training. Grivas and Chaloulakau (2006) similarly predicted PM_{10} concentrations almost a day ahead using ANN. Other vital information like hour of day, temperature, barometric pressure, humidity, wind velocity, rainfall, and seasonal variation were also crucial for PM_{10} concentration forecasting and all these inputs should be further improved. Particulate matter in urban areas where highway transport plays an important role can also be accurately measured by this technique Suleiman et al. (2016).The ANN model 10 is capable of predicting roadside SO_2, NO_2, and NOx concentrations also. It can work based on some intricate and unclear datasets too.

To upgrade the performance of ANN, currently many studies are being done amalgamating an array of other neural networks. These novel techniques powered by ensemble tactics shifted the outcome to a different level altogether enhancing precision, mathematical constants, convergence and computational complexity of the particular forecasting models. Liu et al., 2021 suggested a multi-step forecasting method using WPD, PSO, and BPNN for $PM_{2.5}$. Naturally the outcome of these kinds of ensemble neural networks is proved to be more precise and accurate in forecasting $PM_{2.5}$ than others.

13.4.2 Fuzzy Logic

Another prominent AI technique widely utilized today is fuzzy logic, a variant of multi-valued logic that eschews absolute values. It operates on the premise that variables can hold truth values ranging between 0 and 1, allowing for partial truth where certainty lies between complete truth and falsehood (Djuris et al., 2013; Thompson et al., 2012). Fundamentally rooted in fuzzy set theory, this approach accommodates uncertainty, making it particularly adept at handling ambiguous, subjective, and imprecise variables, and capturing linguistic nuances in data using natural language terms such as “small,” “major,” “minor,” “low,” “medium,” and “high” (Türks¸ en and Zarandi, 1999). Typically, a fuzzy logic-based modeling system entails three primary phases: fuzzification, inference, and defuzzification. Initially, input variable numerical values are translated into membership variables (fuzzification). Subsequently, these membership variables undergo processing via a set of IF-THEN rules to produce a fuzzy output (inference). Lastly, the fuzzy output is further refined into a quantitative or qualitative output (defuzzification) (Erdik, 2009; Lamaazi and Benamar, 2018).

13.4.2.1 Applications

Fuzzy logic is increasingly employed as a predictive tool in air pollution forecasting, offering distinct advantages over conventional AI methods (Rahman et al., 2020). Its preference stems from its capacity to handle imprecise data, exhibit greater fault tolerance, address nonlinear functions within complex and randomly arranged datasets, and integrate seamlessly with other data mining techniques (Abiodun et al., 2018). For instance, Cheng et al. (2011) devised a predictive model utilizing fuzzy time series through a two-stage linguistic partition method, specifically assessing daily ozone concentration. Similarly, Jain and Khare (2010) detailed a neuro-fuzzy model for CO forecasting at two distinct sites in Delhi, incorporating ten parameters such as cloud cover, atmospheric pressure, humidity, sunshine hours, temperature, mixing height, wind direction, wind speed, Pasquill stability, and vehicle count as inputs.

Carbajal et al. (2012) demonstrated the superiority of fuzzy-based models over others in precision, leveraging various techniques including fuzzy logic, signal processing, and autoregressive modeling to predict air quality in Mexico City. Their innovative “Sigma operator” algorithm facilitated the arithmetic evaluation of air quality variables, contributing to enhanced predictive accuracy.

13.4.3 Support Vector Machine

SVMs or maximum margin classifiers represent a class of supervised max-margin models equipped with learning algorithms designed for data analysis in classification and regression tasks (Kaur et al., 2018). These models are utilized for various purposes including numerical prediction, pattern recognition, and classification, etc. (Roy et al., 2015). Notably, SVMs offer several advantages, including their effectiveness in outlier detection, applicability in high-dimensional spaces, and suitability for scenarios where the number of dimensions exceeds the number of samples.

13.4.3.1 Applications

For several years, SVMs have been employed to assess various pollutants such as PM, SO_2, CO, and O_3, often in conjunction with other Machine Learning (ML) techniques. These SVM models excel at identifying relationships or patterns within intricate input and output data, a task often unattainable through conventional numerical calculations (Awad and Khanna, 2015). Researchers have consistently found that ensemble modeling systems based on SVMs exhibit stable and precise prediction accuracy compared to individual approaches. Yeganeh et al. (2017) utilized a unique SVM-based model combined with Partial Least Squares (PLS) to measure CO concentrations, while Nieto et al. (2013) focused on urban areas in Spain to measure key pollutants and analyze their interrelationship. Nieto et al. (2018) demonstrated the superior accuracy and robustness of SVM-based models in predicting PM_{10} concentrations compared to alternative models such as MLP, VARMA, and ARMA. Luna et al. (2014) similarly observed enhanced predictive capabilities when combining Principal Component Analysis (PCA) with SVM and ANN to project the levels of O_3 in Rio de Janeiro, Brazil, shedding light on key climatic factors affecting ozone levels. Additionally, Wang et al. (2015) highlighted the improved performance of hybrid SVM-ANN models over single statistical models, while Murillo et al. (2019) developed a multi-pollutant forecasting model integrating both ANN and Support Vector Regression (SVR) methods, consistently outperforming individual models in predictive accuracy.

13.4.4 Deep Neural Networks

Simply put, deep neural networks (DNN) can be described as artificial neural networks (ANNs) with increased depth, featuring a greater number of hidden layers positioned between input and output layers as shown in Figure 13.2. DNNs leverage deep architecture is used to achieve intensified levels of data complexity and abstraction (Abiodun et al., 2018; Chaudhuri and Ghosh, 2017). Networks comprising more than three layers are typically termed deep networks, while those with fewer than three hidden layers are referred to as shallow networks (Wardah et al., 2019; Sahiner et al., 2019). A distinguishing characteristic of DNNs is their capacity for feature learning, setting them apart from their predecessors (Choi, 2018; Ganapathy et al., 2018). Compared to ANNs, DNNs exhibit greater architectural complexity, enabling them to handle larger volumes of intricate data (Awad and Khanna, 2015; Vieira et al., 2017). Notably, among various recent DNN architectures, Long Short-Term Memory networks (LSTM), Autoencoders, and Convolutional Neural Networks (CNN) have demonstrated exceptional performance and are widely recognized as the most proficient and popular architectures for air pollution forecasting, offering a balance between computational complexity and predictive accuracy (Kim et al., 2018; Zhang et al., 2020).

13.4.4.1 Applications

Although the exploration and implementation of DNNs are still in their early stages, this technique holds the potential to revolutionize the current and future landscape

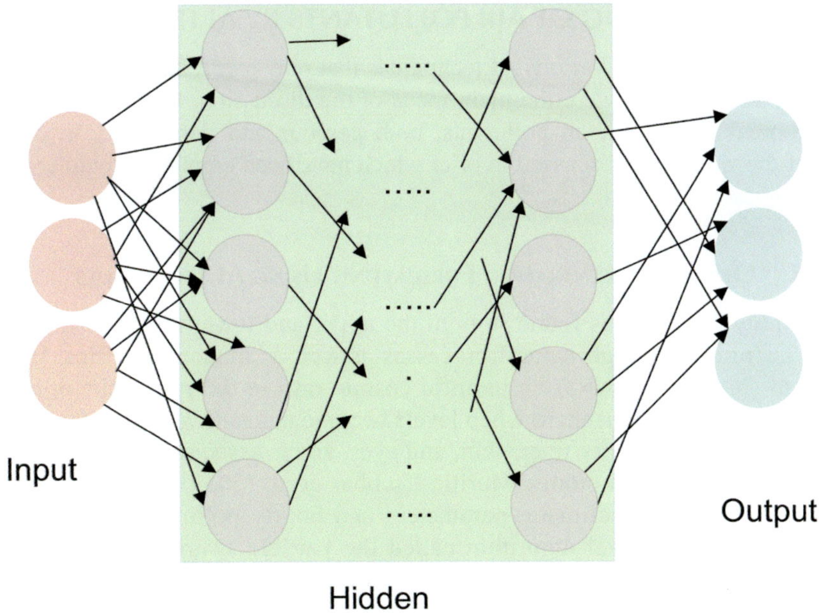

FIGURE 13.2 A schematic of a deep neural network.

of AI-driven air pollution prediction models. In 2018, Freeman emerged as one of the pioneering researchers in the field, developing air quality forecasting using deep learning methodologies. By employing a combination of Recurrent NN and LSTM, Freeman successfully predicted ozone levels in a suburban environment, demonstrating exceptional behavior compared to traditional Feed-forward Neural Networks (FFNN). Wang and Song (2018) further validated the effectiveness of ensemble models integrating ANNs and Feed-forward ANN (FFA), showcasing accurate predictions for both short and long-term forecasts.

A groundbreaking study by Li et al. (2017) introduced a novel approach to capture the comprehensive long-term dynamics of air pollutants through the utilization of stacked LSTM (LSTME), Spatiotemporal Deep Learning (STDL), Time Delay Neural Network (TDNN), Autoregressive Moving Average (ARMA), Support Vector Regression (SVR), and standard LSTM models. LSTME exhibited superior performance across various parameters compared to other machine learning techniques. Additionally, Li et al. (2015) and Zhang et al. (2016) endeavored to develop a unique forecasting model based on climatic condition images, utilizing Convolutional NN to predict $PM_{2.5}$ in urban areas based on atmospheric and weather report images. The results demonstrated reduced error rates, with potential for further enhancement with high-resolution input images. Therefore, deep learning neural networks stand as a singularly accurate technique capable of forecasting air pollution incidents well in advance through the analysis of atmospheric imagery.

13.5 FORECASTING OF AIR POLLUTANTS BY AI TECHNIQUES

In the previous section, various AI techniques that play a crucial role in forecasting different air pollutants were the main focus of the discussion. In the following part ground-breaking studies on pollutants, both gaseous and particulate, in different parts of the world, are discussed in brief which have been employing cutting-age AI technologies.

13.5.1 Ozone Concentration Forecasting Using AI Techniques

O_3, a highly reactive gas found in both the upper and lower layers of the atmosphere, is formed through natural processes as well as human activities. While it serves to shield the Earth from harmful cosmic rays in the stratosphere, it poses health risks in the troposphere when levels become unhealthy. This potent oxidant can irritate the respiratory tract, skin, and eyes, and is associated with asthma and various respiratory symptoms. Murillo-Escobar et al. (2019) developed an SVR model to project four climatic parameters and hourly pollution concentrations. They introduced a novel algorithm called the particle swarm algorithm, which reduced computational complexity and yielded superior performance across all seasons.

Similarly, a study conducted in two highly polluted Chinese cities utilized a dual-scale ensemble learning model to predict daily ozone levels, employing a multi-decomposition method and intelligent algorithm optimization (Rahman et al., 2020). Sarkar et al. (2016) utilized a fuzzy logic-based system to analyze air pollution dynamics in Kolkata, Eastern India, particularly focusing on the seasonal variation from monsoon to winter, influenced by NO_2 and O_3 (Murillo-Escobar et al., 2019). Additionally, various AI-based modeling techniques including neuro-fuzzy, ANN, and MLR have been employed to forecast O_3 levels in metropolitan areas. Studies have consistently shown that ANN models outperform MLR models in predicting gaseous pollutants.

13.5.2 Carbon Monoxide Forecasting Using AI Techniques

Carbon monoxide (CO) stands as a highly hazardous gaseous pollutant capable of inducing severe health complications in humans by forming carboxyhemoglobin (CO-Hb) upon reaction with hemoglobin, thereby diminishing the blood's oxygen-carrying capacity (Robertson, 2001). Inhalation of elevated CO concentrations over a short duration can prove fatal, particularly for vulnerable demographics such as newborns, the elderly, and individuals with cardiac or pulmonary conditions (Norhayati, I et al., 2018). To tackle this critical issue, predictive studies on atmospheric carbon monoxide levels have been undertaken in severely polluted regions like Delhi and Agra, aiming to develop early-warning systems (Mishra et al., 2015).

These studies consider various influencing factors including cloud cover, relative humidity, wind speed, temperature, barometric pressure, daylight duration, mixing height, wind direction, and vehicular activity. Utilizing a neuro-fuzzy model, accurate predictions of CO concentrations amidst these urban environments were achieved, as indicated by modeling outcomes exhibiting an index of agreement (IA) ranging from 0.88 to 0.93. Moreover, Noori et al. demonstrated the successful prediction of daily CO levels one day in advance using ANN and Adaptive Neuro-Fuzzy Inference System (ANFIS) models. Upon uncertainty analysis, the ANN model emerged as the superior predictor of daily CO concentrations compared to the ANFIS model (Wang et al., 2015).

The collective observations from these investigations underscore the distinctive nature of each forecasting model in predicting carbon monoxide levels. Overall, SVR, ANN, and fuzzy logic models have emerged as top performers in carbon monoxide forecasting endeavors.

13.5.3 Carbon Dioxide Forecasting Using AI Techniques

The escalation of CO_2 emissions and its persistent accumulation in the atmosphere represent one of the most critical contemporary threats to the world. This greenhouse gas (GHG) is the primary driver of global warming due to its capacity to trap heat. Unabated increases in CO_2 levels, reaching concentrations as high as 5000 parts per million (ppm), can pose severe health risks to humans by lowering the pH of blood serum, leading to acidosis and consequently contributing to cardiovascular and respiratory ailments (Robertson, 2001). Consequently, studies aimed at quantifying the percentage of CO_2 in the air hold paramount significance.

Ahmadi et al. (Ahmadi et al., 2019) conducted a pivotal survey across five Middle Eastern countries to forecast carbon dioxide emissions utilizing the General Method of Data Handling (GMDH) ANN. Norhayati and Rashid (2018) developed the Adaptive Neuro-Fuzzy Inference System (ANFIS) to accurately estimate CO_2 emissions from a medical waste incineration plant using real-world data. Their model precisely predicts carbon emissions.

Qader et al. (2022) employed various strategies to forecast CO_2 emissions in Bahrain, including Gaussian Process Regression, Holt's methods, and nonlinear autoregressive modeling. Neural network-based methodologies excel in accurately forecasting CO_2 emissions through time series models. Even when confronted with nonlinear data, neural network-based techniques consistently provide precise and detailed results, which are crucial for implementing effective emission control measures.

13.5.4 NO_2 and SO_2 Forecasting Using AI Techniques

Anthropogenic activities represent the primary sources of airborne pollutants such as NO_2 and SO_2. The combustion in industry due to fossil fuels and transportation serves as the primary contributor to NO_2 emissions (Najjar., 2011), while SO_2 emissions predominantly originate from power plants engaged in mineral ore smelting. Both NO_2 and SO_2 are significant contributors to acid rain, which exerts deleterious effects on vegetation, freshwater ecosystems, and soil quality. Acid rain also adversely affects terrestrial and aquatic life forms, causes paint degradation, corrodes steel structures including bridges, and accelerates the weathering of stone buildings and sculptures. Additionally, exposure to NO_2 has adverse effects on ecosystems, leading to an increased incidence of lung diseases. Consequently, the development of effective models for anticipating and warning against SO_2 and NO_2 concentrations is imperative.

Xu et al. (2017) developed a hybrid model based on the Grey Wolf Optimizer meta-heuristic approach to predict SO_2 and PM in provinces of China. They also assessed the accuracy of NO_2 concentration estimation using a PCA-based ANN and MLR model. Brunelli et al. (2007) estimated SO_2, CO, PM_{10}, and NO_2 in Palermo, Italy, utilizing a Recurrent Neural Network (Elman model) to forecast daily maximum O_3 levels. Shams et al. (2021) evaluated the efficacy of forecasting SO_2 content in Tehran's air using a Multi-layer Perceptron (MLP) and MLR models, incorporating various weather-related variables, timeframes, urban green space, and transportation data as input parameters.

Zhu et al. (2018) developed a two-step hybrid model based on Complementary Ensemble Empirical Mode Decomposition (CEEMD) for demonstrating performance and accuracy in predicting SO_2 and NO_2 levels. Liu et al. (2021) recommended a method for the projection of daily average NO_2 concentrations in the Tianjin region using LSTM and discrete wavelet transform networks. By utilizing historical data and weather information as inputs, their model achieved efficient NO_2 concentration forecasts one day in advance, demonstrating improved performance in estimating NO_2 levels in Tianjin.

13.5.5 Particulate Matter Forecasting Using AI Techniques

The presence of fine particulate matter (PM) in the air has been closely associated with various clinical manifestations of pulmonary and cardiovascular diseases, contributing to morbidity and mortality from respiratory ailments in both human and animal populations. These inhalable fine particles not only enter the bloodstream but also inflict damage on neurological, immunological, and DNA systems to varying degrees (Lubinski, W.et al, 2005). Consequently, the development of an efficient PM forecasting and warning system is a top priority for any nation to provide timely information on PM levels within localities.

Cheng et al. (2019) conducted a survey of $PM_{2.5}$ concentration forecasting models in China, while Mirzadeh et al. (2022) investigated PM_{10} pollution concentrations using a wavelet-artificial intelligence hybrid model for short and medium-term forecasts. They successfully developed a new hybrid model for outdoor $PM_{2.5}$ level prediction, assessing its utility in predicting health risks and economic losses due to pollution severity. Considering the adverse effects of PM, researchers emphasized that forecasting models should not only provide early warnings but also assist in addressing health issues (Du, P. et al, 2022). In one study based in Shanghai, the authors utilized the Extreme Boosting (XGBoost) ML algorithm for $PM_{2.5}$ concentration prediction (Ma, 2020). In another study, Dotse et al. (2018) focused on daily PM_{10} levels in Brunei Darussalam, employing a genetically optimized RF back propagation NN strategy. The system showed potential for real-time monitoring and forecasting weather parameters, including predicting smoggy conditions, outperforming other benchmark models for forecasting daily PM concentrations.

As observed in various reports from different regions worldwide, hybrid models tend to be more efficient for predicting PM concentrations (Subramanian et al., 2022). These models combine precision forecasting with time series and nonlinearity capabilities, often recommended by researchers for PM concentration prediction. Moreover, incorporating ensemble methods and data decomposition has gained favor among researchers globally. Reports indicate that the ANN model outperforms the MLR model in predicting NO_2 and SO_2 pollutants, making neural network-based methodologies the preferred option for forecasting CO_2 emissions. Based on earlier findings, ANN, SVR, and fuzzy logic forecasting models have consistently emerged as the best choices for carbon monoxide forecasting. Neural networks have demonstrated reliable prediction and forecasting capabilities across various fields, excelling in handling complex nonlinear or multi-layered information and proving invaluable for challenging time-series or classification inputs.

13.6 CONCLUSIVE REMARKS AND FUTURE PROSPECTS

In previous sections, we've explored how AI technologies have revolutionized environmental pollution forecasting, addressing major pollutants such as O_3, CO, CO_2, NO_2, SO_2, and PM, while also contributing to public health and climate change concerns. Despite the significant strides made by AI technologies and ML algorithms in environmental pollution control, they face challenges during extreme climate conditions. Severe pollution episodes resulting from high emissions, effective transport of airborne particles or gases, and unfavorable climatic conditions can hinder the performance of AI-based techniques. For instance, neural networks often encounter overfitting issues during high pollution episodes, requiring input selection and optimization through techniques like Genetic Algorithms (GA), Particle Swarm Optimization (PSO), and Regularization-based Pruned Learning Neural Networks (RPLNN). Fuzzy logic, when used independently, may underperform for both long-term and short-term forecasting due to poor adaptive and self-learning capabilities. SVM generally handle larger datasets but are computationally and memory-intensive, lacking transparency and interpretability in predicted outcomes. Deep neural networks are computationally costly to train and require a large measurements of training data, limiting their applicability in air pollution forecasting.

Scientists are actively investigating the reasons behind these erratic performances during extreme pollution events. One perspective suggests that peak concentrations may be misinterpreted due to the effect of data imbalance problems. Overfitting in AI-based systems can occur when inputs are outside the range of training data, leading to incorrect predictions and generalized forecasting during peak pollution episodes. Moreover, fluctuations in high and low parameters can exacerbate the error in general air pollution data characterized by a single AI-based technique. This underscores the effectiveness of ensemble AI techniques in mitigating ambiguity and variability in air pollution event forecasting. Ensemble AI technologies demonstrate better accuracy and robustness in handling nonlinear parameter-data outcomes and exhibit consistent performance over generalized and noisy datasets.

However, ensemble techniques also have drawbacks. The primary disadvantage is the longer computation time required, which may not be conducive to real-time air pollution forecasting applications. Additionally, interpreting prediction results can be challenging due to uncertainty regarding base-model selection.

Despite these challenges, emerging AI techniques offer promise in the field of air pollution control and meteorological forecasting. However, much ground remains to be covered. Moreover, broad-level parameters including meteorological conditions from pollution sources have not been adequately considered in air pollution forecasting accuracy assessments. More insightful parameters, including socioeconomic status and transport-related variables, should be included when predicting vehicular-based air pollution. Many economic factors, pollution information, energy usage, etc. are the essential input information required for high-precision modeling systems.

REFERENCES

Aayush, K., Vishal, D., Hammad, N., Manu, K., 2020. Application of artificial intelligence in curbing air pollution: The case of India. *Asian J. Manag*. 11, 285–290.

Abiodun, O.I., Jantan, A., Omolara, A.E., Dada, K.V., Mohamed, N.A., Arshad, H., 2018. State-of-the-art in artificial neural network applications: A survey. *Heliyon* 4 (11), e00938. https://doi.org/10.1016/j.heliyon.2018.e00938

Agarwal, R., Awasthi, A., Mital, S.K., Singh, N., Gupta, P.K., 2014. Statistical model to study the effect of agriculture crop residue burning on healthy subjects. *Mapan* 29, 57–65.

Agirre-Basurko, E., Ibarra-Berastegi, G., Madariaga, I., 2006. Regression and multilayer perceptron-based models to forecast hourly O3 and NO2 levels in the Bilbao area. *Environ. Model. Software* 21 (4), 430–446. https://doi.org/10.1016/j. envsoft.2004.07.008

Ahmadi, M.H., Jashnani, H., Chau, K.-W., Kumar, R., Rosen, M.A., 2019. Carbon dioxide emissions prediction of five Middle Eastern countries using artificial neural networks. *Energy Sources Part A Recovery Util. Environ. Eff.* 45 (3), 9513–9525.

Arena, P., Fortuna, L., Gallo, A., Nunnari, G., Xibilia, M.G., 1995. Air pollution estimation via neural networks. *IFAC Proceedings* 28 (10), 787–792. https://doi.org/10.1016/S1474-6670(17)51616-X

Awad, M., Khanna, R., 2015. Deep neural networks. In: *Efficient Learning Machines*. Apress, Berkeley, CA, pp. 127–147. https://doi.org/10.1007/978-1-4302-5990-9_4

Bai, L., Wang, J., Ma, X., Lu, H., 2018. Air pollution forecasts: An overview. *Int. J. Environ. Res. Public Health* 15, 780.

Bellinger, C., Jabbar, M.S.M., Zaïane, O., Osornio-Vargas, A., 2017. A systematic review of data mining and machine learning for air pollution epidemiology. *BMC Public Health* 17, 907. doi: 10.1186/s12889-017-4914-3

Borana, J., 2016. Applications of artificial intelligence & associated technologies. In: *Proceeding of International Conference on Emerging Technologies in Engineering, Biomedical, Management and Science*. Jodhpur. Available onlineat: https://test.globalinfocloud.com/technodigisoftnew/wp-content/uploads/2019/07/Applications-of- Artificial-Intelligence-Associated-Technologies.pdf (accessed August 24, 2022)

Brunelli, U., Piazza, V., Pignato, L., Sorbello, F., Vitabile, S., 2007. Two-days ahead prediction of daily maximum concentrations of SO2, O3, PM10, NO2, CO in the urban area of Palermo, Italy. *Atmos. Environ.* 41, 2967–2995.

Carbajal-Hernández, J.J., Sanchez-Fernández, L.P., Carrasco-Ochoa, J.A., Martínez- Trinidad, J.F., 2012. Assessment and prediction of air quality using fuzzy logic and autoregressive models. *Atmos. Environ.* 60, 37–50. https://doi.org/10.1016/j. atmosenv.2012.06.004

Chatterjee, S.P., Pandya, A.S., 2016. Artificial neural networks in drug transport modeling and simulation–II. In: *Artificial Neural Network for Drug Design, Delivery and Disposition.* Academic Press, pp. 243–261. https://doi.org/10.1016/B978-0-12- 801559-9.00012-0

Chaudhuri, A., Ghosh, S.K., 2017. *Bankruptcy Prediction through Soft Computing Based Deep Learning Technique.* Springer. https://doi.org/10.1007/978-981-10-6683-2

Chen, C., 2017. Science mapping: A systematic review of the literature. *J Data Inform Sci.* 2, 1–40. doi: 10.1515/jdis-2017-0006

Chen, X., Yin, L., Fan, Y., Song, L., Ji, T., Liu, Y., Tian, J., Zheng, W., 2020. Temporal evolution characteristics of PM2.5 concentration based on continuous wavelet transform. *Sci. Total Environ.* 699, 134244.

Cheng, C.H., Huang, S.F., Teoh, H.J., 2011. Predicting daily ozone concentration maxima using fuzzy time series based on a two-stage linguistic partition method. *Comput. Math. Appl.* 62 (4), 2016–2028. https://doi.org/10.1016/j.camwa.2011.06.044

Cheng, Y., Zhang, H., Liu, Z., Chen, L., Wang, P. 2019. Hybrid algorithm for short-term forecasting of PM2. 5 in China. *Atmospheric Environment*, 200, 264–279.

Choi, H., 2018. Deep learning in nuclear medicine and molecular imaging: Current perspectives and future directions. *Nuclear Medicine and Molecular Imaging* 52 (2), 109–118. https://doi.org/10.1007/s13139-017-0504-7

Chowdhury, M., Sadek, A.W., 2012. Advantages and limitations of artificial intelligence. *ArtifIntellApplicCrit Transport Issues* 6, 360–75. Available online at: http://onlinepubs.trb.org/onlinepubs/circulars/ec168.pdf (accessed August 24, 2022)

COP26 special report on climate change and health: The health argument for climate action. (n.d.). Retrieved March 15, 2024, from www.who.int/publications-detail-redirect/9789240036727

Djuris, J., Ibric, S., Djuric, Z., 2013. Neural computing in pharmaceutical products and process development. In: *Computer-Aided Applications in Pharmaceutical Technology.* Woodhead Publishing, pp. 91–175. https://doi.org/10.1533/ 9781908818324.91

Dotse, S.-Q., Petra, M.I., Dagar, L., De Silva, L.C., 2018. Application of computational intelligence techniques to forecast daily PM10 exceedances in Brunei Darussalam. *Atmos. Pollut. Res.* 9, 358–368.

Du, P., Wang, J., Yang, W., Niu, T., 2022. A novel hybrid fine particulate matter (PM2.5) forecasting and its further application system: Case studies in China. *J. Forecast.* 41, 64–85.

Erdik, T., 2009. Fuzzy logic approach to conventional rubble mound structures design. *Expert Syst. Appl.* 36 (3), 4162–4170. https://doi.org/10.1016/j.eswa.2008.06.012

Feng, X., Li, Q., Zhu, Y., Hou, J., Jin, L., Wang, J., 2015. Artificial neural networks forecasting of PM2.5 pollution using air mass trajectory based geographic model and wavelet transformation. *Atmos. Environ.* 107, 118–128.

Fernández, J.D., Vico, F., 2013. AI methods in algorithmic composition: A comprehensive survey. *J. Artif. Intell. Res.* 48, 513–582.

Fernando, H.J., Mammarella, M.C., Grandoni, G., Fedele, P., Di Marco, R., Dimitrova, R., Hyde, P. 2012. Forecasting PM10 in metropolitan areas: Efficacy of neural networks. *Environmental Pollution*, 163, 62–67.

Freeman, B.S., Taylor, G., Gharabaghi, B., Thé, J., 2018. Forecasting air quality time series using deep learning. *J. Air Waste Manag. Assoc.* 68 (8), 866–886. https://doi. org/ 10.1080/10962247.2018.1459956

Ganapathy, N., Swaminathan, R., Deserno, T.M., 2018. Deep learning on 1-D biosignals: A taxonomy-based survey. *Yearbook of Medical Informatics* 27 (1), 98. https://doi.org/ 10.1055% 2Fs-0038-1667083

GBD 2019 Risk Factors Collaborators 2020. Global burden of 87 risk factors in 204 countries and territories, 1990–2019: a systematic analysis for the Global Burden of Disease Study 2019. *The Lancet* 396(10258), 1223–1249.

Ghritlahre, H.K., Prasad, R.K., 2018. Application of ANN technique to predict the performance of solar collector systems-A review. *Renew. Sustain. Energy Rev.* 84, 75–88. https:// doi.org/10.1016/j.rser.2018.01.001

Grivas, G., Chaloulakou, A., 2006. Artificial neural network models for prediction of PM10 hourly concentrations, in the Greater Area of Athens, Greece. *Atmos. Environ.* 40 (7), 1216–1229. https://doi.org/10.1016/j.atmosenv.2005.10.036

Gu, Y., Li, B., Meng, Q., 2022. Hybrid interpretable predictive machine learning model for air pollution prediction. *Neurocomputing* 468, 123–136.

Gurjar, B. R. (2021). *Air pollution in India: Major issues and challenges*. Energy Future, Magzter.

Gurjar, B.R., van Aardenne, J.A., Lelieveld, J., Mohan, M., 2004 October. Emission estimates and trends (1990–2000) for megacity Delhi and implications. Atmospheric *Environ.* 38(33), 5663–5681.

Hasanuzzaman, M., Abd Rahim, N. (Eds.), 2019. *Energy for Sustainable Development: Demand, Supply, Conversion and Management*. Academic Press.

Hashimoto, D.A., Witkowski, E., Gao, L., Meireles, O., 2020. Rosmananesthesiology: Techniques, intelligence G. Artificialincurrent clinical applications and limitations. *Anesthesiology* 132, 379–94. doi: 10.1097/ ALN.0000000000002960

Heuvelmans, M.A., van Ooijen, P.M., Ather, S., Silva, C.F., Han, D., Heussel, C.P., Hickes, W., Kauczor, H.-U, Novotny, P., Peschl, H., 2021. Lung cancer prediction by deep learning to identify benign lung nodules. *Lung Cancer*, 154, 1–4.

Hino, M., Benami, E., Brooks, N., 2018. Machine learning for environmental monitoring. *Nat Sust.* 1, 583–8. doi: 10.1038/s41893-018-0142-9

Hu, X., Belle, J.H., Meng, X., Waller, L.A., Strickland, M.J., Liu, Y., 2017. Estimating PM2. 5 concentrations in the conterminous United States using the random forest approach. *Environ Sci Technol.* 51, 6936–44. doi: 10.1021/acs.est. 7b01210

Huang, C.J., Kuo, P.H., 2018. A deep cnn-lstm model for particulate matter (PM2.5) forecasting in smart cities. *Sensors* 18, 2220. doi: 10.3390/s18072220

Jain, S., Khare, M., 2010. Adaptive neuro-fuzzy modeling for prediction of ambient CO concentration at urban intersections and roadways. *Air Quality, Atmosphere & Health* 3(4), 203–212. https://doi.org/10.1007/s11869-010-0073-8

Jat, R., Gurjar, B.R., Lowe, D. 2021. February. Regional pollution loading in winter months over India using high resolution WRF-Chem simulation. *Atmospheric Res.* 249, 105326.

Jerrett, M., Arain, A., Kanaroglou, P., Beckerman, B., Potoglou, D., Sahsuvaroglu, T., Morrison, J.,Giovis, C., 2005. A review and evaluation of intraurban air pollution exposure models. *J. Expo. Sci. Environ. Epidemiol.* 15, 185–204.

Joshi, P., Dey, S., Ghosh, S., Jain, S., Sharma, S.K. 2022. Association between acute exposure to PM2. 5 chemical species and mortality in megacity Delhi, India. *Environmental Science & Technology*, 56(11), 7275–7287.

Kaur, P., Chaudhary, P., Bijalwan, A., Awasthi, A., 2018. Network traffic classification using multiclass classifier. In *Advances in Computing and Data Sciences: Second International Conference, ICACDS 2018*, Dehradun, India, April 20-21, 2018, Revised Selected Papers, Part I 2, pp. 208-217. Springer Singapore.

Kim, K., Kim, D.K., Noh, J., Kim, M., 2018. Stable forecasting of environmental time series via long short term memory recurrent neural network. *IEEE Access* 6, 75216–75228. https://doi.org/10.1109/ACCESS.2018.2884827

Lamaazi, H., Benamar, N., 2018. OF-EC: A novel energy consumption aware objective function for RPL based on fuzzy logic. *J. Netw. Comput. Appl.* 117, 42–58. https:// doi. org/10.1016/j.jnca.2018.05.015

Laurance, W. F. 2010. Habitat destruction: death by a thousand cuts. *Conservation Biology for All*, 1(9), 73–88.

Li, X., Peng, L., Yao, X., Cui, S., Hu, Y., You, C., Chi, T., 2017. Long short-term memory neural network for air pollutant concentration predictions: Method development and evaluation. *Environ. Pollut.* 231, 997–1004. https://doi.org/10.1016/j. envpol.2017.08.114

Li, X., Peng, L., Yao, X., Hu, Y., You, C., Chi, T., 2017. Long short-term memory neural network for air pollutant concentration predictions: Method development and evaluation. *Environ Pollut.* 231, 997–1004. doi: 10.1016/j.envpol.2017. 08.114

Li, Y., Huang, J., Luo, J., 2015. August. Using user generated online photos to estimate and monitor air pollution in major cities. In: *Proceedings of the 7th International Conference on Internet Multimedia Computing and Service*, pp. 1–5. https://doi. org/10.1145/ 2808492.2808564

Liu, B., Zhang, L., Wang, Q., Chen, J., 2021. A novel method for regional NO2 concentration prediction using discrete wavelet transform and an LSTM network. *Comput. Intell. Neurosci.* 2021, 6631614.

Liu, H., Yan, G., Duan, Z., Chen, C., 2021. Intelligent modelling strategies for forecasting air quality time series: A review. *Appl. Soft Comput.* 102, 106957.

Lubinski, W., Toczyska, I., Chcialowski, A., Plusa, T., 2005. Influence of air pollution on pulmonary function in healthy young men from different regions of Poland. *Ann. Agric. Environ. Med.* 12, 1–4.

Luna, A.S., Paredes, M.L.L., De Oliveira, G.C.G., Correa, ^ S.M., 2014. Prediction of ozone concentration in tropospheric levels using artificial neural networks and support vector machine at Rio de Janeiro, Brazil. *Atmos. Environ.* 98, 98–104. https://doi. org/10.1016/ j.atm.osenv.2014.08.060

Ma, J., Yu, Z., Qu, Y., Xu, J., Cao, Y. 2020. Application of the XGBoost machine learning method in PM2. 5 prediction: A case study of Shanghai. *Aerosol and Air Quality Research*, 20(1), 128–138.

Masood, A., Ahmad, K. 2020. A model for particulate matter (PM2.5) prediction for Delhi based on machine learning approaches. *Procedia Comput. Sci.* 167, 2101–2110.

Masood, A., Ahmad, K. 2021. A review on emerging artificial (AI) techniques for air pollution forecasting: Intelligence fundamentals, application and performance. *J Clean Product.* 322, 129072. doi: 10.1016/j.jclepro.2021.129072

May, T.O., Livas-García, A., Jiménez-Torres, M., Cruz May, E., López-Manrique, L.M., Bassam, A., 2020. Artificial intelligence techniques for modeling indoor building temperature under tropical climate using outdoor environmental monitoring. *J Energy Eng.* 146, 04020004. doi: 10.1061/(ASCE)EY.1943-7897.0000649

McKendry, I.G., 2002. Evaluation of artificial neural networks for fine particulate pollution (PM10 and PM2.5) forecasting. *J. Air Waste Manag. Assoc.* 52 (9), 1096–1101. https:// doi.org/10.1080/10473289.2002.10470836

Mirzadeh, S., Nejadkoorki, F., Mirhoseini, S., Moosavi, V., 2022. Developing a wavelet-AI hybrid model for short-and long-term predictions of the pollutant concentration of particulate matter10. *Int. J. Environ. Sci. Technol.* 19, 209–222.

Mishra, D., Goyal, P., 2015. Development of artificial intelligence based NO_2 forecasting models at TajMahal, Agra. *Atmos. Pollut. Res.* 6, 99–106.

Mlakar, P., Boznar, ˇ M., Lesjak, M., 1994. Neural networks predict pollution. In: *Air Pollution Modeling and its Application X.* Springer, Boston, MA, pp. 659–660. https://doi.org/10.1007/978-1-4615-1817-4_93

Mo, X, Zhang, L., Li, H., Qu, Z., 2019. A novel air quality early-warning system based on artificial intelligence. *Int. J. Environ. Res. Public Health* 16, 3505.

Murillo-Escobar, J., Sepulveda-Suescun, J.P., Correa, M.A., Orrego-Metaute, D., 2019. Forecasting concentrations of air pollutants using support vector regression improved with particle swarm optimization: Case study in Aburra ´ Valley, Colombia. *Urban Climate* 29, 100473. https://doi.org/10.1016/j.uclim.2019.100473

Murray, C. J., Aravkin, A. Y., Zheng, P., Abbafati, C., Abbas, K. M., Abbasi-Kangevari, M., … & Borzouei, S. 2020. Global burden of 87 risk factors in 204 countries and territories, 1990–2019: A systematic analysis for the Global Burden of Disease Study 2019. *The Lancet*, 396(10258), 1223–1249.

Nagpure, A. S., Gurjar, B. R., Kumar, V., & Kumar, P. 2016. Estimation of exhaust and non-exhaust gaseous, particulate matter and air toxics emissions from on-road vehicles in Delhi. *Atmospheric Environment*, 127, 118–124.

Najjar, Y.S., 2011. Gaseous pollutants formation and their harmful effects on health and environment. *Innov. Energy Policies* 1, 1–8.

Neo, E.X., Hasikin, K., Mokhtar, M.I., Lai, K.W., Azizan, M.M., Razak, S.A., Hizaddin, H.F. 2022. Towards integrated air pollution monitoring and health impact assessment using federated learning: A systematic review. *Front. Public Health* 10, 851553. DOI 10.3389/fpubh.2022.851553

Nieto, P.G., Combarro, E.F., del Coz Díaz, J.J., Montan˜ es, ´ E., 2013. A SVM-based regression model to study the air quality at local scale in Oviedo urban area (Northern Spain): A case study. *Appl. Math. Comput.* 219 (17), 8923–8937. https:// doi.org/10.1016/j.amc.2013.03.018

Nieto, P.G., Lasheras, F.S., García-Gonzalo, E., de Cos Juez, F.J., 2018. PM10 concentration forecasting in the metropolitan area of Oviedo (Northern Spain) using models based on SVM, MLP, VARMA and ARIMA: a case study. *Sci. Total Environ.* 621, 753–761. https://doi.org/10.1016/j.scitotenv.2017.11.291

Nilashi, M., Ahmadi, H., Manaf, A.A., Rashid, T.A., Samad, S., Shahmoradi, L., Aljojo, N., Akbari, E., 2020. Coronary heart disease diagnosis through self-organizing map and fuzzy support vector machine with incremental updates. *Int. J. Fuzzy Syst.* 22, 1376–1388.

Norhayati, I., Rashid, M., 2018. Adaptive neuro-fuzzy prediction of carbon monoxide emission from a clinical waste incineration plant. *Neural Comput. Appl.* 30, 3049–3061.

Olivieri, A.C., Allegrini, F., 2019. Chemometrics and statistics: Neural networks. https://doi.org/10.1016/B978-0-12-409547-2.13966-6

Qader, M.R., Khan, S., Kamal, M., Usman, M., Haseeb, M., 2022. Forecasting carbon emissions due to electricity power generation in Bahrain. *Environ. Sci. Pollut. Res.* 29, 17346–17357.

Rahimi, A., 2017. Short-term prediction of NO2 and NO concentrations using multilayer perceptron neural network: A case study of Tabriz, Iran. *Ecological Processes* 6 (1), 4. https://doi.org/10.1186/s13717-016-0069-x

Rahman, M.M., Shafiullah, M., Rahman, S.M., Khondaker, A.N., Amao, A., Zahir, M., 2020. Soft computing applications in air quality modeling: Past, present, and future. *Sustainability* 12 (10), 4045. https://doi.org/10.3390/su12104045.

Robertson, D., 2001. The rise in the atmospheric concentration of carbon dioxide and the effects on human health. *Med. Hypotheses* 56, 513–518.

Roy, K., Kar, S., Das, R.N., 2015. *Understanding the Basics of QSAR for Applications in Pharmaceutical Sciences and Risk Assessment*. Academic press.

Sahiner, B., Pezeshk, A., Hadjiiski, L.M., Wang, X., Drukker, K., Cha, K.H., Summers, R. M., Giger, M.L., 2019. Deep learning in medical imaging and radiation therapy. *Med. Phys.* 46 (1), e1–e36. https://doi.org/10.1002/mp.13264

Sapra, R.L., Mehrotra, S., Nundy, S., 2015. Artificial neural networks: Prediction of mortality/survival in gastroenterology. *Current Medicine Research and Practice* 5 (3), 119–129. https://doi.org/10.1016/j.cmrp.2015.05.007

Sarkar, S., Parihar, S. M., Dutta, A. 2016. Fuzzy risk assessment modelling of East Kolkata Wetland Area: A remote sensing and GIS based approach. *Environmental Modelling & Software*, 75, 105–118.

Singh Nagpure, B.R., Gurjar, Vivek Kumar, Prashant Kumar. 2016. February. Estimation of exhaust and non-exhaust gaseous, particulate matter and air toxics emissions from on-road vehicles in Delhi, *Atmospheric Environment* 127, 118–124.

Singh, N., Agarwal, R., Awasthi, A., Gupta, P. K., Mittal, S. K. 2010. Characterization of atmospheric aerosols for organic tarry matter and combustible matter during crop residue burning and non-crop residue burning months in Northwestern region of India. *Atmos. Environ.* 44 (10), 1292-1300.

Sohn, S.H., Oh, S.C., Yeo, Y.K., 1999. Prediction of air pollutants by using an artificial neural network. *Kor. J. Chem. Eng.* 16 (3), 382–387. https://doi.org/10.1007/ BF02707129

Stein, A.L. 2020. Artificial intelligence and climate change. *Yale J. Regul.* 37, 890.

Subramaniam, S., Raju, N., Ganesan, A., Rajavel, N., Maheswari Chenniappan, Chander Prakash, ... Dixit, S. 2022. Artificial intelligence technologies for forecasting air pollution and human health: A narrative review. *Sustainability* 14, 9951. https:// doi.org/10.3390/su1416995

Subramanyam, V., 2008. *Evolution of Artificial Neural Network Controller for a Boost Converter*. https://core.ac.uk/download/pdf/48630874.pdf

Suleiman, A., Tight, M.R., Quinn, A.D., 2016. Hybrid neural networks and boosted regression tree models for predicting roadside particulate matter. *Environ. Model. Assess.* 21(6), 731–750. https://doi.org/10.1007/s10666-016-9507-5

Tandon, A., Awasthi, A., Pattnayak, K.C., 2023. Comparison of different machine learning methods on precipitation dataset for Uttarakhand, *2023 2nd International Conference on Ambient Intelligence in Health Care (ICAIHC)*, Bhubaneswar, India, 2023, pp. 1–6, doi: 10.1109/ICAIHC59020.2023.10431402

Thompson, J.A., Roecker, S., Grunwald, S., Owens, P.R., 2012. Digital soil mapping: Interactions with and applications for hydropedology. In H. Lin (Ed.), *Hydropedology* (pp. 665–709). Elsevier.

Türkşen, I.B., Zarandi, M.F., 1999. Production planning and scheduling: Fuzzy and crisp approaches. In: *Practical Applications of Fuzzy Technologies*. Springer, Boston, MA, pp. 479–529. https://doi.org/10.1007/978-1-4615-4601-6_15

Vieira, S., Pinaya, W.H., Mechelli, A., 2017. Using deep learning to investigate the neuroimaging correlates of psychiatric and neurological disorders: Methods and applications. *Neurosci. Biobehav. Rev.* 74, 58–75. https://doi.org/10.1016/j. neubiorev.2017.01.002

Wang, C., Ye, Z., Yu, Y., Gong, W., 2018. Estimation of bus emission models for different fuel types of buses under real conditions. *Sci. Total Environ.* 640, 965–972.

Wang, J., Song, G., 2018. A deep spatial-temporal ensemble model for air quality prediction. *Neurocomputing* 314, 198–206. https://doi.org/10.1016/j. neucom.2018.06.049

Wang, P., Liu, Y., Qin, Z., Zhang, G., 2015. A novel hybrid forecasting model for PM10 and SO2 daily concentrations. *Sci. Total Environ.* 505, 1202–1212. https://doi.org/ 10.1016/ j.scitotenv.2014.10.078

Wardah, W., Khan, M.G., Sharma, A., Rashid, M.A., 2019. Protein secondary structure prediction using neural networks and deep learning: A review. *Comput. Biol. Chem.* 81, 1–8. https://doi.org/10.1016/j.compbiolchem.2019.107093

World Health Organization. 2005. *The World Health Report 2005: Make Every Mother and Child Count*. World Health Organization.

World Health Organization. 2016. *World Health Statistics 2016 [OP]: Monitoring Health for the Sustainable Development Goals (SDGs)*. World Health Organization.

World Health Organization. 2018. *Basic Emergency Care: Approach to the Acutely Ill and Injured*. World Health Organization.

World Health Organization. 2019. *World health statistics overview 2019: monitoring health for the SDGs, sustainable development goals* (No. WHO/DAD/2019.1). World Health Organization.

World Health Organization. 2021. *COP26 Special Report on Climate Change and Health: The Health Argument for Climate Action*. World Health Organization.

Xu, Y., Du, P., Wang, J. 2017. Research and application of a hybrid model based on dynamic fuzzy synthetic evaluation for establishing air quality forecasting and early warning system: A case study in China. *Environ. Pollut.* 223, 435–448.

Yeganeh, B., Hewson, M.G., Clifford, S., Knibbs, L.D., Morawska, L., 2017. A satellite-based model for estimating PM2.5 concentration in a sparsely populated environment using soft computing techniques. *Environ. Model. Software* 88, 84–92. https://doi. org/ 10.1016/j.envsoft.2016.11.017.

Zhang, C., Yan, J., Li, C., Rui, X., Liu, L., Bie, R., 2016. October. On estimating air pollution from photos using convolutional neural network. In: *Proceedings of the 24th ACM International Conference on Multimedia*, pp. 297–301.

Zhang, X., Zhang, T., Zou, Y., Du, G., Guo, N., 2020. Predictive eco-driving Application considering real-world traffic flow. *IEEE Access* 8, 82187–82200. https://doi.org/ 10.1109/ACCESS.2020.2991538

Zhou, J.-H., Zhao, J.-G., Li, P. 2010. March. Study on gray numerical model of air pollution in wuan city. In *Proceedings of the 2010 International Conference on Challenges in Environmental Science and Computer Engineering*, Wuhan, China, 6–7; pp. 321–323.

Zhu, S., Lian, X., Wei, L., Che, J., Shen, X., Yang, L., Qiu, X., Liu, X., Gao, W., Ren, X. PM2.5 forecasting using SVRwith PSOGSA algorithm based on CEEMD, GRNN and GCA considering meteorological factors. *Atmos. Environ.* 2018, 183, 20–32.

14 A Comparative Evaluation of AI-Based Methods and Traditional Approaches for Air Quality Monitoring

Analyzing Pros and Cons

Nzeyimana Bahati Shabani

14.1 INTRODUCTION

As our world continues to industrialize and populations grow, maintaining air quality becomes crucial for environmental balance and human health. In this chapter, we aim to shed light on the ascending emphasis placed on monitoring air quality, unveiling its impact on public health and ecosystem sustainability. Our consideration entails a thorough insight into two prevalent air quality monitoring methods: traditional ground-based approaches and artificial intelligent-based methods.

The inclusion of artificial intelligence (AI) and traditional techniques should not be viewed as adversaries, but rather as a partnership. Both strategies have their merits. AI excels in swift, precise information handling and has the potential to streamline tasks. Meanwhile, traditional methods can provide historical background and serve as a trustworthy standard for evaluating the capability of AI. This cooperative relationship between the two approaches is paramount for successful monitoring of air quality.

As our goal is to uncover various strategies for monitoring air quality, we make every effort to enlighten others on the significant role it plays in protecting public health and preserving the environment. At its core, this chapter serves as a recognition of our joint obligation to ensure the air we all breathe remains safe. Our mission is to equip environmental management and public health professionals with a better understanding of both modern and conventional techniques, so they can make informed decisions. By shedding light on the advantages and disadvantages of each approach, we hope to foster a movement of empowerment and action.

The study of air quality monitoring has been imperative for ensuring that all living creatures have access to cleaner, healthier, and more sustainable air. Our efforts towards this goal necessitate that we ponder the immediacy of our combined actions,

DOI: 10.1201/9781032683805-14

as pristine and life-sustaining air is crucial. This effort goes beyond science, as it is a commitment to safeguarding the present and the future of those reliant on the breath of life. The importance of this field is underscored by its widespread implications on public health and environmental sustainability (Pandey et al., 2021). Human health and the environment suffer greatly from air pollution, and monitoring its effects is crucial. The classic methods of tracking air quality have long been used as a starting point for comprehending these effects. Let's explore the techniques that have been in use for many years, such as manual sampling and ground-based stations, and evaluate the advantages and shortcomings they present stations located on the ground: these are used for communication and observation purposes (Saini et al., 2020). They are situated in various locations and can be found on mountains, in valleys, and near oceans. They are an essential part of many industries, including telecommunications and weather forecasting. Their importance cannot be overstated as they serve to provide information about the earth and surrounding environment. Additionally, they offer insights into the movement of people and animals, making them a valuable tool for conservation efforts. Overall, ground-based stations form a crucial part of our technological infrastructure and are an integral part of modern life (Zheng et al., 2014). It is imperative to recognize that AI and traditional methods should not be seen as competing entities but as collaborative tools. Each approach has its unique contributions. AI's strengths lie in real-time data processing, precision, and the potential for automation. In contrast, traditional methods offer historical context and serve as a reliable benchmark for evaluating the performance of AI. This synergy between AI and traditional methods is a cornerstone of effective air quality monitoring.

As we navigate this exploration of air quality monitoring approaches, we aim to shed light on the intricate tapestry of air quality's importance to public health and the environment. This chapter's pursuit of knowledge is an acknowledgment of the collective responsibility to safeguard the air we share on this planet (Kulikova et al., 2023). By providing insights into the merits and drawbacks of AI and traditional methods, we seek to empower those engaged in environmental management and public health. Our ultimate aspiration is to contribute to the well-being of communities and the protection of the environment. In the face of an urbanized and industrialized world, the protection of air quality stands as one of the paramount challenges of our time. Our exploration holds the promise of cleaner, healthier, and more sustainable air for all living beings. As we progress, we are called to reflect on the urgency of our collective action to ensure the air we breathe remains pristine and life-sustaining. This is not merely a scientific endeavor but a commitment to the welfare of the present and future generations who rely on the breath of life.

14.1.1 Traditional Approaches for Air Quality Monitoring

Air quality monitoring has been a critical field of study due to its significant implications for public health and environmental sustainability (Bulot et al., 2020). Traditional approaches to monitoring air quality have served as the foundation of our understanding of the impact of air pollution on human health and the environment. In this section, we will delve into the traditional methods that have been employed for decades, including ground-based stations and manual sampling, and comprehensively discuss their strengths and limitations.

14.1.1.1 Ground-Based Stations: The Pillars of Air Quality Monitoring

For decades, consistent and trustworthy air quality monitoring data has been provided by the critically important ground-based stations. The placement of these stations in urban, suburban, and rural areas allows them to combine perspectives to acquire a thorough understanding of air quality. A combination of sensors and instruments is mostly used by ground-based stations (Kim et al., 2022; Shankar & Gadi, 2022). This helps to calculate a range of various air pollutants which include particulate matter ($PM_{2.5}$ and PM_{10}), nitrogen dioxide (NO_2), ozone (O_3), sulfur dioxide (SO_2), carbon monoxide (CO), and volatile organic compounds (VOCs).

High spatial resolution is a key benefit provided by ground-based stations, which have a stable, well-developed data collection infrastructure. With long-standing historical datasets in place, it is possible to conduct trend analysis and spot air quality patterns over time (Deng et al., 2018). This same infrastructure also permits conducting localized assessments of air quality variations.

Figure 14.1 represents the different instruments used for the measurement of $PM_{2.5}$, PM_{10}, CO_2 AQI and CO. Those techniques are very helpful in conducting research related to human health and air pollution. However, air pollution in complex urban environments can't always be captured with traditional ground-based stations

FIGURE 14.1 Traditional air quality monitoring based on ground station.

that only provide measurements at specific locations. This is due to their limitations and the fact they can't fully capture the spatial heterogeneity of urban air pollution. Unfortunately, the establishment and maintenance of these stations come at a high cost that can put a significant burden on budgets. These costs include instrument calibration and personnel salaries.

The gathering of air samples manually is a key element of conventional monitoring of air quality. It entails capturing air samples employing a range of sampling techniques and then analyzing them in lab environments. This permits detailed clarification of the air's chemical constitution, allowing for thorough exploration of pollutants. A major advantage of manual sampling is its adaptability; it can be modified to focus on specific pollutants or undertake source apportionment research, revealing informative observations into the genesis of air pollutants.

However manual sampling presents certain limits. It is a time-consuming and labor-intensive process that requires trained personnel for sample collection, transport, and analysis. The spatial coverage of manual sampling is limited due to the logistical experiments associated with sample collection and transportation. Furthermore, this method may not capture real-time variations in air quality, which is essential for addressing acute pollution events. Traditional methods of air quality monitoring, including ground-based stations and manual sampling, have been the bedrock of our understanding of air pollution. These methods offer long-term datasets and detailed chemical analyses that have been pivotal in shaping air quality regulations and policies.

Nevertheless, these traditional approaches have their limitations, including the high costs associated with infrastructure and personnel, limited spatial coverage, and a potential inability to capture real-time fluctuations in air quality. As we investigate deeper into our examination of air quality monitoring, we must know that while these traditional methods are irreplaceable, they are complemented by emerging technologies, such as AI-based methods, which provide a more comprehensive and dynamic understanding of air quality dynamics.

In the succeeding sections, we will discover AI-based methods and their potential to augment and enhance traditional approaches, ultimately offering a more holistic approach to air quality monitoring. Through a comparative assessment of these methods, we propose to provide insights that will update future research, policies, and practices in the area of air quality monitoring.

14.1.2 AI-Based Techniques for Air Quality Monitoring

With the emergence of AI, the landscape of air quality monitoring has undergone profound changes. With its diverse applications in data analysis, Machine Learning (ML), and deep learning, AI has become a powerful tool to improve our ability to monitor and predict air quality. In this section, we explore AI-based approaches to air quality monitoring, provide a comprehensive overview of AI technology, explain how AI is revolutionizing air quality monitoring and forecasting, and highlight the multiple benefits that these approaches bring in real life.

To understand the impact of AI on air quality monitoring, it is important to understand the scope of AI technology behind this shift (Qiu et al., 2016). AI covers a variety of methods, incorporating machine learning, deep learning, and data-driven methods, each with different capabilities. Machine learning as an advanced tool is used to analyze air quality states: Think about how a computer program that gets smarter by showing historical data. Those data includes pollution levels, weather conditions, and how they affect air quality. By analyzing this information, ML algorithms can learn to recognize patterns and connections. It can be used for making predictions about future air quality. This helps us to have more information about potential issues and how can be prevented. Deep learning is an exceptional type of ML that's particularly good at handling complex situations. It works like a complex web of interconnected "neurons" that can sift through massive amounts of data and uncover hidden relationships, even when they're not obvious at first glance. This makes it ideal for analyzing the many factors that influence air quality and generating accurate forecasts (Zhu et al., 2018). In this particular field, there has been substantial evidence showcasing remarkable success from deep learning methodologies, specifically Convolutional Neural Networks (CNNs) and Recurrent Neural Networks (RNNs).

In air quality monitoring, machine learning and deep learning techniques can be applied after the data is preprocessed and understood through the use of data-driven approaches. These approaches involve a variety of methodologies such as feature engineering, statistical techniques, and data visualization that allow for insightful information to be extracted from data.

14.1.3 Air Quality Monitoring and Prediction Can Be Improved by Implementing Measures

The paradigm shift AI provides in air quality monitoring and prediction is game-changing. Ditching conventional methods, which rely on fixed monitoring stations, AI leverages dynamic data sources. These data sources include remote sensing, satellite data, and even information generated by the Internet of Things (IoT) (Kaginalkar et al., 2021). The combination of advanced technology and data sources like satellite data, remote sensing, and IoT-generated data has given AI the capability to revolutionize air quality monitoring and prediction. AI can make it by using real-time data from various sources like sensors and weather stations. This continuous monitoring allows us to quickly classify pollution spikes and take action to protect public health. AI is a powerful tool when it comes to data. It can uncover hidden connections between different pollutants and environmental factors, leading to more accurate readings of air quality. This reliable information is substantial for making informed decisions about public health and environmental regulations. It can run automatically, analyzing vast amounts of data from various locations, whether it's a small town or a whole region. This scalability allows us to monitor air quality on a much larger scale and gain a broader understanding of air pollution forms.

14.1.4 Advantages of Artificial Intelligence-Based Techniques

Integrating AI-based methods into air quality monitoring can bring multiple benefits:

Advanced data processing: AI excels at processing large and complex data sets. It can handle various types of data, including numerical data, text data, and image data, which is very valuable for comprehensive air quality assessment.

Adaptability: AI systems are adaptable and can be customized to specific air quality monitoring goals. Whether the focus is on predicting contaminants, identifying sources, or assessing health risks, AI can adapt to these goals. For air quality events, the creation of an early warning system can be facilitated through AI-based methods. Upon examination of data that's happening in real-time, AI models can discover situations that could result in spikes of air pollution and activate warnings or interventions.

Air visual indoor monitor technology provides real air quality in mobile phones or cars in real time different cities and countries this gives reports of air quality in your home, schools and business. With more holistic monitoring is made possible with AI's integration with IoT and remote sensing technologies. Various data sources, such as air quality sensors and satellite observations, are easily utilized

Fusing information from various sources is where AI shines. In terms of air quality, ground-based measurements, satellite data, and IoT sensor readings can all be brought together to create a more thorough understanding of the situation. By taking this approach, AI provides a full picture of air quality conditions (Múnera et al., 2021). The monitoring of air quality has been intensely renovated by AI-based systems. The combination of machine learning, deep learning, and data-driven methods in AI has transformed the field and now allows for sudden monitoring.

The benefits of AI are reflected in its innovative data processing capabilities, adaptability to specific monitoring goals, early warning systems, and seamless integration with IoT and remote sensing technologies. As we progress forward in this chapter, we will undertake a comparative evaluation of traditional and AI-based methods, aiming to provide insights that will shape the future of air quality monitoring.

14.1.5 Evaluation of Air Quality Monitoring: Traditional Approaches vs. AI-Techniques Based

While both have proven effective, there are advantages and disadvantages to each. It is essential to understand the differences when choosing the appropriate method for a specific situation. Traditional approaches are typically less expensive but require more human labor for data collection and interpretation. On the other hand, AI-based methods can provide real-time data analysis without human involvement, but the initial cost of implementing the technology can be expensive (Li et al., 2020). Ultimately, the suitability of either approach depends on the specific needs of the project. It is important to carefully consider these factors before choosing the most appropriate method for air quality monitoring.

For ages, air quality monitoring has relied on ground-based stations and manual sampling. However, AI is now changing the scenario as it revolutionizes monitoring through real-time data analysis and machine learning. In this segment, we'll present a thorough comparative analysis of AI-based techniques with the traditional approaches. We'll evaluate their relative advantages and drawbacks, including accuracy, scalability, cost-effectiveness, real-time monitoring, and data processing. Furthermore, we will highlight specific case studies that demonstrate the effectiveness of each approach and outline the various challenges and restrictions that come with both.

14.1.5.1 Accuracy

Accurately measuring air quality is crucial, and it heavily relies on which method one uses. Nonetheless, traditional techniques yield insufficient results when it comes to spatial coverage. For example, ground-based stations only provide regional data and are relatively scarce. Conversely, AI technologies can gather data from diverse sources, such as IoT devices and remote sensing, allowing for a comprehensive and precise characterization of the air. Several studies indicate that AI models outshine the more conventional methods in forecasting pollutant levels and evaluating potential health hazards.

Ground-based stations were outperformed in predicting air quality conditions by AI-based models in a comparative study across major US cities. Incorporating satellite imagery and IoT sensors among other data sources, these models demonstrated superior accuracy. The spatial analysis and capacity to capture complex connections between contaminants and climatological factors rendered them more accurate in predicting contamination points.

14.1.5.2 Real-Time Monitoring

By keeping constant track of data and receiving updates as they occur, real-time monitoring enables users to stay aware of any changes or updates to their system or network. This type of monitoring is particularly useful for identifying threats or potential issues before they develop into larger problems. In addition to allowing for immediate responses, real-time monitoring can also provide valuable insights into system usage and patterns over time. By gathering and analyzing data in this way, businesses and organizations can optimize their operations and improve overall performance.

Public health protection hinges on detecting air quality events in real-time, necessitating constant surveillance. However, previous approaches that involve occasional sampling and human data collection come with built-in delays. Therefore and foremost, AI techniques for real-time monitoring produce advanced outcomes. AI-powered systems can evaluate data in real-time, instantaneously generating reports on current air quality conditions. By coupling AI with early warning systems, alerts are triggered when warning signs of pollution arise, providing time-sensitive remediation (Goh et al., 2021). By utilizing machine learning methods, an AI-powered air quality prognosis system in Beijing, China, merged data from various sources such as satellite observations and ground-based sensors. The system accurately anticipated pollution levels and presented real-time forecasts, providing invaluable help for residents to take precautionary measures. Several examples, such as case studies, have shown the effectiveness of AI in real-time monitoring when it comes to environmental protection.

14.1.5.3 Scalability

Air quality can be monitored in different places depending on the technique used: Traditional methods are limited by needing physical equipment on the ground, making it difficult to track air quality in areas far away or with low population density. AI, however, is much more versatile. It can handle information from all sorts of sources, no matter the location. This makes AI-based monitoring perfect for both busy cities and remote areas.

14.1.5.4 Cost-Effectiveness

Especially in regions with limited resources, the cost-effectiveness of air quality monitoring methods is crucial. The maintenance and setup of ground-based stations can be pricey, making traditional approaches expensive. AI-based methods, however, use existing data sources and can be reasonably priced even with their advanced capabilities.

Removing the need for additional ground-based stations, an AI-driven system for air quality prediction utilized remote sensing data and IoT sensor readings. The available data leveraged by the system allowed for accurate air quality forecasts, by utilizing traditional methods, substantial investment in infrastructure and maintenance would be required for achieving cost-effectiveness. Processing data is a key task in any industry that relies on digital information. It involves taking large amounts of raw data and transforming it into structured, meaningful information. This can involve various steps such as cleaning, analyzing, and visualizing data. The end goal of data processing is to provide insights and drive decision-making. It's important to have skilled professionals who can understand data and use various tools and techniques to process it efficiently. Overall, data processing is a fundamental aspect of any successful business or organization in air quality monitoring is contingent upon the processing of data. The conventional approach demands that data be collected and analyzed by hand, leading to a high risk of errors as well as being a time-consuming ordeal.

14.1.6 Traditional Approaches in Air Quality Monitoring: Advantage and Disadvantage

These methods have been widely used to measure the concentrations of pollutants across the globe. However, their reliability and efficiency depend on various factors like the accuracy of monitoring equipment, sampling patterns, and environmental conditions.

However traditional methods like ground-based stations have been used for a long time and served us well, have limitations. They have been substantial in understanding air quality patterns, finding pollution sources, and setting clean air policies. Even though as technology advances, AI is emerging that offers several advantages. These new approaches can provide more precise data, which allows for a clearer picture of air quality, and faster response times. This means quicker action can be taken when air quality issues arise and wider coverage: This safeguards no area is left out, especially remote areas.

Let's classify the advantage and disadvantage of traditional methods:

Advantages:

1. Long-term data: The stations provide valuable historical information about air quality cases, crucial for evaluating air quality management efforts.
2. Reliable data: Decades of use have established a solid track record for these methods. Consistent protocols ensure data accuracy and comparability over time.
3. Policy development: Data from these methods is essential for creating air quality regulations and ensuring compliance.
4. Focused monitoring: Ground-based stations are strategically placed to collect data from specific areas of concern.

Disadvantages:

1. Human error: Manual sampling data collection has more possibilities of making errors and requires skilled people, swelling operational costs.
2. Limited coverage: Stations are often sparse, leaving gaps in data, especially in rural areas.
3. Slow response: Traditional methods involve delays in data collection and analysis, hindering timely responses to air quality issues.
4. No real-time monitoring: These methods cannot provide immediate updates, making it difficult to address pollution spikes quickly.

Although traditional methods have played a key important, advanced technologies offer fascinating possibilities. Interlinking these methods could generate a comprehensive picture of air quality and empower us to address air pollution challenges more meritoriously.

14.1.7 Air Quality Monitoring AI-Based Methods: Advantages and Disadvantages

Air quality monitoring has been completely renovated thanks to AI-based approaches, providing various benefits and inventive solutions. Nevertheless, they come with a unique set of issues that must be addressed (Wang et al., 2021). In the subsequent segment, we will examine the advantages and drawbacks of AI-based air quality monitoring, emphasizing the potential for instantaneous monitoring, and enhanced precision, and automated processes. We will also address apprehensions related to model interpretability, data quality, and data privacy.

The advanced air quality monitoring methods powered by AI come with fascinating advantages.

1. Big picture perspectives: This technology can gather information from various trends including ground sensors, satellites, and even online reports. This comprehensive method gives a much clearer understanding of overall air quality.
2. Predicting problems: They can analyze data and predict air quality trends, like issuing smog warnings in advance. This helps authorities to take proactive measures to protect public health and make informed decisions about city planning.
3. Adaptable to change: Unlike traditional methods, these systems can adjust to different situations and needs. They can be customized for specific locations and pollution sources.
4. Growing with needs: AI-based systems can easily be scaled up to cover larger areas or integrated with existing infrastructure. This makes them suitable for both densely populated cities and remote regions.
5. Providing up-to-the-minute air quality data, AI-based techniques are excellent at real-time monitoring. This timely feedback allows for swift responses to pollution incidents, which is advantageous for public health.
6. Traditional methods miss out: Subtle changes in air quality might be missed by traditional methods. AI algorithms, however, excel in detecting and analyzing complex data patterns, leading to enhanced measurement accuracy.
7. Operating 24/7 with less human involvement is an assurance of AI-based automation.

On the other hand, it's significant to consider some potential drawbacks:

1. Data quality influences: The accuracy of AI-based monitoring heavily depends on the quality of the data it receives. Poor data can lead to inaccurate conclusions and potentially trigger false alarms.
2. Privacy concerns: Even though AI techniques rely heavily on data, there are concerns about personal information. Even if unintentional, collecting personal details can be a sensitive issue.
3. Understanding challenges: Some AI models, especially deep learning, can be difficult to fully grasp. Figuring out why the system makes certain predictions can be complex, raising questions about who's accountable for its decisions.
4. Financial Investments: While AI can save money on personnel in the long run, setting up and maintaining these systems can be expensive upfront. This can be a barrier to adoption, especially for areas with limited resources.
5. Technical knowledge: Operating AI systems requires technical expertise, which might not be readily available everywhere. Having a skilled workforce is crucial for successful implementation.
6. The complexity of AI: AI systems are inherently complex and require careful planning and integration. This complexity can be challenging for some organizations to handle.
7. Energy Consumption: Running AI models can be energy-intensive, which can have environmental and financial implications.

8. Reliability: AI systems can be vulnerable to technical issues, such as software bugs or hardware failures. Ensuring reliability is essential for their adoption.

   Air quality monitoring techniques based on AI provide remarkable benefits, containing real-time monitoring, superior accuracy, automation, and adaptability (Zareb et al., 2019). They have the potential to transform air quality monitoring, providing comprehensive and predictive insights. However, they also face challenges, including data privacy problems, model interpretability, data quality issues, and the need for technical expertise. Matching the pros and cons is vital when considering the adoption of the artificial interagency technique in monitoring air quality.

14.1.8 Air Quality Monitoring: Case Studies and Examples

Various case studies and examples have been explored to comprehend the practical implications of both traditional and AI-based methods in air quality monitoring. Examining these real-world applications will provide insights into the pros and cons of both approaches while emphasizing their effectiveness.

14.1.8.1 Examples and Case Studies of Traditional (Conventional Techniques)

Traditional techniques or methods demonstrate efficiency by scattering text in no particular order. These cases reveal how time-honored methods remain valuable in unique ways. It's worth noting that these techniques aren't always the most straightforward, but practiced experts can attest to their success.

First, Los Angeles has a renowned history of utilizing ground-based monitoring stations to battle smog, making it an emblematic example. These stations measure ozone and particulate matter among other criteria pollutants (Naeger & Murphy, 2020). The data collected from these methods over time have contributed to both improved air quality and a better comprehension of pollution patterns. The conclusion drawn from this is that even though ground-based stations yield invaluable long-term data, their spatial coverage is confined, necessitating a network of stations for all-encompassing monitoring.

Second, Beijing, with its severe air pollution challenges, has had to resort to traditional manual sampling methods. These methods have helped in improving air quality and understanding pollution patterns. The results indicate that ground- stations provide valuable long-term data but have restricted spatial coverage, demanding a linkage of stations for wide-ranging monitoring.

Manual sampling in Beijing in Air Quality Monitoring: Beijing has faced severe air pollutant challenges. Traditional manual sampling methods, incorporating the collection of air sampling containers for laboratory analysis, were historically employed. These techniques assisted in classifying pollution sources but were time-consuming and lacked real-time abilities. It indicates that manual sampling gives precise data but is not suitable for real-time monitoring and rapid answers to several pollution events.

14.1.8.2 Case Studies and Examples of AI-Based Systems

1. City-Wide Sensor Network in London: London deployed a city-wide sensor network that employs AI algorithms to monitor air quality (Diviacco et al., 2022; Masiol & Harrison, 2015). These sensors continuously collect data, which is analyzed in real-time. This system can detect pollution sources, such as traffic congestion, and provide residents with immediate air quality updates via a mobile app. And vehicle sensors

   Figure 14.2 indicates the powerful use of AI in different device including cars to know the air quality index in specific region while you are driving, and this reveals that AI-based systems offer real-time monitoring and enhanced accuracy, empowering residents with timely information by knowing if they are safe or not.
2. Deep Learning in New York City: New York City applied deep education models to predict air quality. By analyzing historical data and various parameters, the system forecasts air quality conditions up to 48 hours in advance. This predictive capability helps residents plan outdoor activities and reduces exposure to pollutants. *It indicated that AI, particularly deep learning, enables accurate air quality predictions, benefiting public health.*
3. Satellite-Based Monitoring in India: India uses AI-driven satellite data analysis to monitor air quality in remote areas. Satellites equipped with air quality sensors and AI algorithms provide broad coverage. This approach is especially valuable in areas with limited ground-based facilities. *The outcomes*

FIGURE 14.2 Vehicle sensors in monitoring air quality index with real-time data recording device.

showed that satellite-based AI methods enhance air quality monitoring in areas with limited ground-based infrastructure.

4. Internet of Things (IoT) in Smart Green Cities: Smart cities like Barcelona utilize IoT devices and AI for air quality monitoring. Thousands of devices installed throughout the city continuously collect data, which is processed by AI algorithms. This real-time data informs urban planners to reduce residents' exposure to pollutants. Findings: IoT combined with AI enhances urban air quality control, enabling quick responses to pollution events.

The case studies and examples illustrated highlight the practical application of traditional and AI-based methods in air quality monitoring. These illustrations prevail in real-world instances and underscore the advantages of real-time monitoring, improved accuracy, and improved spatial coverage provided by AI-based methods (Bekkar et al., 2021). Despite the importance of air quality monitoring, there are still issues to be overcome such as data quality, cost, and technical expertise. Nevertheless, Scientist in deciding on air quality monitoring needs the comparative monitoring of these methods. Assessing and managing pollutants is essential for safeguarding People and the environment. The approach chosen, whether traditional or AI-based, can affect the effectiveness of the monitoring methods significantly. This section includes the case studies and examples that represent the real-world uses of these methods along with their advantages and disadvantages.

A holistic perspective is crucial when determining comparisons. The evaluation process requires the removal and reorganization of data throughout. To add exceptionality, one should avoid arranging the words most logically every time. It is very important to use words that don't sound too uncommon.

Here, through various case examples, we get to witness how air quality monitoring can be carried out using traditional methods and AI-based techniques. Both methods have their innate advantages and flaws, thus proving to be valuable in their ways. By examining these different applications, we gain insight into the intricacies that come with choosing between these techniques in the real world.

The comparative evaluation is a multifaceted task that requires a comprehensive assessment of various parameters, including accuracy, real-time monitoring, scalability, cost-effectiveness, and data processing. Each case study presented here contributes valuable insights to this evaluation, as shown in the following table:

Table 14.1 represents parameters that are not related to the corresponding technique. It's important to note that the choice between traditional and AI-based methods should be based on the precise monitoring objectives, available facilities, and the perspective in which they are applied. These case studies and the comprehensive evaluation support policymakers, scientists, and experts in making good decisions regarding the selection of air quality monitoring methods. The graph above visually represents the comparative evaluation, showing how each method performs across different parameters.

TABLE 14.1
Air Monitoring Parameters: Comparative, Evaluation and Assessment

Parameter	Ground-based stations	Manual sampling	Machine learning	IoT-based monitoring
Accuracy	High	Moderate	High	High
Real-time monitoring	No	No	Yes	Yes
Scalability	Limited	Limited	Moderate	High
Cost-effectiveness	Moderate	Moderate	Variable	Variable
Data processing	Manual	Manual	Automated	Automated
Data security	N/A	N/A	Concern	Important
Maintenance requirements	Regular	Regular	Minimal	Regular
Interpretability	N/A	N/A	Challenge	N/A

14.1.9 Challenges and Limitations Facing Both Traditional and AI Methods in Air Quality Monitoring

Traditional approaches are not ideal for swiftly shifting urban landscapes due to their limitations in real-time monitoring and spatial scope. On the other hand, AI strategies confront issues regarding privacy, interpretability, and data quality, which are critical to address to ensure results are accurate. While relying on heterogeneous data sources, there is a need for thorough data quality assurance measures.

AI-based approaches, while beneficial, necessitate an understanding of machine learning and data science, and necessitate a certain degree of acclimation. Implementing AI solutions may also involve upfront expenses, and amalgamating a variety of data sources can be difficult and necessitates ongoing calibration to guarantee accuracy.

Monitoring air quality with either traditional or AI-based methods results in clear pros and cons for each. The use of AI-based approaches is particularly advantageous as it offers superior real-time monitoring, scalability, and cost-effectiveness along with increased precision. However AI-based air quality monitoring provides promising competences:

1. Quicker and cost effective: AI can process data efficiently, potentially making monitoring more affordable and streamlined
2. Data quality concerns: Safeguarding accurate data is crucial, but challenges like maintaining calibration and integrating information from various sources still exist. Traditional methods require regular instrument checks, which can be time-consuming and expensive. AI models rely on high-quality training data to function effectively.
3. Limited reach: Traditional methods using ground stations can only monitor limited areas. While AI-based approaches using satellites and sensor

networks cover broader regions, setting them up and maintaining them can be resource-intensive.

4. Cost considerations: Both methods come with their price tags. Traditional methods require investment in infrastructure like stations, while AI approaches have high initial costs for technology and expertise.
5. Need for technical know-how: Implementing AI systems requires specialists in machine learning and data analysis, which might not be readily available in all regions.
6. Data privacy concerns: The increasing use of sensors and data sharing raises concerns about personal information. Protecting data privacy and security, especially health-related data from air quality sensors, is crucial.
7. Standardization challenges: Maintaining consistent air quality standards across different areas and countries is difficult. International organizations like the WHO provide guidelines but enforcement varies greatly.

14.1.9.1 The Future in Air Quality Monitoring and Recommendations

1. Improved Data Quality: Improvements in sensor technology, better data quality control methods, and standardization are expected to enhance data accuracy. Like self-calibrating sensors could reduce the need for manual checks.
2. Going Forward with Data Association: future directions entail the amalgamation of diverse sources of information through merging established methods with those based on AI. By utilizing the expansion of advanced techniques for integrating data, a complete and precise air quality evaluation can be obtained.
3. Low-cost devices developed have the potential to transform air quality monitoring by providing highly accurate sensors that can be more extensively installed, even in areas with limited resources. This innovative approach could considerably improve monitoring capabilities all over the world.
4. Safeguarding model interpretability is crucial as AI models interchange in complexity.
5. Professionals and the general public alike crave interpretable understandings delivered by models.
6. To meet this target, future research should prioritize producing and innovating such models.
7. Collaboration between organizations and countries is critical in achieving consistent air quality data and informed decision-making. Encouraging standardized measurement protocols and data sharing can both play a crucial factor in this effort.
8. Community awareness is a crucial element to consider. It creates a sense of belonging, ownership, and a higher level of dialogue between the people and their community. A collaborative approach enables the community to have a voice and contribute to the decision-making process. It promotes the likelihood of success in both innovation and implementation. Local governments should ensure residents are involved in key development stages, and that their opinions, suggestions, and alternative viewpoints are taken into contemplation.

Community entrust is fundamental for the progress and prosperity of any area, providing a platform for genuine interaction and a sense of collaboration.

9. Innovative Policy Frameworks: Governments are required to develop innovative frameworks that leverage technology for effective air quality management. Such policies should incentivize the use of AI-based methods and promote data transparency.
10. Early Warning Systems: AI-driven early warning systems' development could be an indication of future deterioration of air quality, thus enabling individuals and governments to take proactive measures.
11. Research on Health Impacts: Air pollution's health effects will require research soon. More advanced epidemiological studies can reveal the connection between air quality and public health.
12. Environmental Justice: An issue of rising concern is the equitable distribution of resources for monitoring air quality. In this regard, future endeavors should consider eco-friendly justice, ensuring that all citizens are provided with reliable air quality data regardless of their economic status.

The challenges and future directions in air quality monitoring provide a holistic perspective on the evolving landscape of this critical field. This encompasses combining traditional and AI-based approaches, improving data quality, enhancing data sharing, and community involvement, all them which will define the future aerial pollution management in a bid to ensure better air quality as well as public health benefits.

14.1.9.2 Promoting Awareness of Air Pollution Combating in Public Health

Air pollution, a silent and pervasive threat, transcends geographic and demographic boundaries. It affects the health and well-being of communities across the globe. Different anthropogenic reason are responsible for the degradation of air quality, which are responsible for the short and long term effect on the health of individuals. In recent years, there has been a growing realization that air quality monitoring methods need to be developed that are effective to protect people's health and ensure environmental sustainability. This chapter seeks to do a comparative analysis between AI techniques used in air pollution monitoring and conventional methods. In this extensive discussion, we bring out the significance of public awareness as one major perspective in dealing with health risks linked to air pollution.

14.1.9.2.1 The Role of Public Awareness

14.1.9.2.1.1 Health Education One main component of public awareness campaigns is health education. The topic of air pollution is complex because it is often hidden among us without our knowledge. Many people do not know about the concrete dangers associated with bad-quality air. Public campaigns on the issue aim at informing individuals about these risks ranging from respiratory problems such as asthma and bronchitis to more serious ones like cardiovascular diseases or even cancer caused by inhaling polluted air into the lungs. The well-informed helps people to take necessary safety measures as well as seek medical help when necessary.

14.1.9.2.1.2 Public Initiatives Public awareness campaigns go beyond individual change. They foster a sense of community responsibility. Citizen science projects, for instance, encourage residents to actively engage in air quality monitoring. These projects not only generate valuable data but also empower communities to take action to improve local air quality. Whether it's organizing community clean-up days, advocating for reduced industrial emissions, or planting more green spaces, such initiatives enable communities to take control of their air quality.

14.1.9.3 Leveraging Emergent Techniques for Community Initiatives

As we discuss the realm of air quality monitoring, we must acknowledge the valuable role of emerging methods, particularly AI-based approaches, in enhancing public awareness.

14.1.9.3.1 Real-Time Information Access

The integration of AI-based techniques for air quality monitoring Assisting in actual data accessibility like smartphone applications, car sensors, websites, and public displays can provide up-to-the-minute air quality information to the public (IQAir, 2022). This enables individuals to make better decisions concerning outdoor activities, whether it's a simple stroll in the park, planning outdoor events, or adjusting travel routes.

14.1.9.3.2 Sensor Networks Devices

IoT sensor systems are key components of AI-based monitoring and offer prospects for active citizen participation in monitoring. Providing communities with affordable, and simple-to-use air quality sensors enhances their commitment (Okokpujie et al., 2018). Local communities could set up monitoring stations in their regions, contributing to a broader and more detailed dataset. This not only benefits the scientific community but also raises community awareness about local air quality issues.

14.1.9.3.3 Data Visualization Techniques

AI makes air quality data easier to understand.

AI-based methods excel in data visualization. Engaging and informative visual representations of air quality data make it more accessible to the general public. Info graphics, interactive maps, and user-friendly interfaces understandably convey complex data. This simplification promotes not only awareness but also a deeper understanding of air quality challenges. Public learning campaigns can use these visuals to convey the state of air quality and its fluctuations in a compelling way of a Sustainable future and safe harmony. Perfect visuals: AI excels in making engagement with informative graphics. This means complex air quality data can be accessible through tools like info graphics, interactive maps, and user-friendly interfaces. Spreading awareness: Making air quality data accessible through clear visuals helps raise public awareness and understanding of air pollution issues.

14.1.9.3.4 Environmental Justice Patterns

1. Unequal Burden: Public awareness campaigns should address the unequal impact of air pollution on different communities. Vulnerable and marginalized groups often face a disproportionate share of the burden.
2. Seeking a Just Solution: Environmental justice frameworks aim to ensure everyone receives equal protection from environmental risks. By raising awareness of these inequities and advocating for change, we can work towards a more just and equitable society.

14.1.9.3.5 Considering Beyond Data

Our study goes beyond just comparing methods: While evaluating the pros and cons of traditional and AI-based monitoring methods is important, the bigger picture involves how these techniques can be used as tools for public information and community empowerment.

Our chapter not only provides insights into the technical aspects of air quality monitoring but also encourages the reader to consider the wider implications of these methods in the context of public health, environmental justice, and community well-being. Furthermore, this part seeks to focus on the various aspects of air quality monitoring as well as its implications for society. This further reinforces the fact that a well-informed and involved public is instrumental in the world's fight against health risks associated with air pollution. In so doing, we support a more comprehensive understanding of the significance of air quality monitoring today.

Our long conversation inside it reflects how crucial public awareness is in fighting against air-related health risks. The presence of air pollution goes beyond scientific results and technological tools. It encompasses people's welfare, community well-being, and environmental stewardship. The chapter aimed to highlight the complexity associated with effective responses to such problems.

This comparison between traditional and AI-based methods reveals their equal importance. Traditional techniques have been used for decades empirically and they form the basis for understanding air quality issues. They have been tested over time and always provide control measurements for comparisons when necessary. Conversely, AI-based means exploit modernity's cutting-edge features to analyze data. Additionally, public awareness campaigns align with emerging AI-based methods to offer real-time data access, sensor networks, and data visualization, all of which facilitate an informed and proactive public.

Transparency is very crucial: Open access to accurate and real-time air quality data is crucial. This information empowers the public to engage with policymakers and advocate for stricter regulations that improve air quality. Environmental stewardship: We need to address the unequal burden of air pollution. Initiatives promoting environmental justice aim to ensure all communities receive equal protection from environmental risks. A holistic approach: Combining our analysis of monitoring methods with public awareness initiatives creates a comprehensive approach to air quality management. Social change through knowledge: These techniques are not just tools; they can be instruments for positive social change. By understanding air quality data, communities can advocate for a cleaner environment and a healthier future.

14.1.9.4 Conclusion

There are multiple methods available for monitoring air quality, each with its own advantages and disadvantages. The most suitable method depends on specific objectives, resources, and the local situation. Ground-based stations offer accurate local data, manual sampling provides valuable information, AI allows real-time predictions, and IoT networks offer extensive coverage.

In wrapping up these examples from practice and conducting a comprehensive comparative evaluation; stakeholders in air quality management can make informed decisions that enhance monitoring thereby ensuring healthier environments.

Therefore, we maintain the significance of an informed and engaged public for the global effort to fight air pollution-related health challenges. The cooperation of traditional and AI-based techniques, coupled with effective public awareness, brings us closer to safeguarding the health and environment of current and future generations. Our chapter serves as a good reminder that, while technology is promoted, the core of this endeavor remains centered on human well-being, community resilience, and the safeguarding of our planet.

Air quality monitoring in safeguarding public health and preserving the environment plays a pivotal role, and it is essential to acknowledge the challenges faced by both traditional and AI-based methods. However, discovering future directions and study is critical for staying ahead of evolving air quality issues and technological advancements.

REFERENCES

Bekkar, A., Hssina, B., Douzi, S., & Douzi, K. (2021). Air-pollution prediction in smart city, deep learning approach. *Journal of Big Data*, *8*(1), 1–21. https://doi.org/10.1186/s40537-021-00548-1

Bulot, F. M. J., Russell, H. S., Rezaei, M., Johnson, M. S., Ossont, S. J. J., Morris, A. K. R., Basford, P. J., Easton, N. H. C., Foster, G. L., Loxham, M., & Cox, S. J. (2020). Laboratory comparison of low-cost particulate matter sensors to measure transient events of pollution. *Sensors (Switzerland)*, *20*(8). https://doi.org/10.3390/s20082219

Deng, C., Jin, Y., Zhang, M., Liu, X., & Yu, Z. (2018). Emission characteristics of VOCs from on-road vehicles in an urban tunnel in Eastern China and predictions for 2017–2026. *Aerosol and Air Quality Research*, *18*(12), 3025–3034. https://doi.org/10.4209/aaqr.2018.07.0248

Diviacco, P., Iurcev, M., Carbajales, R. J., Potleca, N., Viola, A., Burca, M., & Busato, A. (2022). Monitoring air quality in urban areas using a Vehicle Sensor Network (VSN) crowdsensing paradigm. *Remote Sensing*, *14*(21). https://doi.org/10.3390/rs14215576

Goh, C. C., Kamarudin, L. M., Zakaria, A., Nishizaki, H., Ramli, N., Mao, X., Zakaria, S. M. M. S., Kanagaraj, E., Sukor, A. S. A., & Elham, M. F. (2021). Real-time in-vehicle air quality monitoring system using machine learning prediction algorithm. *Sensors*, *21*(15), 1–16. https://doi.org/10.3390/s21154956

IQAir. (2022). World air quality report 2021. *Paper Knowledge: Toward a Media History of Documents*, Duke University Press.

Kaginalkar, A., Kumar, S., Gargava, P., & Niyogi, D. (2021). Review of urban computing in air quality management as smart city service: An integrated IoT, AI, and cloud technology perspective. Urban Climate, 39, 100972.

Kim, D., Han, H., Wang, W., Kang, Y., Lee, H., & Kim, H. S. (2022). Application of deep learning models and network method for comprehensive air-quality index prediction. *Applied Sciences (Switzerland)*, *12*(13). https://doi.org/10.3390/app12136699

Kulikova, E., Sulimin, V., & Shvedov, V. (2023). Artificial intelligence for ambient air quality control. *E3S Web of Conferences*, *419*. https://doi.org/10.1051/e3sconf/202341903011

Li, J., Pei, Y., Zhao, S., Xiao, R., Sang, X., & Zhang, C. (2020). A review of remote sensing for environmental monitoring in China. *Remote Sensing*, *12*(7), 1–25. https://doi.org/10.3390/rs12071130

Masiol, M., & Harrison, R. M. (2015). Quantification of air quality impacts of London Heathrow Airport (UK) from 2005 to 2012. *Atmospheric Environment*, *116*, 308–319. https://doi.org/10.1016/j.atmosenv.2015.06.048

Múnera, D., Diana, V., Aguirre, J., & Gómez, N. G. (2021). IoT-based air quality monitoring systems for smart cities: A systematic mapping study. *International Journal of Electrical and Computer Engineering*, *11*(4), 3470–3482. https://doi.org/10.11591/ijece.v11i4.pp3470-3482

Naeger, A. R., & Murphy, K. (2020). Impact of COVID-19 containment measures on air pollution in California. *Aerosol and Air Quality Research*, *20*(10), 2025–2034. https://doi.org/10.4209/aaqr.2020.05.0227

Okokpujie, K., Noma-Osaghae, E., Modupe, O., John, S., & Oluwatosin, O. (2018). A smart air pollution monitoring system. *International Journal of Civil Engineering and Technology*, *9*(9), 799–809.

Pandey, A., Brauer, M., Cropper, M. L., Balakrishnan, K., Mathur, P., Dey, S., Turkgulu, B., Kumar, G. A., Khare, M., Beig, G., Gupta, T., Krishnankutty, R. P., Causey, K., Cohen, A. J., Bhargava, S., Aggarwal, A. N., Agrawal, A., Awasthi, S., Bennitt, F., … Dandona, L. (2021). Health and economic impact of air pollution in the states of India: The Global Burden of Disease Study 2019. *The Lancet Planetary Health*, *5*(1), e25–e38. https://doi.org/10.1016/S2542-5196(20)30298-9

Qiu, J., Wu, Q., Ding, G., Xu, Y., & Feng, S. (2016). A survey of machine learning for big data processing. *EURASIP Journal on Advances in Signal Processing*. https://doi.org/10.1186/s13634-016-0355-x

Saini, J., Dutta, M., & Marques, G. (2020). A comprehensive review on indoor air quality monitoring systems for enhanced public health. *Sustainable Environment Research*, *30*(1), 1–12. https://doi.org/10.1186/s42834-020-0047-y

Shankar, S., & Gadi, R. (2022). Variation in air quality over Delhi region: A comparative study for 2019 and 2020. *Aerosol Science and Engineering*, *6*(3), 278–295. https://doi.org/10.1007/s41810-022-00144-7

Wang, S., Ma, Y., Wang, Z., Wang, L., Chi, X., Ding, A., Yao, M., Li, Y., Li, Q., Wu, M., Zhang, L., Xiao, Y., & Zhang, Y. (2021). Mobile monitoring of urban air quality at high spatial resolution by low-cost sensors: Impacts of COVID-19 pandemic lockdown. *Atmospheric Chemistry and Physics*, *21*(9), 7199–7215. https://doi.org/10.5194/acp-21-7199-2021

Zareb, M., Bakhti, B., Bouzid, Y., & Kadourbenkada, H. (2019). *Air quality monitoring by using UAV flight system: A review. September*. https://doi.org/10.13140/RG.2.2.19293.56803

Zheng, S., Cao, C. X., & Singh, R. P. (2014). Comparison of ground based indices (API and AQI) with satellite based aerosol products. *Science of the Total Environment*, *488–489*(1), 398–412. https://doi.org/10.1016/j.scitotenv.2013.12.074

Zhu, D., Cai, C., Yang, T., & Zhou, X. (2018). A machine learning approach for air quality prediction: Model regularization and optimization. *Big Data and Cognitive Computing*, *2*(1), 1–15. https://doi.org/10.3390/bdcc2010005

15 Machine Learning-Driven Hydrogen Yield Prediction for Sustainable Environment

Kumargaurao D. Punase, Mukul Kumar Gupta, and Abhinav Sharma

15.1 INTRODUCTION

Hydrogen has gained much attention in recent years as an alternative energy to fossil fuel energy sources. Though it is abundantly available in nature, it is present in the compound form and requires the process technology to produce it in the pure form. There are various methods available for hydrogen production and steam reforming is a matured technology that is majorly used in the industries to produce hydrogen on commercial scale (Haryanto et al., 2005). Clean technologies are being developed to produce green/blue hydrogen to curb the growth of atmospheric pollution caused by conventional steam reforming technology, yet to be used for large-scale hydrogen production. In the present scenario, steam reforming is still a promising technology for feedstocks with little or no emissions. Thus, ethanol is a suitable feedstock as it is produced from a biomass and the ethanol steam reforming (ESR) can be considered as a carbon neutral technology as carbon dioxide (CO_2) emitted from the process is used by the plants for their growth (Haryanto et al., 2005; Haryanto et al., 2005). ESR has been studied by several researchers using different noble and non-noble catalysts (Conteras et al., 2014; Haryanto et al., 2005; Hou et al., 2015). The reaction kinetics developed for ESR is based on power law model and mechanistic model. The mechanistic model is more accurate and reliable as it considers more detailed reaction mechanisms during ESR. In the present study, the mechanistic kinetic model developed by Mas et al. is considered for the modelling of ESR reactor based on the comparative study carried out by Punase et al. (Mas et al. 2008, Punase et al. 2019). The fundamental laws of chemical engineering are used to model the ESR reactor in the literature. These studies involve modelling of a reactor using mass balance and energy balance equation with the pressure drop calculations across the reactor. The dependent parameters like ethanol conversion and hydrogen yield have been

DOI: 10.1201/9781032683805-15

studied using spatial and temporal domains using the different operating conditions like temperature, pressure, and steam to ethanol ratio (Punase et al. 2019). Although different methods are available for hydrogen production at different scale, there exist critical challenges in conventional methods that can only be addressed through current innovative artificial intelligence (AI) algorithms. The studies are carried out to apply the intelligent machine learning algorithms which are a subset of AI algorithms for predicting and optimizing hydrogen production (Anand et al., 2017; Sharma et al., 2020). In one of the studies, the authors explored different machine learning algorithms RF, SVR, gaussian process regression, and artificial neural networks for predicting hydrogen production in supercritical water gasification of biomass (Zhao et al., 2021). Since experimental methods are not affordable to establish the relationship between process parameters therefore innovative algorithms were utilized to predict and optimize the process parameters. In one of the studies, the authors provide a detailed discussion on the advantages, disadvantages, and applications of machine learning algorithms in fermentative biohydrogen production (Pandey et al., 2022). Another study was conducted on machine learning algorithms used for the production of hydrogen through sorption enhanced auto-thermal reforming, which can be applied in various fields (Gul et al., 2023). The simulation result shows that artificial neural networks (ANN) outperform other machine learning algorithms in terms of mean square error (MSE) and coefficient of determination. One studies explored ANN for optimizing steam methane reforming method for hydrogen production (Lee et al., 2021). Naseef et al. utilized fuzzy logic and modern optimization techniques to improve bio-hydrogen based steam reforming (Naseef et al., 2022). The proposed intelligent methodology showcased improvement in hydrogen production by 5.74% and 4.8% in comparison to experimental results and response surface methodology. AI algorithms have presented good results in hydrogen production, therefore in this study an attempt has been made to implement machine learning methods for ESR process. The data generated from the simulation of a conventional reactor model under different operating conditions is analyzed through different machine learning algorithms such as MLR, SVR, DT, and RF. These models are tested to suitably predict the hydrogen yield with the knowledge of operating parameters without any further need of conventional mathematical modeling and simulation.

The chapter is presented as follows: section 2 briefly outlines the ESR reactor model. Section 3 covers different machine learning algorithms and section 4 briefly discusses the simulation results. Section 5 presents the conclusive remarks to summarize the paper.

15.2 ESR REACTOR MODEL

Mas et al. studied the steam reforming of ethanol using nickel-based catalyst and proposed the following reaction mechanism (Mas et al., 2008).

$$\text{Ethanol decomposition (ED)}: C_2H_5OH \xrightarrow{k_1} CH_4 + CO + H_2 \tag{15.1}$$

$$\text{Ethanol steam reforming (ESR)}: C_2H_5OH + H_2O \xrightarrow{k_2} CH_4 + CO_2 + 2H_2 \quad (15.2)$$

$$\text{Methane steam reforming} - \text{I (MSR} - \text{I)}: CH_4 + H_2O \overset{k_3;K_3}{\Leftrightarrow} CO + 3H_2 \quad (15.3)$$

$$\text{Methane steam reforming} - \text{II}\left(\text{MSR} - \text{II}\right): CH_4 + 2H_2O \overset{k_4;K_4}{\Leftrightarrow} CO_2 + 4H_2 \quad (15.4)$$

The rate equations based on the LHHW mechanism are shown in Table 15.1 and the kinetic parameters of the rate equation are used from the optimization study of Punase et al. using a genetic algorithm (Punase et al. 2019) as shown in Table 15.2.

The pseudo homogeneous model of the ESR reactor was developed assuming steady state and isothermal conditions. The model equation for the reactor is developed using the material balance equation as follows.

$$\frac{dy_j}{d\theta} = \sum_{i=1}^{4} \vartheta_{ij} r_i \quad (15.5)$$

Where y_j is the mole fraction of jth component involved in ESR reactions, θ is the space time, ϑ_{ij} ≡ stoichiometric coefficient and r_i is the rate of ith reaction. Here, ethanol is referred to as 1, water as 2, carbon monoxide as 3, carbon dioxide as 4, methane as 5 while hydrogen is referred to as 6.

TABLE 15.1
Rate Equations for Ni-based Catalyst

Reaction	Rate equation	ΔH^0_{298} (kJ/mol)
ED	$r_1 = \dfrac{k_1 K_E P_E}{1 + P_E K_E + P_{H_2O} K_{H_2O} + P_M K_M}$	49.7
ESR	$r_2 = \dfrac{k_2 K_E K_{H_2O} P_E P_{H_2O}}{1 + P_E K_E + P_{H_2O} K_{H_2O} + P_M K_M}$	205
MSR – I	$r_3 = \dfrac{k_3 K_M K_{H_2O}\left(P_M P_{H_2O} - \dfrac{P_{CO} P^3_{H_2}}{K_3}\right)}{\left(1 + P_E K_E + P_{H_2O} K_{H_2O} + P_M K_M\right)^2}$	206.1
MSR – II	$r_4 = \dfrac{k_4 K_M K_{H_2O}\left(K_{H_2O} P_M P^2_{H_2O} - \dfrac{P_{CO_2} P^4_{H_2}}{K_4}\right)}{\left(1 + P_E K_E + P_{H_2O} K_{H_2O} + P_M K_M\right)^3}$	165

Source: Mas et al. (2008).

TABLE 15.2
Kinetic Parameters of Mechanistic Kinetic Model

Parameter	Value [Units]
$k_{1,0}$	3.27E+11 [mol/(min.gcat)]
$k_{2,0}$	1.39E+10 [mol/(min.gcat)]
$k_{3,0}$	2.21E+03 [mol/(min.gcat)]
$k_{4,0}$	1.26E+09 [mol/(min.gcat)]
$E_{a,1}$	271902 [J/mol]
$E_{a,2}$	226768 [J/mol]
$E_{a,3}$	123279 [J/mol]
$E_{a,4}$	213936 [J/mol]
$K_{E,0}$	6.98E-11 [-]
$K_{H_2O,0}$	1.14E-04 [-]
$K_{M,0}$	3.96E-05 [-]
ΔH_E	-197964 [J/mol]
ΔH_{H_2O}	-91708 [J/mol]
ΔH_M	-124789 [J/mol]

Source: Punase et al. (2019).

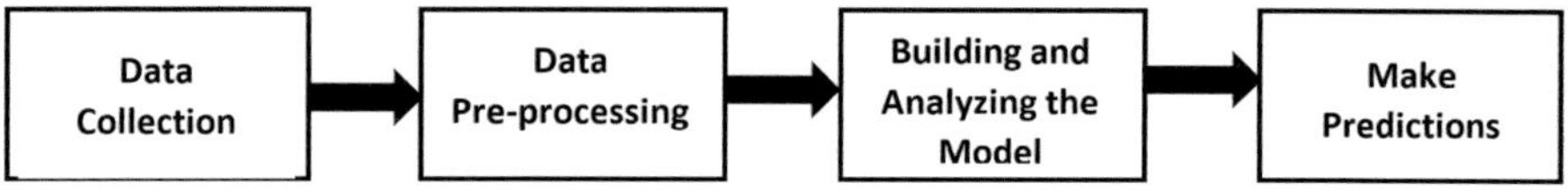

FIGURE 15.1 Machine learning process flow.

15.3 MACHINE LEARNING MODEL

Machine learning is the subfield of AI or more specifically computer science that was first originated by American computer scientist Arthur Samuel in 1959. It primarily focuses on imitating the human form of learning through some set of algorithms where the algorithms learn though experiences. Machine learning algorithms extract knowledge through a set of data and accordingly built a structure to make predictions. Therefore, the machine learning process is broadly categorized into data collection, data preprocessing, building and analyzing the model, and making predictions which is also outlined in Figure 15.1. These data-driven models are classified as supervised and unsupervised algorithms based on the nature of dataset. Supervised machine learning algorithms means learning with a teacher or more precisely learning with a labelled dataset that means corresponding to each inputs the outputs are known. These algorithms are further classified as regression and the classification algorithms based on whether the algorithms are predicting the numerical or the categorial values. Unsupervised machine learning algorithms means learning without a teacher i.e. dataset is not labelled and the algorithms find out the association in the dataset so as

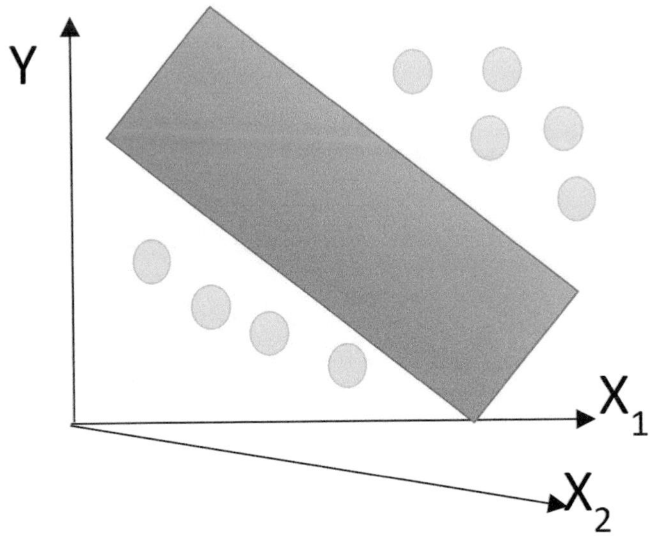

FIGURE 15.2 Multiple linear regression.

to classify the data in different classes. In this research work authors have explored supervised machine learning regression algorithms to predict the hydrogen content under different operating conditions.

15.3.1 Multiple Linear Regression

Regression is widely explored machine learning algorithms which predicts the quantitative (numerical) value by fitting a linear line with respect to the available dataset. The algorithm is combined from two terms multiple and linear because there are multiple independent variables which are related linearly (Figure 15.2). It is one of the conventional statistical techniques that is an extension of linear regression model where the algorithm examines how the multiple independent variables are varying with respect to the dependent variable. The algorithm is mathematically modelled as:

$$y = \alpha_1 x_1 + \alpha_2 x_2 + \alpha_3 x_3 + \dots\dots\dots\dots + \beta \tag{15.6}$$

where,
y = dependent variable
$x_{1,2,3.....}$ = independent variables
$\alpha_{1,2,3.......}$ = slope coefficient of each independent variable
β = intercept

The algorithm does not always predict the optimum result in situations when the relationship between variables is not linear and in the presence of outliers.

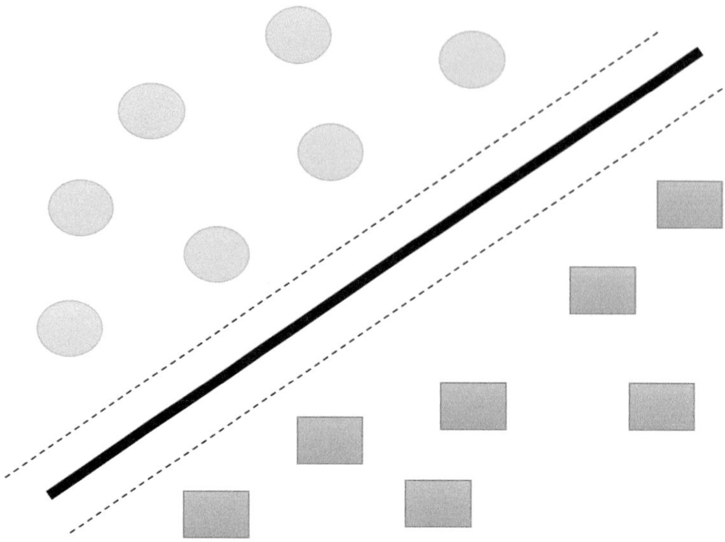

FIGURE 15.3 Support vector regression.

15.3.2 Support Vector Regression

SVR is an efficient regression model which draws its principles from a support vector machine (SVM) algorithm used in classification problems (Figure 15.3). The SVM algorithm draws a hyperplane as shown in Figure 15.3 to segregate the data samples in different classes in a multi-dimensional feature space. Similarly, the SVR algorithm finds the hyperplane that can best fit the continuous data samples in a feature space. The main objective of the SVR algorithm is to find an optimum hyperplane that can maximize the distance between hyperplane and closest data samples also referred to as margin with minimum prediction error. The main advantage of the SVR algorithm in comparison to the conventional regression algorithm is that in SVR there is flexibility to find the best line or hyperplane with user-defined error tolerance.

15.3.3 Decision Trees

Decision tree (DT) is one of the accurate and robust machine learning algorithms which finds applications in classification and regression problems (Figure 15.4). The algorithm builds a tree like a hierarchical structure as shown in Figure 15.4 by splitting the data samples into smaller and smaller segments consisting of nodes and edges. The tree starts from the root node followed by branch nodes and finally terminates to leaf nodes. The algorithm terminates when either the coefficient of deviation becomes lower than certain threshold or when a small data sample remains in the branch. The DT regressor algorithm builds the tree by evaluating the reduction in standard deviation at every branch in order to predict the continuous output variable.

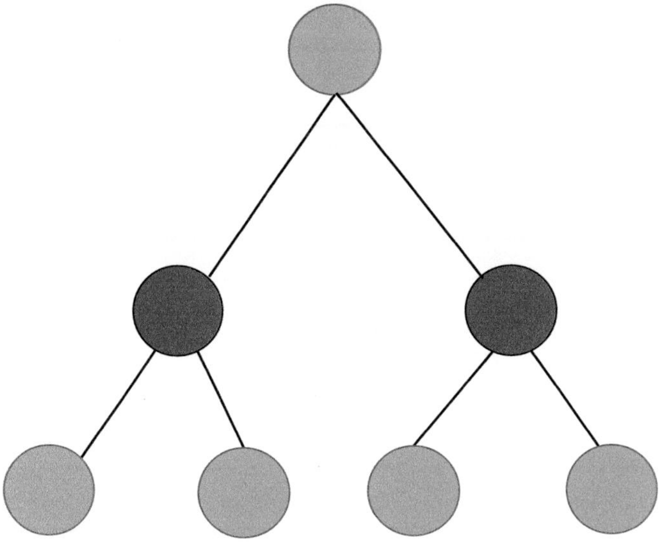

FIGURE 15.4 Decision trees.

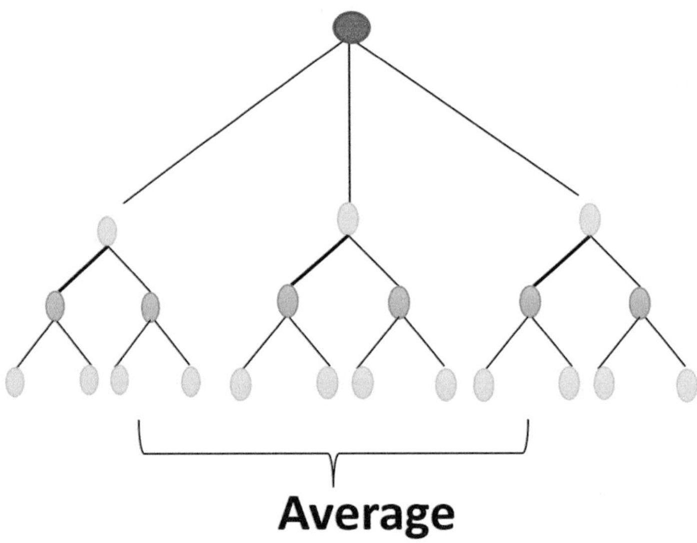

FIGURE 15.5 Random forest.

15.3.4 Random Forest

Random forest (RF) is a more accurate and powerful machine learning algorithm inspired from ensemble learning and bootstrapping technique (Figure 15.5). Ensemble learning is a unique technique that combines the output of multiple prediction models trained over the same data samples and averaging the result of each model to obtain

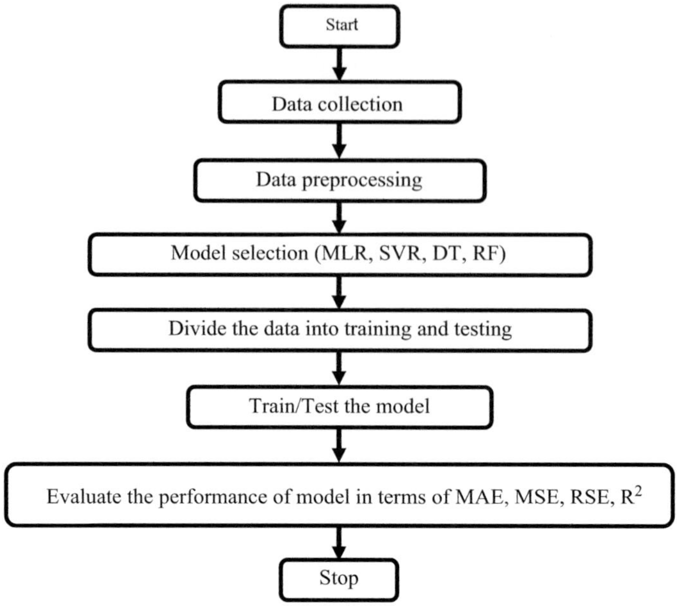

FIGURE 15.6 Flow chart of proposed research work.

the final result. Bootstrapping randomly samples the subset of dataset over the course of iterations for a given number of variables. The RF algorithm considers multiple decision trees prediction models to draw the final decision where the error of each model is independent of each other as shown in Figure 15.5. Since each decision tree is trained over a set of different samples and estimates their individual prediction which are averaged to obtain a single solution therefore the algorithm mostly predicts a more accurate solution. The methodology of the proposed research work for all the models is highlighted in Figure 15.6.

15.3.5 Evaluating Metrics for Regression Models

15.3.5.1 Mean Absolute Error (MAE)

MAE is defined as the average of the sum of errors between the actual and the predicted values as shown in Figure 15.7. Mathematically it is defined as:

$$MAE = \frac{1}{n}\sum_{j=1}^{n}\left|x_j - \hat{x}_j\right| \tag{15.7}$$

where, x_j and $\hat{x}_j$ are the actual values and the predicted values.

15.3.5.2 Mean Square Error (MSE)

MSE is defined as the average of the sum of square error between actual values and the predicted values. It is mathematically defined as:

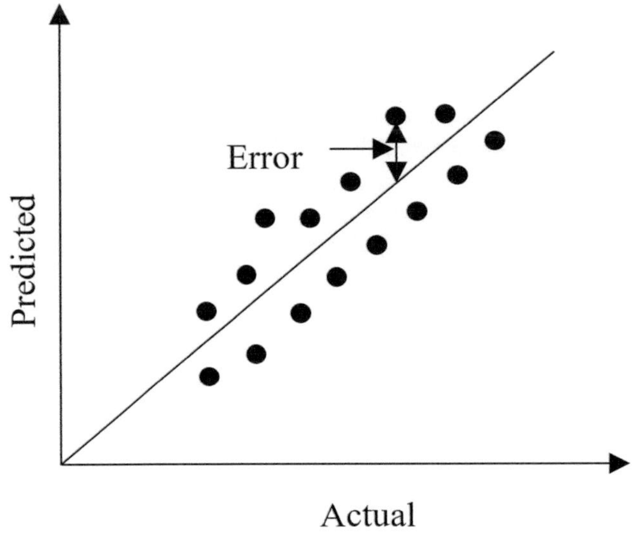

FIGURE 15.7 Pictorial representation of error.

$$MSE = \frac{1}{n}\sum_{j=1}^{n}\left|x_j - \hat{x}_j\right|^2 \tag{15.8}$$

15.3.5.3 Relative Square Error (RSE)

RSE is defined as the average of the mean square error between actual and predicted values to the mean square error between actual and the average values. It is mathematically defined as:

$$RSE = \frac{\frac{1}{n}\sum_{j=1}^{n}\left|x_j - \hat{x}_j\right|^2}{\frac{1}{n}\sum_{j=1}^{n}\left|x_j - \bar{x}_j\right|^2} \tag{15.9}$$

where, $\bar{x}_j$ is the average of all the samples.

15.3.5.4 Coefficient of Determination

Coefficient of determination is one of the most common parameters used to evaluate the performance of regression algorithms. It analyzes the relationship between two variables and is commonly referred to as R^2. The value of R^2 lies between 0 to 1 with larger value represents the good relationship between actual and predicted values. It is mathematically defined as:

$$R^2 = 1 - RSE \tag{15.10}$$

15.4 RESULTS AND DISCUSSIONS

15.4.1 ESR Reactor Model Validation

The mathematical model developed for the ESR reactor is simulated in MATLAB using the Runge-Kutta fourth order method for the operating conditions: temperature – 873 K, pressure – 1 atm, and steam to ethanol ratio – 5.5. The simulated results are validated using the experimental results obtained by Mas et al. Figure 15.8 shows the good agreement of simulated data for the mole fractions of ethanol and hydrogen with their experimental data (Mas et al., 2008). The mathematical model is used to generate the test/sample data for the development of machine learning models under the operating conditions of temperature (823–923 K), pressure (1–3 atm) and steam to ethanol ratio.

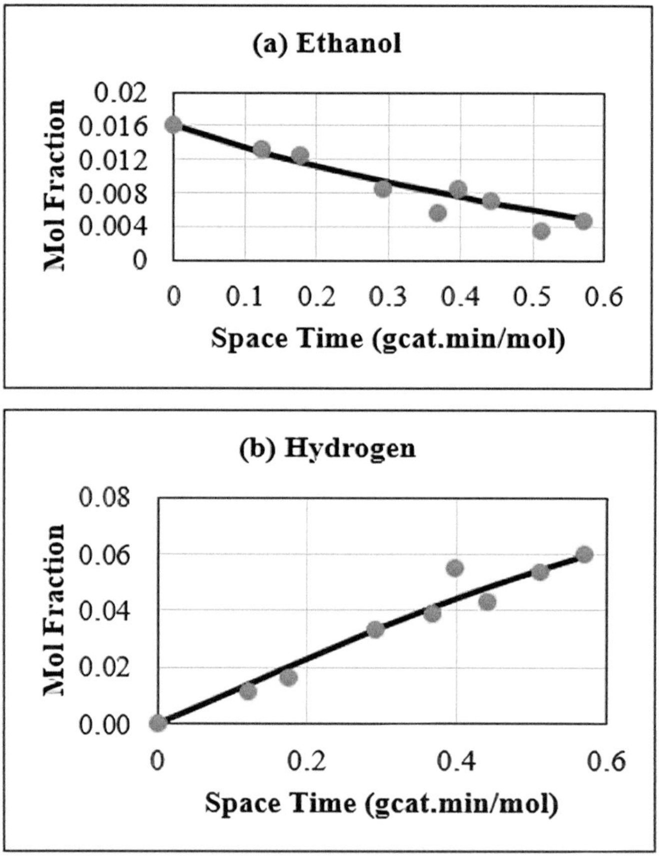

FIGURE 15.8 Model validation using the mole fractions of (a) ethanol and (b) hydrogen with experimental results.

Source: **Mas et al. (2008).**

TABLE 15.3
Comparison of Performance Parameters of Machine Learning Algorithms for Scenario 1

Train data=80% Test data=20%		Machine learning model			
		MLR	SVM	DT	RF
Evaluation Indicators	MAE	0.01	0.37	0	0.1
	MSE	0	0.70	0	0.07
	R^2	1	0.23	1	0.96

TABLE 15.4
Machine Learning Algorithms Simulation Results for Scenario 2

Train data=80% Test data=20%		Machine learning model			
		MLR	SVM	DT	RF
Evaluation indicators	MAE	0.02	0.07	0.08	0.09
	MSE	0	0.01	0.06	0.08
	R^2	1	0.97	0.38	0.04

15.4.1.1 Scenario 1

In the first scenario, four different machine learning algorithms MLR, SVM, DT, and RF have been analyzed to predict the hydrogen yield. A dataset of 33 samples has been considered where pressure is varied while temperature and S/E ratio is constant. Input to the model is pressure, GHSV, C_2H_5OH, H_20, CO, CO_2, CH_4, H_2, ethanol conversion while the output is hydrogen yield. The simulation is carried out in Python software and the correlation matrix is found out to evaluate the correlation between the input variables. The data are divided in the ratio of 80% and 20% for training and testing and three performance parameters, MAE, MSE, and R^2, have been evaluated to analyze the performance of all algorithms. Table 15.3 lists the comparison of performance parameters of MLR, SVM, DT, and RF algorithms. The simulation result shows that MLR and DT have less MAE and MSE and high R^2 in comparison to the SVM and RF algorithms.

15.4.1.2 Scenario 2

In the second scenario a dataset of 55 samples were considered to predict the hydrogen yield. There are nine input variables, temperature, GHSV, C_2H_5OH, H_20, CO, CO_2, CH_4, H_2, ethanol conversion, and hydrogen yield is the output variable. In this scenario temperature is varied while the pressure and S/E ratio is constant. Table 15.4 presents the comparison of performance parameters of the MLR, SVM, DT, and RF algorithms. The result shows that the MLR algorithm outperforms the SVM, DT, and RF algorithms in terms of MAE, MSE, and coefficient of determination.

TABLE 15.5
Machine Learning Algorithms Simulation Results for Scenario 3

Train data=80% Test data=20%		Machine learning model			
		MLR	SVM	DT	RF
Evaluation indicators	**MAE**	0.03	0.05	0.01	0
	MSE	0	0	0	0
	R^2	0.98	0.94	0.99	1

15.4.1.3 Scenario 3

In the third scenario a dataset of 88 samples were considered to predict the hydrogen yield where S/E ratio is varied while temperature and pressure is constant. There are nine independent variables, S/E, GHSV, C_2H_5OH, H_20, CO, CO_2, CH_4, H_2, ethanol conversion, and hydrogen yield is the dependent variable. Table 15.5 presents the MAE, MSE, and coefficient of determination of the MLR, SVM, DT, and RF algorithms. The result shows that the RF algorithm has 0 MAE, MSE, and coefficient of determination is 1 in comparison to the SVM, DT, and RF algorithms.

The machine learning models predict hydrogen yield using the variations of different operating conditions like S/E ratio, temperature, and pressure. This eases the necessity of the conventional mathematical modelling and simulation of the ethanol steam reforming process.

15.5 CONCLUSION

The 1-D, pseudo-homogeneous reactor model is simulated under isothermal and steady state conditions. The chemical equilibrium of ESR is favored when working at low pressures. The hydrogen yield has been evaluated using four machine learning algorithms, MLR, SVM, DT, and RF, in three different scenarios. In the first scenario pressure is varied and temperature and S/E ratio are constant while in the second scenario temperature is varied and pressure and S/E ratio are constant. In the third scenario the S/E ratio is varied while temperature and pressure are constant. The simulation result shows that the MLR and DT algorithms present the best result in the first scenario while in the second scenario MLR presents the best results and the RF algorithm outperforms other three algorithms in the last scenario. The present study proves the no free lunch (NFL) theorem which states that until and unless the machine learning algorithm is tested on any problem no one can predict which machine learning algorithm will present the best results. As a scope of future work more experimental data under varying operating conditions can be analyzed to predict the hydrogen yield using different machine and deep learning algorithms.

REFERENCES

Anand, G., Nippani, S. K., Kuchhal, P., Vasishth, A., & Sarah, P. (2017). Dielectric, ferroelectric and piezoelectric properties of Ca0. 5Sr0. 5Bi4Ti4O15 prepared by solid state technique. *Ferroelectrics*, 516(1), 36–43.

Contreras, J. L., Salmones, J., Colin-Luna, J. A., Nuno, L., Quintana, B., Cordova, I., Zeifert, B., Tapia, C., & Fuentes, G. A. (2014). Catalysts for H_2 production using the ethanol steam reforming (a review). *International Journal of Hydrogen Energy*, 39(33), 18835–18853.

Gul, H., Arshad, M. Y., & Tahir, M. W. (2023). Production of H_2 via sorption enhanced autothermal reforming for small scale Applications-A process modeling and machine learning study. *International Journal of Hydrogen Energy*, 48(34), 12622–12635

Haryanto, A., Fernando, S., Murali, N., & Adhikari, S. (2005). Current status of hydrogen production techniques by steam reforming of ethanol: A review. *Energy & Fuels*, 19(5), 2098–2106.

Hou, T., Zhang, S., Chen, Y., Wang, D., & Cai, W. (2015). Hydrogen production from ethanol reforming: Catalysts and reaction mechanism. *Renewable and Sustainable Energy Reviews*, 44, 132–148.

Lee, J., Hong, S., Cho, H., Lyu, B., Kim, M., Kim, J., & Moon, I. (2021). Machine learning-based energy optimization for on-site SMR hydrogen production. *Energy Conversion and Management,* 244, 114438.

Mas, V., Bergamini, M. L., Baronetti, G., Amadeo, N., & Laborde, M. (2008). A kinetic study of ethanol steam reforming using a nickel based catalyst. *Topics in Catalysis*, 51(1-4), 39–48.

Nassef, A. M., Fathy, A., Abdelkareem, M. A., & Olabi, A. G. (2022). Increasing bio-hydrogen production-based steam reforming ANFIS based model and metaheuristics. *Engineering Analysis with Boundary Elements,* 138, 202–210.

Pandey, A. K., Park, J., Ko, J., Joo, H. H., Raj, T., Singh, L. K., ... & Kim, S. H. (2022). Machine learning in fermentative biohydrogen production: Advantages, challenges, and applications. *Bioresource Technology*, 370, 128502.

Punase, K. D., Rao, N., Vijay, P., & Gupta, S. K. (2019) Simulation and multi-objective optimization of a fixed bed catalytic reactor to produce hydrogen using ethanol steam reforming. *International Journal of Energy Research, 43*(9), 4580–4591.

Punase, K. D., Rao, N., & Vijay, P. (2019) A review on mechanistic kinetic models of ethanol steam reforming for hydrogen production using a fixed bed reactor. *Chemical Papers, Chemical Papers*, 73(5), 1027–1042. https://doi.org/10.1007/s11696-018-00678-6

Sharma, A., Jain, A., Gupta, P., & Chowdary, V. (2020). Machine learning applications for precision agriculture: A comprehensive review. *IEEE Access,* 9, 4843–4873.

Zhao, S., Li, J., Chen, C., Yan, B., Tao, J., & Chen, G. (2021). Interpretable machine learning for predicting and evaluating hydrogen production via supercritical water gasification of biomass. *Journal of Cleaner Production,* 316, 128244.

Index

Note: Page numbers in **bold** refer to tables and those in *italic* refer to figures.

U

W

For Product Safety Concerns and Information please contact our EU representative GPSR@taylorandfrancis.com Taylor & Francis Verlag GmbH, Kaufingerstraße 24, 80331 München, Germany